Industrial Production Management in Flexible Manufacturing Systems

Industrial Production Management in Flexible Manufacturing Systems

Contributors

Don Ajith Rohana Dolage, Abu Bakar Sade et al.

AURIS
Reference

www.aurisreference.com

Industrial Production Management in Flexible Manufacturing Systems

Contributors: Don Ajith Rohana Dolage, Abu Bakar Sade et al.

Published by Auris Reference Limited

www.aurisreference.com

United Kingdom

Industrial Production Management in Flexible Manufacturing Systems

ISBN: 978-1-78154-934-6

British Library Cataloguing in Publication Data
A CIP record for this book is available from the British Library

Printed in the United Kingdom

Exclusively distributed by CBS Publishers & Distributors Pvt. Ltd.

Sales & Distribution Rights only for India, Pakistan, Bangladesh, Sri Lanka, Nepal and Bhutan. This book is not to be sold outside these territories.

Contents

List of Abbreviations

ASMI	Annual Surveys of Manufacturing Industries
CAD	Computer Aided Design
CIM	,Computer-integrated manufacturing
CNC	Control Machine Tools
DEA	Data envelopment analysis
DAA	Deadlock Avoidance Algorithm
DMU	Decision making units
DOS	Department of Statistics Malaysia
FMM	Federation of Malaysian Manufacturers
FAS	Flexible assembly systems
FMS	Flexible manufacturing system
FMT	Flexible Manufacturing Technology
GLO	Geographic Location
GCI	Gini concentration index
GRE	Global routing efficiency
GRR	Global routing range
GRV	Global routing versatility
GRO	Growth Rate of Output
INLP	Integer nonlinear programming
LAN	Local Area Networks
MIDA	Malaysian Industrial Development Authority
MFP	Multifactor productivity
PCM	Price cost margin
PCA	Principal Component Analysis
PCs	Principal Components
PLC	Programmable Logic Controllers
SME	Sequential mutual exclusionsSME
SFPF	Stochastic Frontier Production Function
TE	Technical efficiency
TP	Technological progress
TFPG	Total productivity growth
WIP	Work in process

List of Contributors

Don Ajith Rohana Dolage
International Graduate School of Business, University of South Australia, Adelaide, Australia

Abu Bakar Sade
Faculty of Management, Multimedia University, Cyberjaya, Malaysia

Don Ajith Rohana Dolage
International Graduate School of Business, University of South Australia, Adelaide, Australia

Abu Bakar Sade
Faculty of Management, Multimedia University, Cyberjaya, Malaysia

Prasenjit Chatterjee
Mechanical Engineering Department, MCKV Institute of Engineering, Howrah - 711 204, West Bengal, India

Shankar Chakraborty
Department of Production Engineering, Jadavpur University, Kolkata - 700 032, INDIA

Jingshan Li
Department of Electrical and Computer Engineering and Center for Manufacturing, University of Kentucky, Lexington, KY 40506, USA

Ningjian Huang
Manufacturing Systems Research Lab, General Motors Research and Development Center, General Motors Corporation, Warren, MI 48090-9055, USA

Ranbir Singh
Research Scholar, Deptt. of Mech. Engg., DCRUST Murthal, Sonipat (Haryana), 131039, India

Rajender Singh
Professor, Deptt. of Mech. Engg., DCRUST Murthal, Sonipat (Haryana), 131039, India \

B.K. Khan
Director, MSIT Jagdishpur, Sonipat (Haryana), India

S.R. Singh
Department of Mathematics, D. N. College, Meerut, 250103,India

Leena Prasher
Centre for Mathematical Sciences, Banasthali University, Banasthali, Rajasthan 304022, India

F. Zammori
Dipartimento di Ingegneria Meccanica, Nucleare e della Produzione Via Bonanno, 25/A 56126 Pisa, Italia

M. Braglia
Dipartimento di Ingegneria Meccanica, Nucleare e della Produzione Via Bonanno, 25/A 56126 Pisa, Italia

M. Frosolini
Dipartimento di Ingegneria Meccanica, Nucleare e della Produzione Via Bonanno, 25/A 56126 Pisa, Italia

Mussa I. Mgwatu
Department of Mechanical and Industrial Engineering, University of Dar es Salaam, P.O. Box 35131, Dares Salaam, Tanzania

Paul E. Deering
Department Engineering Technology and Management, Ohio University, Athens, OH, USA.

S.R. Singh
Department of Mathematics, D. N. College, Meerut, India

Leena Prasher
Department of Mathematics, QIFGOI, Mohali, India

Neha Saxena
Department of Mathematics, D. N. College, Meerut, India

Ranbir Singh
Research Scholar, Department of Mechanical Engineering, DCRUST Murthal, Sonepat, India

Rajender Singh
Professor, Department of Mechanical Engineering, DCRUST Murthal, Sonepat, India

B. K. Khan
Technical Advisor, MSIT, Sonepat, India

Yigang Xu
School of Management and Economics, University of Electronic Science and Technology of China, Chengdu, China

Yifei Du
School of Management and Economics, University of Electronic Science and Technology of China, Chengdu, China

Yong Zeng
School of Management and Economics, University of Electronic Science and Technology of China, Chengdu, China

Shiming Li
School of Management and Economics, University of Electronic Science and Technology of China, Chengdu, China

J. R. Jadhav
Department of Mechanical Engineering, Sardar Patel College of Engineering, Bhavan's Campus, Munshi Nagar, Andheri (West), Mumbai 400 058, India

S. S. Mantha
All India Council for Technical Education (AICTE), 7th Floor, Chanderlok Building, Janpath, New Delhi 110 001, India

S. B. Rane
Department of Mechanical Engineering, Sardar Patel College of Engineering, Bhavan's Campus, Munshi Nagar, Andheri (West), Mumbai 400 058, India

Preface

The text *Industrial Production Management in Flexible Manufacturing Systems* addresses the present discussions surrounding flexible production systems based on automation, robotics and cybernetics as they continue to replace the traditional production systems. It also covers issues related to the use of multi-servicing in the operational management of the industrial production and its scheduling systems. The aim of first chapter is to examine the impact of the degree of adoption of flexible manufacturing technology (FMT) on the technical efficiency of the manufacturing industry of Malaysia. Second chapter explores the impact of the adoption of flexible manufacturing technology (FMT) on the Malaysian manufacturing industry. Third chapter focuses on the applications of six preference ranking methods for effectively solving the FMS selection problems. A Markovian approach on quality evaluation in flexible manufacturing systems has been presented in fourth chapter. The aim of fifth chapter is to solve a real-world machine loading problem with an objective of balancing the workload of the flexible manufacturing system (FMS) with decreased computational time. The objective of sixth chapter is to determine the optimal policy for production system, which maximizes the total profit subject to some constraints under consideration. Seventh chapter focuses on routing flexibility, which is the ability to manufacture a part type via several routes and/or to perform different operations on more than one machine. Eighth chapter demonstrates the importance of incorporating and solving the machining optimization problem jointly with part selection and machine loading problems in order to avoid unbalanced workload in the FMS. Ninth chapter presents the sufficient conditions for deadlock to exist in a flexible manufacturing technology. A centralized reverse channel structure with flexible manufacturing under the stock out situation has been presented in tenth chapter. A critical review of machine loading problem in flexible manufacturing system has been proposed in eleventh chapter. Twelfth chapter discusses about flexible manufacturing technology of continuous process (CP) enterprise. The purposes of last chapter are to study and compare the UNIDO–ACMA model of lean implementation and validate the ISM model through comparison with UNIDO–ACMA model.

Chapter 1

A FRONTIER APPROACH TO MEASURING IMPACT OF ADOPTION OF FLEXIBLE MANUFACTURING TECHNOLOGY ON TECHNICAL EFFICIENCY OF MALAYSIAN MANUFACTURING INDUSTRY

Don Ajith Rohana Dolage[1], Abu Bakar Sade[2]

[1]International Graduate School of Business, University of South Australia, Adelaide, Australia

[2]Faculty of Management, Multimedia University, Cyberjaya, Malaysia

ABSTRACT

This paper examines the impact of the adoption of Flexible Manufacturing Technology (FMT) on the Technical Efficiency of Malaysia Manufacturing Industry. Owing to the potential multicollinearity, the Principal Component Analysis has been adopted to extract the most appropriate underlying dimensions of FMT in an effort to substitute the eight FMT variables. The study has been conducted within FMT intensively adopted 16 three-digit industries that encompass 50 five-digit industries covering the years 2000-2005. The results obtained from the two situations, one, including the industry fixed effects dummy variables and the other without these, are contrasted. It is found that the model that included the industry fixed effect dummy variables possesses a greater explanatory power. The two principal components that account for the greater variation in FMT show positive and moderately significant relationship with TE. The study concludes with sufficient evidence that FMT has a direct and moderately significant relationship with TE.

INTRODUCTION

Over the last two decades many researchers conducted studies that examined the factors attributing to the performance of the East Asian Economies. According to Mahadevan [1] in particular, the GDP growth of these economies was found to be driven by the perspiration factor of input accumulation rather than the inspiration factor of total productivity growth (TFPG). Owing to the increasing concern regarding the continued growth of these economies, new endogenous growth theories have been adopted in an effort to explain the performance of these economies. The significance of the manufacturing sector in these economies is so much that in each economy it accounts for well over 30 percent of the overall GDP. Therefore, more focused studies looking into the inspiration factor of TFPG, which also means factor productivity should be undertaken.

According to the conventional growth accounting methodology, TFPG is considered synonymous with technological progress (TP) which is also known as technical change. According to Coelli Rao, O'Donnell and Battese [2], TP represents advances made in technology that may be represented by an upward shift in the production frontier. This methodology which is a non frontier approach, is based on the assumption that all industries are fully realising their capacity in the production process and thus, are technically efficient. In this approach, no distinction is made between TP and changes in technical efficiency (TE) with which a known technology is adopted in production. As a result, this approach does not separately account for the technological improvement embodied in labour or the capital stock (change in efficiency) contained in the TFPG. Under a newer approach named Stochastic Frontier Production Function (SFPF), originally proposed by Aigner, Lovell and Schmidt [3], output growth is decomposed into input growth and TFPG and TFPG is in turn decomposed into TP and TE. The word frontier signifies the idea of maximality and represents the "best practice" approach to production. Mahadevan [4] showed that unlike the growth accounting approach which provided a shape of an average industry, the estimation of a frontier function was heavily influenced by the best performing industries. According to Mahadevan and Kalirajan [5], TE can be due to the accumulation of knowledge in the learning-by-doing process, improvements in the instructions for mixing together raw materials, diffusion of new technology, improved managerial practices or R&D undertaken by government or profit maximising agents, or can be affected by overall market structure of industry as it affects the methods used for acquiring, developing or modifying technology.

It is widely believed that intensive regimes of contemporary manufacturing paradigms such as mass customisation, customerisation and instant customerisation can pave the way for a competitive manufacturing industry. The studies show that mass customisation is the core manufacturing paradigm. The studies also showed that the crucial determinant of the successful implementation of mass customisation is the abundant use of Flexible Manufacturing Technology (FMT) [6,7]. Kumar and Desmukh [8] remark that today's customer not only expects quality, reliability and competitive pricing but also customised products with timely delivery, it is desirable that an organisation is as flexible as possible,. According to Sinha and Noble [9], FMT can represent a huge cost for adopting firms, but may also offer the chance to achieve competitive advantage through superior manufacturing. Therefore, it would be important to examine the causal link between the degree of FMT adoption and TE.

The average GDP growth of Malaysia during 2000- 2007 (5.5 percent) is lower than that during 1990-2000 (7.0 percent). Malaysian Manufacturing sector GDP during 2000-2007 (13.0 percent) is much lower than the same for the period 1990-2000 (4.8 percent). These are some of the key indicators to the declining competitiveness of the Malaysian manufacturing industry over the period 2000-2007. In the Malaysian manufacturing industry, FMT is widely adopted and has received the due attention from the industry policy makers. The Malaysian Industrial Development Authority (MIDA) [10] has recognised a number of promoted activities and products (for the development and production) with regard to high technology establishments. The engagement in these activities will make them entitled to pioneer status or investment tax allowance under the promotion of Investment Act 1986. This includes FMT products such as, Computer process control systems/equipment, Process instrumentation, and Robotic equipment and Computer numerical control machine tools. The Ninth Malaysia Plan which is aimed at achieving changes in the structure and improved performance of the economy with every economic sector achieving higher value added and total factor productivity. The "Thrust 1" of this Plan states that, "Application of high technology and production of higher value added products will be given emphasis. Measures will be undertaken to migrate the electrical and electronics (E&E) industry towards high-technology and higher value added activities". Hence, the Malaysian experience is used as a case study given its suitability.

Mahadevan [1] adopted SFPF in a study on TFPG of Malaysia's Manufacturing Industries. Sun and Kalirajan [11] found that 2.5 percent average annual rate of TP during this period was the major contributor to TFPG in the Korean manufacturing industry whereas TE grew by a modest 1.1 percent per annum. Zhang and Zhang [12] adopted SFPF to estimate the TE of China's large and medium sized iron and steel enterprises. Lee, Kim and Heo [13] in their empirical study on TP versus TE Gains in Manufacturing Sector of Korea, adopted the nonparametric Malmquist productivity index to break down the productivity growth into two components; technological change (innovation) and efficiency change (catching up). According to Battese and Broca [14], SFPF involves an unobservable random variable associated with the technical inefficiency of production of individuals, in addition to the random error in a traditional regression model. These studies show that TE is adopted extensively to measure efficiency of the manufacturing sector.

As shown above, evidently, only a few studies have examined the impact of specific technologies on the measures of competitiveness such as productivity, technical efficiency and profitability at industry level using less aggregated data. Berndt and Morrison [15] examined the impact of high-tech investments on multifactor productivity (MFP) and three profitability measures. While the study found only limited evidence of a positive relationship between profitability and the share of high-tech capital in the total physical capital stock, it establish that they were negatively correlated with MFP. Dolage, Sade and .Elsadig [16] investigated to impact of FMT on the TFPG of Malaysian manufacturing industry and established the significance of certain types of FMT on the TFPG.

Amato and Amato [17], Dolage and Sade [18] have investigated the impact of high-tech or FMT investments on Price Cost Margin. This study established that there was a positive impact from high-tech investments regardless of whether or not the specification includes industry effects dummy variables to account for the differences in technological opportunity among industries. Hence, according to productivity literature evidently no empirical study has been undertaken to investigate the impact of FMT adoption on TE of Malaysian manufacturing industry.

METHODOLOGY

The basic research hypothesis of the study is:

A high degree of FMT adoption enhances TE of the manufacturing industry of Malaysia.

Estimation of TE Using SFPF

The SFPF adopted in this study to compute the industrywise technical efficiency is based on the same adopted by Mahadevan [4] in the study on "A frontier approach to measuring TFPG in Singapore Manufacturing Industry". The derivation of the function to evaluate TE, based on SFPF is described below in Equation (1):

$$\ln Y_{it} = \delta_{1i} + \sum_{j=1}^{2} \delta_{ij} \ln W_{ijt}$$

$$(1)$$

where

$i = 1,2,3,\cdots,16$ (no of three-digit manufacturing industries);

$j = K$, and L (K-capital and L-labour);

$t = 1,2,3$ (no of years from 2000-2005);

Y = Value added output measured in 2000 prices;

W_K = Capital expenditure measured in 2000 prices;

W_L = Number of workers employed;

δ_{1i} =Intercept term of the ith three-digit manufacturing industry;

δ_{ij} = Actual response of output to the method of application of the jth input used by the ith manufacturing industry.

Mahadevan [19] in the study on "Is There a Real Growth Measure for Malaysia's Manufacturing Industries?" incorporated time dummies in the SFPF to capture the effects of time on the TE. Therefore, time dummies are incorporated in this study too and accordingly Equation (1) shown above can be modified to accommodate this effect; the revised model is represented in Equation (2) shown below:

$$\ln Y_{it} = \delta_{1i} + \sum_{j=1}^{2} \delta_{ij} \ln W_{ijt} + \sum_{t=1}^{6} \delta_t \tag{2}$$

Kalirajan and Shand [20] in their study on Frontier Production Functions and Technical Efficiency Measures, explained that the efficient use of inputs due to various industry specific characteristics contributed individually to the technical efficiency of the industry and the contributions can be measured by the magnitudes of the random slope coefficients. All other production characteristics are captured by the varying random intercept term.

Since intercepts and slope coefficients can vary across industries they can be represented as:

$$\delta_{ij} = \delta_j^1 + u_{ij} \tag{3}$$

$$\delta_{1i} = \delta_1^1 + v_{ij} \tag{4}$$

where δ_j^1 is the mean response coefficient of output with respect to the jth input, and u_{ij} and v_{ij} are random disturbance terms.

Combining Equations (2), (3) and (4), SFPF can be presented in Equation (5) shown below:

$$\ln Y_{it} = \delta_1^1 + \sum_{j=1}^{2} \delta_j \ln W_{ijt} + \sum_{j=1}^{2} u_{ij} \ln W_{ijt} + v_{1i} + \sum_{t=1}^{6} \delta_t \tag{5}$$

Mahadevan [21] showed that by adopting Aitken's generalised least squares method proposed by Hildredth and Houck and the estimation procedure by Griffith the industry-specific and input-specific response coefficient estimates of the above model could be obtained. The highest values of each response coefficient and the intercept determine the frontier coefficient of the potential production function. If δ^* denotes the parameter estimates of the frontier production function, then

$\delta_j^* = \max\{\delta_{ij}\}$ where i = 1, 2, $\cdots$16 and j = K, L

$\delta_1^* = \max\{\delta_{1i}\}$ where i = 1, 2, $\cdots$16

$\delta_t^* = \max\{\delta_t\}$ where t = 1, 2, $\cdots$ 6 Since $\delta_1^* = \delta_1^1 + \max\{v_t\}$ $\delta_j^* = \delta_j^1 + \max\{u_{ij}\}$, and j represents both K and L, the above equations can be rearranged as follows:

$\delta_K^* = \delta_K^1 + \max\{u_{iK}\}$

$\delta_L^* = \delta_L^1 + \max\{u_{iL}\}$

$\delta_1^* = \delta_1^1 + \max\{v_t\}$

$\delta_t^* = \max\{\delta_t\}$

The potential output of each industry can be realised when each industry adopts the "best practice" techniques and the maximum potential output for each industry is given by Equation (6):

$$\ln Y_{it}^* = \delta_1^* + \sum_{j=1}^{2} \delta_j^* \ln W_{ijt} + \delta_t^*$$

(6)

In Equation (6), since j represent both capital (K) and labour (L) it can be expanded to form a new equation (Equation (7)) which is given below:

$$\ln Y_{it}^* = \delta_1^* + \delta_K^* \ln W_{iKt} + \delta_L^* \ln W_{iLt} + \delta_t^*$$

(7)

The industry-specific TE can be represented by Equation (8) shown below as the ratio of the industry's actual realised output to that of its potential output:

$$TE_{it} = \frac{Y_{it}}{\exp(\ln Y_{it}^*)}$$

(8)

The numerator Y_{it} can be computed simply as the "gross output" take away "cost of inputs", values for which were obtained from ASMI.

Equation (5) in which j is representative of both capital (K) and labour (L) can be expanded and presented as follows:

$$\ln Y_{it} = \delta_1^1 + \delta_K \ln W_{iKt} + \delta_L \ln W_{iLt} + u_{iK} \ln W_{iKt}$$

$$+ u_{iL} \ln W_{iLt} + v_{1t} + \sum_{t=1}^{6} \delta_t \tag{9}$$

All the values relating to the six years for the 50 MSIC five-digit industries (altogether 300 cases) are to be substituted in Equation (9) given above:

The technique of multiple regression analysis was adopted to ascertain the respective response coefficients for capital and labour and the constants ($\delta_i^*, \delta_k^*, \delta_i^*$ and δ_i^*). The widely used statistical software, Statistical Package for Social Sciences (SPSS) was adopted to run the above multiple regression and from its output, the maximum values for u_{iK}, u_{iL} and v_i were obtained to compute δ_i^*, δ_K^*, δ_L^* and δ_i^*. Once these values were ascertained, as the next step, they were substituted in the Equation (7) in order to compute the maximum frontier outputs for each industry for all the 6 years with the use of EXCEL.

Review of Factors Affecting TE in the Manufacturing Industry

Geographic Location (GLO): According to ASMI reports, Malaysia shows a significant diversity in the number of firms located within different provinces. Certain regions in a country could be disadvantaged by its distal location from the major markets due to less access to physical and human capital and technologies. Sun and Kalirajan [11] stated that firms applied their production technology and inputs differently as a result of the differences in location, experience, and firm size. Zhang and Zhang [12] stated that geographic location affected the efficiency of an enterprise. Vu [22], Margono and Sharma [23], Zhang and Zhang [12], Sun and Kalirajan [11] have considered GLO as an explanatory variable of TE.

Malaysia does not publish indicators which show the favourability of a particular province for an industry. The ASMI contains the number of establishments located within a province; usually Table 5 of ASMI depicts this information. The researcher assumes that the major reason why a large number of establishments are concentrated in a particular province is that it has a conducive environment for industries with respect to many aspects namely, access to skilled labour, technology, support industries, raw materials, close proximity to markets and ports etc. So, the corollary is that the number of firms in a province indicates the location advantage of the province. Accordingly, the provinces were ranked based on the number of establishments located in each. The directory published by Federation of Malaysian Manufacturers (FMM) contained the addresses of establishments including the province in which each establishment is situated [24]. In the FMM directory 2007, the manufacturing industries have been categorised under MSIC four-digit level. Thereafter, each establishment in a particular province was multiplied by its respective ranking. This ranking for a province was not a constant over the six year period considered since the number of establishments located within provinces has evidently changed yearly, though of course marginally. Finally, GLO was computed for each MSIC five-digit industry as the summation of the products of the number of establishment and the ranking of the respective province.

Ownership (OWN): It has long been established that generally foreign owned establishments are more efficient than locally owned establishments. Vu [22], Margono and Sharma [23], Zhang and Zhang [12], Sun and Kalirajan [11] and Mahadevan [25] considered OWN as an explanatory variable of TE. Since Malaysia is known to have a long history as a well considered recipient of FDI since mid seventies, the variable OWN was incorporated to account for the "ownership" impacts on TE.

Malaysia does not publish the status of establishments with respect to their ownership. Nevertheless, the DOS from its Economic Census data provided the number of establishments belonging to the Non Malaysian and Malaysian categories for each MSIC five-digit industry for all the years under review. Mahadevan [25] used a dummy variable "1" for industries in which more than 45 percent of the total number of establishments were either wholly foreign owned or joint ventures which were more than half foreign owned; otherwise "0". However, information provided by the DOS in this respect was not detailed enough to assess the degree of foreign ownership regarding the Non Malaysian firms. A dummy variable (OWN) was used to distinguish between industries that had higher and lower percentages of Non-Malaysian firms. In this study, dummy variable "1" was used if in a particular five-digit industry more than 30% of the establishments were Non-Malaysian otherwise, the dummy variable "0" was used.

Firm Age (FAGE): The length of existence of a firm can indicate advancement of different factors favourable to TE namely, propensity to employ skilled workers and degree of learning by doing. According to Zhang and Zhang [12], the maturity of a firm represented by its age is a factor contributing towards inefficiency but the firm size coefficients are expected to affect inefficiency negatively. Alternatively, vintage of capital can be used to detect technical progress in production, especially the type of technical progress embodied in capital (Zhang and Zhang [12]. However, measuring the vintage of capital is not straightforward owing to the lack of data on the past capital investments at the firm level. Since it is reasonable to assume that FAGE indicates the vintage of capital and the experience of the firm, it is considered as a control variable in the TE model of this study. Sun and Kalirajan [11] as well as Margono and Sharma [23] have considered FAGE as an explanatory variable of TE in their studies.

The DOS Malaysia does not publish data pertaining to FAGE. The FMM directory gives the year of incorporation of each establishment listed under MSIC four-digit level industry. Using this value, the average FAGE for each industry was computed as at year 2000. The average firm age (FAGE) for the later years was computed by increasing the average firm age as of 2000 by one for each subsequent year.

***Incentive Payments (INC)*:** It is rational to consider that incentive schemes and bonus payments can motivate employees to work hard resulting in efficiency gains for the industry. Mahadevan [25] highlights differential incentive systems as one of the factors causing inefficiency. However, in this study INC has not been incorporated in the TE model as an explanatory variable. Vu [22] has included bonus payments as one of the hypothesised influencing factors in the TE model adopted to study technical efficiency of industrial state-owned enterprises. In this study, INC was measured as the ratio of incentive payment to salary. The DOS Malaysia does not publish data for INC industry-wise. The researcher computed INC from the Economic Census data maintained by the DOS Malaysia.

Firm Size (FSIZE): This variable is a crude proxy for scale of entry barrier. Theoretically, the minimum efficient plant size is a better proxy but could not be included due to the non availability of data. Higher productivity gains can be expected in the presence of oligopolistic competition. Therefore, researchers include average plant size or firm size in TE models to take account of such effects: Chandrasiri [26], Amato and Amato [17] and McGuckin [27]. However, the direction of the link is ambiguous. In two separate studies, Amato and Amato [17] and Round [28] have incorporated FSIZE in their PCM model to account for the entry barriers. In this study, FSIZE was measured as the average firm size of the eight largest firms in each industry. As these data are not annually published in Malaysia, they had to be computed using the data obtained from the Economic Census data maintained by the DOS.

The eight types of FMT considered in this study are given below:

Computer Numerical Control Machine Tools (CNC): Measured as the percentage of firms in each MSIC fivedigit industry using microprocessor based numerical control technologies, referred to as computer numerical control machine tools. Numerical Controlled Machine Tools (NC): Measured as the percentage of firms in each MSIC five-digit industry using numerical controlled machine tools.

Robotics (ROB) Measured as the percentage of firms in each MSIC five-digit industry using robotics.

Programmable Logic Controllers (PLC): Measured as the percentage of firms in each MSIC five-digit industry using programmable logic controllers.

Automated Inspections (INS): Measured as the percentage of firms in each MSIC five-digit industry using automated sensor-based inspection, either during the production process or final product.

Automated Storage and Retrieval Systems (ASR): Measured as the percentage of firms in each MSIC fivedigit industry using automated storage and retrieval systems.

Computer Aided Design (CAD): Measured as the percentage of firms in each MSIC fivedigit industry using computer aided design to control manufacturing machinery.

Local Area Networks (LAN): Measured as the percentage of firms in each MSIC five-digit industry using local area networks.

Industry Fixed Effects Dummy Variables (IND_j): The study involved 50 five-digit industries included in 16 three-digit industries. It is logical to assume that industry characteristics among these 16 three-digit industries can be diverse and need to be captured by a variable. Therefore, 16 dummy variables (IND_j) were incorporated into the TE model to capture the industry fixed effects. The model representing the relationship among TE explanatory variables and FMT variables can be specified as given below:

Technical Efficiency Model

$$
\begin{aligned}
TE = {} & \lambda_0 + \lambda_1 GLO + \lambda_2 OWN + \lambda_3 FSIZE + \lambda_4 FAGE \\
& + \lambda_5 INC + \lambda_6 CNC + \lambda_7 NC + \lambda_8 ROB + \lambda_9 PLC \\
& + \lambda_{10} INS + \lambda_{11} ASR + \lambda_{12} CAD + \lambda_{13} LAN \\
& + \lambda_{13+j} IND_j + \mu
\end{aligned}
\tag{10}
$$

DATA AND ESTIMATION

Inclusion Criteria

According to the Malaysian Standard Industrial Classification 2000 (MSIC 2000) there are 53 three-digit industries. In order to obtain a rational outcome, the study needs to be conducted only within industries in which FMT is intensively adopted. On account of this, inclusion criteria were formulated in an effort to select FMT intensively adopted MSIC three-digit industries for the sample, which is shown below:

Industries with high "capital/labour" ratio.

Industries in which product variation is a marketing strategy.

Industries in which products are susceptible to demand fluctuation.

Using the above criteria a sample of 16 MSIC threedigit industries which together comprise 50 five-digit industries was selected.

Primary Data

The data that indicate the degree of adoption of FMT is not published by any organisation in Malaysia. Hence, a questionnaire survey was conducted to gather information necessary to compute the percentage of establishments adopting each specific type of FMT in a given year, within a given MSIC five-digit industry. The questionnaires were sent to all the establishments, listed under the 50 MSIC five-digit industries that appeared in the directory of Federation of Malaysian Manufacturers.

Secondary Data

In order to compute TE, industry-wise data is required for output, intermediate input, capital input and labour input. The closest indicators for these values were obtained from of the ASMI published for the years 2000 through 2005 by the DOS of Malaysia. The variables GLO was computed using the data obtained from this table. OWNFISZE and INC were computed using the data obtained from the Economic Censes conducted by the DOS Malaysia. The information required to calculate FAGE was obtained from the Directory published by the Federation of Malaysian Manufacturers.

EMPIRICAL RESULTS

Multicollinearity of FMT

Since only FMT intensively used industries were included in the sample, naturally some similarity in the sequence and characteristics of the production processes could be expected even amongst different five-digit industries. Hence, there could be a tendency for a similarity in the technology adopted amongst these industries. Due to the similarities in technologies, a high prevalence of multicollinearity among the eight types of FMT could be anticipated. In this study, bivariate Pearson productmoment correlation analysis has been conducted using SPSS to test for multicollinearity amongst FMT. The output that reveals potential multicollinearity among FMT variables is displayed in Table 1. According to Coakes, Steed and Price [29] and Field [30], when a considerable number of correlations are exceeding 0.3, the matrix is suitable for Principal Component Analysis (PCA).

PCA was performed using SPSS in order to obtain underlying dimensions (Principal Components) of FMT as a remedy for multicollinearity. As per both standard methods of (i.e. screen test and eigen values greater than one) extracting the optimal number of components, three Principal Components (PCs) were extracted that account for 67 percent of the variation in the FMT. According to Table 2, the loadings of variables onto the three PCs obtained from both types of rotations (Orthogonal and Oblique) are quite similar. Hence, due to simplicity, PCs obtained from orthogonal rotation was used in the rest of the analysis.

Once the most appropriate type of rotation and the resultant PCs were decided, the variables loading onto each of these PCs were examined as the next step. An examination of the component loadings depicted in Table 2 indicates that LAN, CAD, PLC and CNC load onto PC1; ASR, INS and ROB load onto PC2 while only NC loads onto PC3. Usually it is difficult to give clear cut themes or names to PCs that only relate to or encompass particular variables that are loading onto it. Hence, only the best possible names have been assigned to the PCs extracted from this analysis. The technologies LAN, CAD, PLC and CNC are used in the manufacturing set up as process control technologies. Since these load onto PC1,

Table 1. Correlations among FMT

		CNC	NC	ROB	PLC	INS	ASR	CAD	LAN
CNC	Pearson Correlation	1.000	0.160**	0.351**	0.634**	0.307**	0.237**	0.248**	0.322**
	Sig. (2-tailed)		0.005	0.000	0.000	0.000	0.000	0.000	0.000
	N	300	300	300	300	300	300	300	300
NC	Pearson Correlation	0.160**	1.000	0.012	0.164**	0.177**	0.126*	0.141*	0.171**
	Sig. (2-tailed)	0.005		0.836	0.005	0.002	0.030	0.014	0.003
ROB	Pearson Correlation	0.351**	0.012	1.000	0.368**	0.250**	0.427**	0.391**	0.236**
	Sig. (2-tailed)	0.000	0.836		0.000	0.000	0.000	0.000	0.000
PLC	Pearson Correlation	0.634**	0.164**	0.368**	1.000	0.302**	0.257**	0.394**	0.380**
	Sig. (2-tailed)	0.000	0.005	0.000		0.000	0.000	0.000	0.000
INS	Pearson Correlation	0.307**	0.177**	0.250**	0.302**	1.000	0.564**	0.115*	0.186**
	Sig. (2-tailed)	0.000	0.002	0.000	0.000		0.000	0.046	0.001
ASR	Pearson Correlation	0.237**	0.126*	0.427**	0.257**	0.564**	1.000	0.308**	0.129*
	Sig. (2-tailed)	0.000	0.030	0.000	0.000	0.000		0.000	0.025
CAD	Pearson Correlation	0.248**	0.141*	0.391**	0.394**	0.115*	0.308**	1.000	0.609**
	Sig. (2-tailed)	0.000	0.014	0.000	0.000	0.046	0.000		0.000
LAN	Pearson Correlation	0.322**	0.171**	0.236**	0.380**	0.186**	0.129*	0.609**	1.000
	Sig. (2-tailed)	0.000	0.003	0.000	0.000	0.001	0.025	0.000	

**Correlation is significant at the 0.01 level (2-tailed);*Correlation is significant at the 0.05 level (2-tailed).

Table 2. Comparison of components obtained from two types of rotations

	Component One			Component Two			Component Three		
	Orthogonal	Oblique		Orthogonal	Oblique		Orthogonal	Oblique	
		Pattern	Structure		Pattern	Structure		Pattern	Structure
LAN	0.816	0.861	0.811						
CAD	0.816	0.841	0.801						
PLC	0.666	0.640	0.722			0.445			
CNC	0.555	0.517	0.621			0.467			
ASR				0.845	0.858	0.851			
INS				0.816	0.844	0.826			
ROB	0.477	0.412	0.542	0.526	0.460	0.573			
NC							0.883	0.871	0.883

so can be named as "process control" technologies. The technologies ASR, INS and ROB load onto PC2, so can be named as "production and quality control" technologies. PC3 has only one variable i.e. NC, loading onto it so can be called the "general control" technology.

As the next step, the eight FMT variables were substituted with the three PCs namely, PC1, PC2 and PC3. Therefore, the TE model was reformulated as follows:

$$TE = \lambda_0 + \lambda_1 GLO + \lambda_2 OWN + \lambda_3 FSIZE$$
$$+ \lambda_4 FAGE + \lambda_5 INC + \lambda_6 PC1 + \lambda_7 PC2$$
$$+ \lambda_8 PC3 + \lambda_{8+i} IND_i + \mu$$

Multiple Regression Analysis of TE

As described, the model contains a set of 16 industry fixed dummy variables (IND_j) to account for the differences of technological opportunity among industries. Although it is theoretically desirable to include IND_j, the consequent impact of adding these 16 extra variables needs to be examined by comparing and contrasting the results obtained without considering the IND_j in the model. A separate regression was performed for this scenario and the tables of Model Summary, ANOVA and Coefficients were obtained. In order to facilitate the easy comparison of the results, the tables of output obtained from regression analysis for the two situations, one with the IND_j included and the other without the IND_j have been combined into one. The tables of Model Summary, ANOVA and Coefficients contained in the SPSS output for these situations have been reproduced in Tables 3-5 respectively.

According to **Table 3**, Adjusted R square is considerably high (0.983) when IND_j has been included. This indicates that the explanatory variables together explain 51.8 percent of the variance in TE. However, the explanatory power of the model has decreased significantly when the IND_j has been excluded; Adjusted R square (0.271) has decreased.

According to ANOVA, the F statistics for both situations of including and excluding IND_j in the models are 10.905 and 14.884 respectively. They both are larger than the critical value (1.569) of the F distribution, obtained from the F distribution calculator for $\alpha = 0.05$ level of significance when degrees of freedom are 23 and 276.

The statistical test for the existence of a linear relationship between dependent variable and the independent variables is:

$$H_0 : \lambda_1 = \lambda_2 = \lambda_3 = \cdots = \lambda_k = 0$$

$H_1 :$ Not all $\lambda_i \left(i = 1, 2, \cdots, 24 \right)$ are zero

As the F statistic is in the rejection region, H_0 was rejected and H_1 was accepted. Since "p < 0.000", it can be concluded that there is strong evidence of TE having a linear regression relationship with any of the explanatory variables in the model with a probability of less than 0.1 percent of making an error in this conclusion.

IND_j Included

According to **Table 5**, the variables namely, GLO (0.000) and INC (0.000) are very highly significant at "p < 0.001". This implies that chances of making an error by assuming that these variables correlate with TE is less than 0.1 percent. Also both variables show a positive relationship with the dependent variable. FSIZE (0.136) is marginally significant at "0.10 < p < 0.15" and positively correlated whereas OWN (0.199) is insignificant and positively correlated. However, FAGE (0.889) is very highly insignificant. Since the main focus of this model is to test the

Table 3. Model summary[b]

R		R Square		Adjusted R Square		Std. Error of the Estimate	
IND. Included	IND. Excluded	IND. Included	IND. Excluded	IND. Included	IND. Excluded	IND. Included	IND. Excluded
0.645	0.418	0.416	0.175	0.367	0.152	0.062700	0.072548

[b]Dependent Variable: TE.

Table 4. ANOVA[b]

	Sum of Squares		df		Mean Square		F		Sig. (p-value)	
	Indj included	Indj excluded	Indj included	Indj excluded	Indj included	Indj excluded	Indj included	Indj excluded	Indj included	Indj excluded
Regression	8.784	5.357	23	8	0.382	0.670	10.905	14.884	0.000	0.000
Residual	9.666	13.093	276	291	0.035	0.045				
Total	18.450	18.450	299	299						

[b]Dependent Variable: TE.

Table 5. Coefficients[a]

Variable	IND, included		IND, excluded	
	B	Sig. (p-value)	B	Sig. (p-value)
(Constant)	0.526	0.000	0.567	0.000
GLO	0.000	0.000	0.000	0.000
OWN	0.054	0.199	-0.082	0.012
FSIZE	2.702E-11	0.136	-2.943E-11	0.035
FAGE	0.000	0.889	-0.002	0.565
INC	1.955	0.000	1.669	0.000
PC1	0.022	0.069	0.036	0.003
PC2	0.015	0.097	0.024	0.008
PC3	-0.038	0.005	-0.058	0.000

[a]Dependent Variable: TE.

significance of the correlation of FMT with TE, an examination of the correlation of the three PCs with TE becomes necessary. Both PC1 (0.069) and PC2 (0.097) are moderately significant at "$0.1 < p < 0.05$" and positively correlated with TE. Although PC3 (0.005) is highly significant, its relationship with TE is negative.

IND$_j$ Excluded

A separate regression was performed for the scenario in which 16 IND$_j$ variables were excluded from the TE model and the tables of Model Summary, ANOVA and Coefficients contained in the respective SPSS output have been reproduced in the respective tables for the scenario of IND excluded. The Adjusted R square (0.271) has decreased considerably. According to the output of the model that excluded the IND$_j$, the significance of explanatory variables, GLO, OWN, FSIZE, FAGE, INCPC1, PC2 and PC3 are 0.000, 0.012, 0.035, 0.565, 0.000, 0.003, 0.008 and 0.000 respectively. One crucial difference in this model is contrary to a priori expectations, both OWN and FSIZE have negative coefficients. However, PC1 and PC2 are highly significant at "$0.001 < p < 0.01$" which are only moderately significant according to the model in which IND$_j$ variables were included. However, after considering all attributes it is inferred that the reliability of the first TE model which included IND$_j$ variable is higher.

For all the cases, Mahalanobis distance and Cooks distance which indicate the impact of outliers had been saved in the SPSS data editor. The critical chi-square value of 51.179 at $\alpha = 0.001$ level of significance was taken as the critical value for the Mahalanobis distance. There were 19 cases which exceeded the critical value indicating that there were 19 multivariate outliers among the 300 cases. The critical value considered for the Cooks distance was one and only in two cases the critical value was exceeded.

The variables, PC1 and PC2 of the FMT which represent two important themes (dimensions), namely "process control technologies" and "production and quality control technologies" which together account for 53 percent of the variance in FMT is significant. The third PC which represents "general control technology" is insignificant and it only accounts for 13 percent of the variance in FMT. According to both TE models, the null hypotheses that PC1 and PC2 have no partial correlation with TE (i.e. $\lambda_6 = 0$, and $\lambda_7 = 0$) can be rejected. Moreover, the TE model which included IND_j (the more reliable model), the relationship both PC1 and PC2 have with the TE is moderately significant. Therefore, the alternative hypotheses can be accepted which means that FMT has a significant correlation with TE which is positive (since in both models λ_6 and λ_7 are positive). This leads to the acceptance of the research hypothesis: A high degree of FMT adoption enhances TE of the Manufacturing Industry of Malaysia.

CONCLUSIONS

The aim of this paper was to examine the impact of the degree of adoption of FMT on the technical efficiency of the manufacturing industry of Malaysia. The types of FMT considered were namely, Computer Numerical Control machine tools (CNC), Numerical Controlled Machine Tools (NC), Robotics (ROB), Programmable Logic Controllers (PLC), Automated Inspections (INS), Automated Storage and Retrieval Systems (ASR), Computer Aided Design (CAD) and Local Area Networks (LAN). In order to remove the effects of multicollinearity among the eight types of FMT they were substituted in the regression model with the three PCs. The FMT variables load onto PCs as follows: LAN, CAD, PLC and CNC load onto PC1; ASR, INS and ROB load onto PC2 and NC only loads onto PC3. The three PCs were labelled so that they best describe their respective constituents; PC1—"process control" technologies, PC2— "production and quality control" technologies and the PC3—"general control" technology.

The most important finding of the study is that both PC1 and PC2 show moderately significant and positive relationships with TE. This indicates that the increasing adoption of process control technologies and production and quality control technologies have direct impact on TE of the FMT intensively adopted sub sector of the manufacturing industry. In contrast, PC3 shows a very highly significant and negative relationship with TE. Since both PC1 and PC2 together account for greater variation (53 percent) and PC3 account for relatively smaller variation (12 percent) among the eight FMT, it can be concluded that a high degree of FMT adoption enhances TE of the manufacturing industry of Malaysia. This is consistent with the a priori expectations regarding FMT.

This shows that the degree of adoption of FMT has a positive relationship with embodied technological change which captures the effects of learning by doing (experience), advances in applied technology, managerial efficiency and industrial organisation which affords better methods and organisations that improve the efficiency of both new and old factor inputs. In other words, empirical findings of the present study postulate a correlation between the degree of FMT adoption and the above stated effects. Although the empirical findings suggest prevalence of a greater efficiency within the FMT intensively adopted industries, it does not indicate which specific effect is causing the efficiency. In the light of the deductions made in this section it can be stated that the higher the degree of adoption of FMT, the greater will be the ability to manufacture more output with the same factor inputs, in other words the ability to produce cost effectively.

In its usual call for future research, the authors recommend studies that investigate the relationship of investments in FMT rather than the degree of FMT adoption have with the TE of the manufacturing industry. It can be safely admitted that the accuracy of findings can be increased considerably by considering investments in FMT rather than the degree of adoption of FMT. Hence, it is proposed that future studies need be undertaken in collaboration with the statuary bodies established to oversee and facilitate the manufacturing industry which makes establishments obligatory to divulge investments made in FMT in order to evaluate the impact of investments in FMT on TE.

REFERENCES

1. R. Mahadevan, "Perspiration versus Inspiration: Lessons from a Rapidly Developing Economy," Journal of Asian Economics, Vol. 18, No. 2, 2007, pp. 331-347.doi:10.1016/j.asieco.2007.02.009

2. T. J. Coelli, D. S. P. Rao, C. J. O'Donnell and G. E. Battese, "An Introduction to Efficiency and Productivity Analysis," Springer, New York, 2005.

3. D. J. Aigner, C. A. K. Lovell and P. J. Schmidt, "Formulation and Estimation of Stochastic Frontier Production Models," Journal of Econometrics, Vol. 6, No. 1, 1977, pp. 21-37.doi:10.1016/0304-4076 77 90052-5

4. R. Mahadevan, "A Frontier Approach to Measuring Total Factor Productivity Growth in Singapore's Services Sector," Journal of Economic Studies, Vol. 29, No. 1, 2002, pp. 48-58. doi:10.1108/01443580210414111

5. R. Mahadevan and K. P. Kalirajan, "On Measuring Total Factor Productivity Growth in Singapore's Manufacturing Industries," Applied Economics

Letters, Vol. 6, No. 5, 1999, pp. 295-298. doi:10.1080/135048599353267

6.   J. Wind and A. Rangaswamy, "Customisation: The Next Revolution in Mass Customization," Journal of Interactive Marketing, Vol. 15, No. 1, 2001, pp. 13-31.

7.   G. Da Silveria and F. S. Fogliatto, "Effects of Technology Adoption on Mass Customisation Ability of Broad and Narrow Market Firms," Gestao & Producao, Vol. 12 No. 3, 2005, pp. 347-359. doi:10.1590/S0104-530X2005000300006

8.   P. Kumar and S. G. Deshmukh, "A Model for Flexible Supply Chain through Flexible Manufacturing," Global Journal of Flexible Systems Management, Vol. 7, No. 3-4, 2006, pp. 17-24.

9.   R. Sinha and C. H. Noble, "The Adoption of Radical Manufacturing Technologies and Firm Survival," Strategic Management Journal, Vol. 29, No. 9, 2008, pp. 943- 962.doi:10.1002/smj.687

10. Malaysian Industrial Development Authority (MIDA), "Malaysia-Investment in the Manufacturing Sector," MIDA, 2007.

11. C. Sun and K. L. Kalirajan, "Gauging the Sources of Growth of High-Tech and Low-Tech Industries: The Case of Korean Manufacturing," Blackwell Publishing Ltd., Hoboken, 2005.

12. X. Zhang and S. Zhang, "Technical Efficiency in China's Iron and Steel Industry: Evidence from the New Census Data," International Review of Applied Economics, Vol. 15, No. 2, 2001, pp. 199-211. doi:10.1080/02692170151137078

13. J. D. Lee, T. A. Kim and E. Heo, "Technological Progress versus Efficiency Gain in Manufacturing Sectors," Review of Development Economics, Vol. 2, No. 3, 1998, pp. 268-281. doi:10.1111/1467-9361.00041

14. G. Battese and S. S. Broca, "Functional Forms of Stochastic Frontier-Production Functions and Models for Technical Inefficiency Effects: A Comparative Study for Wheat Farmers in Pakistan," Journal of Productivity Analysis, Vol. 8, No. 4, 1997, pp. 395-414.

15. E. R. Brendt and C. J. Morrsison, "High Tech Capital Formation and Economic Performance in USA Manufacturing Industries: An Exploratory Analysis," Journal of Econometrics, Vol. 65, No. 1, 1995, pp. 9-43.

16. D. A. R. Dolage, A. B. Sade and M. A. Elsadig, "The Influence of Flexible Manufacturing Technology Adoption on Productivity of Malaysian Manufacturing Industry," Economic Modelling, Vol. 27, No. 1, 2010, pp. 395- 403. doi:10.1016/j.econmod.2009.10.005

17. L. H. Amato and C. H. Amato, "The Impact of High Tech Production

Techniques on Productivity and Profitability in Selected US Manufacturing Industries," Review of Industrial Organization, Vol. 16, No. 4, 2000, pp. 327-342.doi:10.1023/A:1007800121100

18. D. A. R. Dolage and A. B. Sade, "The Impact of Adoption of Flexible Manufacturing Technology on Price Cost Margin of Malaysian Manufacturing," Technology and Investment, Vol. 3, No. 1, 2012, pp. 26-35. doi:10.4236/ti.2012.31005

19. R. Mahadevan, "Is There a Real TFP Growth Measure for Malaysia's Manufacturing Industries?" ASEAN Economic Bulletin, Vol. 19, No 2, 2002, pp. 178-190.doi:10.1355/AE19-2E

20. K. P. Kalirajan and R. T. Shand, "Frontier Production Functions and Technical Efficiency Measures," Journal of Economic Surveys, Vol. 13, No. 2, 1999, pp. 149-172.doi:10.1111/1467-6419.00080

21. R. Mahadevan, "Assessing the Output and productivity Growth of Malaysia's Manufacturing Sector," Journal of Asian Economics, Vol. 12, No. 4, 2001, pp. 587-597.doi:10.1016/S1049-0078 01 00104-X

22. Q. N. Vu, "Technical Efficiency of Industrial StateOwned Enterprises in Vietnam," Asian Economic Journal, Vol. 17, No. 1, 2003, pp. 87-101.

23. H. Margono and S. C. Sharma, "Efficiency and Productivity Analysis of Indonesian Manufacturing Industries," Journal of Asian Economics, Vol. 17, No. 6, 2006, pp. 979-995.doi:10.1016/j.asieco.2006.09.004

24. Federation of Malaysian Manufacturers (FMM), "Directory -2007," FMM, 2007.

25. R. Mahadevan, "How Technically Efficient Are Singapore's Manufacturing Industries?" Applied Economics, Vol. 32, No. 15, 2000, pp. 2007-2014.doi:10.1080/00036840050155931

26. S. Chandrasiri, "Productivity and Technology in Sri Lankan Manufacturing Industry, Human Development in a Knowledge-Based Society: Sri Lankan Scene," The Sri Lanka Economic Association, Colombo, 2005.

27. R. H. McGuckin and M. L. Streitwieser, "The Effect of Technology Use on Productivity Growth," Centre of Economic Studies, 1996.

28. D. K. Round, "Price Cost Margins in Australian Manufacturing Industries, 1971-1972," University of Adelaide, Adelaide, 2001.

29. S. J. Coakes, L. Steed and J. Price, "SPSS 15.0, Analysis without Anguish," John Wiley & Sons Australia, Ltd., Melbourne, 2008.

30. A. Field, "Discovering Statistics Using SPSS," SAGE Publications Ltd., London, 2005.

Chapter 2

THE IMPACT OF ADOPTION OF FLEXIBLE MANUFACTURING TECHNOLOGY ON PRICE COST MARGIN OF MALAYSIAN MANUFACTURING INDUSTRY

Don Ajith Rohana Dolage[1], Abu Bakar Sade[2]

[1]International Graduate School of Business, University of South Australia, Adelaide, Australia

[2]Faculty of Management, Multimedia University, Cyberjaya, Malaysia

ABSTRACT

This paper explores the impact of the adoption of Flexible Manufacturing Technology (FMT) on the Malaysian Manufacturing Industry. The Principal Component Analysis has been adopted to extract the most appropriate underlying dimensions of FMT to use in place of the eight FMT variables owing to the potential multicollinearity. The study has been conducted within FMT intensively adopted 16 three-digit industries that encompass 50 five-digit industries covering the years 2000-2005. The results obtained from the two scenarios, one, including the industry fixed effects dummy variables and the other without these, are contrasted. It is established that the model that included the industry fixed effect dummy variables has a greater explanatory power. The two principal components that account for the greater variation in FMT show positive and moderately significant relationship with PCM. The study provides sufficient evidence to conclude that FMT has a direct and moderately significant relationship with PCM.

INTRODUCTION

It is widely accepted that FMT has the potential to bring about impressive cost savings while at the same time affording facilities to manufacture high quality products. In today's highly competitive world, a customer not only expects quality, reliability and competitive pricing but also customised products with timely delivery [1]. FMT refers to computer-based manufacturing technologies that make automation programmable rather than fixed. When fully developed, FMT organizations use the computer to integrate the functional areas of marketing, design, manufacturing, inventory control, materials handling and quality control into a continuous, sometimes unattended, round-the-clock operation. Actual installations today range from single, flexible manufacturing machining systems to "islands of automation" to fully computer-integrated manufacturing (CIM) operations [2]. The ability to alter the production of diverse products rapidly can provide manufacturers with a distinct competitive advantage. Companies adopting FMT rather than conventional manufacturing technology can react more quickly to market changes, provide certain economies, enhance customer satisfaction and increase profitability. FMT affords manufacturing flexibility that comprises three components: 1) the flexibility to produce a variety of products using the same machines and to produce the same products on different machines; 2) the flexibility to produce new products on existing machines; and 3) the flexibility of the machines to accommodate changes in the design of products. The ability to alter the production of diverse products rapidly can provide manufacturers with a distinct competitive advantage. Firms adopting FMT rather than conventional manufacturing technology can react more quickly to market changes, provide certain economies, enhance customer satisfaction and increase profitability. FMT is sometimes referred to in literature as high technology or high tech.

Nonetheless, the degree to which such potential benefits have been derived and reflected in the market performance has not been adequately explored. The review of literature shows empirical studies on FMT have been carried out in the following areas; types of flexibility, types of FMT, procedure bias on investment appraisal of FMT, operational problems, market structure and competitiveness. However, it is observed that the influence of FMT adoption on the competitiveness of the Malaysian manufacturing industry has not been adequately explored to give conclusive findings. The extant empirical studies have revealed that due to the potential operational problems of FMT implementation, derivation of potential benefits of FMT might be impeded [3-5]. MoreoverSlagmulder and Bruggemen [6] and Fine and Freund [7] showed that due to the "procedure bias on investment appraisal of FMT, investments in FMT" did not take place smoothly or effectively. Sinha and Noble [8] found strong support for the research hypothesis: Adoption of advanced manufacturing technologies (i.e. those that changed competitive dynamics in the industry) was a significant predictor of survival. Hence, there is a necessity for further studies that focus on measurement of FMT contribution towards the profitability in the manufacturing industry.

In the Malaysian manufacturing industry, FMT is widely adopted and has received the due attention from the industry policy makers. The Malaysian Industrial Development Authority (MIDA) [9] has recognised a number of promoted activities and products (for development and production) with regard to high technology establishments. The engagement in these activities will make them entitled to pioneer status or investment tax allowance under the promotion of Investment Act 1986. This includes FMT products such as, computer process control systems/equipment, process instrumentation and robotic equipment and computer numerical control machine tools. The Ninth Malaysia Plan which is aimed at achieving changes in the structure and improved performance of the economy with every economic sector achieving higher value added and total factor productivity. The "Thrust 1" of this plan states that, "Application of high technology and production of higher value added products will be given emphasis. Measures will be undertaken to migrate the electrical and electronics (E & E) industry towards high-technology and higher value added activities".

It is known that adherence to intensive regimes of contemporary manufacturing paradigms namely mass customisation, customerisation and instant customerisation is essential to stay competitive in the manufacturing industry. The studies show that mass customisation is the core manufacturing paradigm. The studies also showed that the crucial determinant of the successful implementation of mass customisation is the abundant use of Flexible manufacturing Technology (FMT) [10,11].

There are three widely adopted measures of competitiveness namely productivity technical efficiency and profitability. Both productivity and technical efficiency provide insights into particular facets of the competetiveness of the manufacturing industry but do not reflect the profitability within industries. Invariably these measures adopt value of "all assets" as that of the capital due to non availability of segregated data for land, building and machinery. Since this value is always very large and changes in the value of "all assets" due to machinery alone may not be discernible, a measure devoid of capital is needed to be incorporated in the study. Besides, sustainability of a particular industry depends mostly on the industry profitability and so a measure capable of evaluating this dimension of manufacturing industry should also be incorporated in any study that focuses on manufacturing industry competitiveness. When a new and cost efficient technology such as FMT emerges, the first to grab such technology are the top performing firms followed by the others. With the passage of time, once all the firms could gain access to the FMT, the advantage the top performing firms enjoyed purely due to adoption of FMT diminishes and as a result the profit margin of a unit product would squeeze. However, due to factors such as lowering of product prices, more customised products and larger assortments of models from the same products, new custromers would be attracted to the market. This results in a situation where the sales volume has increased without commensurate additional costs to the firm. Therefore, owing to the economies of scale, Industry-wide PCM could change in response to the degree of adoption of FMT. Hence, it is rational to anticipate that profit within a particular industry would increase despite squeezing unit profits.

Evidently, only a few studies have examined the impact of specific technologies on the industry level profitability using less aggregated data. Berndt and Morrison [12] examined the impact of high-tech investments on profitability using three profitability measures. The study found a significant and negative relationship between profitability and the share of high-tech capital in the total physical capital stock. Amato and Amato [13] investtigated the impact of high-tech investments on Price Cost Margin. Okada [14] adopts PCM to measure the product market competition on productivity in Japanese manufacturing industries. Hence, this study adopts price cost margin (PCM) which has the ability of addressing both these requirements.

This study established that prior findings of a negative relationship between profitability and high-tech might result from omitting industry fixed effect dummy variables (to account for the differences in technological opportunity among industries) from regression models.

Berndt and Morrison [12] found limited evidence of a positive relationship between profitability and the share of high-tech capital in the total physical capital. Amato and Amato [13] established a negative and significant relationship between the price cost margin and high-tech methods when industry fixed variable was excluded; relationship became insignificant when industry fixed variable was included. Hence there is a need for a separate study that incorporates differences in technologies considered, differences in the industries considered, differences in countries considered and the differences in the explanatory variables considered.

The main aim of this empirical study is to investigate the impact of FMT adoption on PCM in selected manufacturing industries of Malaysia. The study developed inclusion criteria to identify industries in which FMT is intensively adopted. As such 16 three-digit industries have been selected which consist of 50 five-digit industries. The study depends on the Annual Surveys of Manufacturing Industries (ASMI) during 2000-2005 and Economic census data maintained by the Department of Statistics Malaysia (DOS), for the secondary data [15]. The present study, unlike prior similar studies that have been conducted at the four-digit level, is carried out at the five-digit level.

Also the present study contributes to the previous studies by considering less aggregated data and also by considering PCM being computed using data obtained from ASMI.

This study also takes into consideration a higher number of specific FMT variables such as, Computer Numerical Control Machine Tools (CNC), Numerical Controlled Machine Tools (NC), Robotics (ROB), Programmable Logic Controllers (PLC), Automated Inspections (INS), Automated Storage and Retrieval Systems (ASR), Computer Aided Design (CAD) and Local Area Networks (LAN). In order to overcome multicollinearity among FMT variables, the study extracts three underlying dimensions of FMT by adopting Principal Component Analysis. They are namely; "process control" technologies, "production and quality control" technologies and "general control" technology. The study adopts a questionnaire survey to compute the degree of adoption of FMT among the selected 50 five-digit industries. The present study considers eight types of FMT instead of five specific technologies, evidently the maximum number considered in a prior study. The study covers only six years from 2000 to 2005 due to the limitation of data availability.

METHODOLOGY

The basic research hypothesis of the study is: A high degree of FMT adoption enhances PCM of the manufacturing industry of Malaysia.

Estimation of FMT

The methodology of this study comprises two stages. The first stage is to compute PCM for all the industries considered in the sample. The second stage is to identify the explanatory variables of PCM. The PCM approach to measuring profitability is widely adopted in the manufacturing industry. Siraz [16], Go, Kamerschen and Delorme [17], Yean [18] and Lee [19] adopted PCM to measure profitability. Collins and Preston [20] defined PCM as the percentage gross return (before taxes) on sales. Also PCM can be defined as the difference between price (p) and marginal cost (mc) as a fraction of price that is $(p-mc)/p$. The price cost margin is usually taken as an indicator of market power which means the ability of buyers or sellers to exert influence over the price or quantity of a good, service, or commodity exchanged in a market. Siraz [16] and Round [21] in their investigation to the influence of major market structure elements on performance in manufacturing industries in the United Kingdom and Australia respectively, used PCM to measure the competitiveness. The formula of PCM i.e. $(p-mc)/p$ can be expanded and presented in Equation 1 in order to measure PCM industry-wise:

PCM =

$$\frac{[\text{Value of sales} + \text{Inventory} - \text{Labour} - \text{Cost of materials}]}{\text{Value of Sales}} \qquad (1)$$

In a study, Go, Kamerschen and Delorme [17] used PCM to measure the profitability of Philippine manufacturing industry where the explanatory variables were sellers' concentration, capital output ratio, industry growth rate, import share, export share and degree of foreign participation. Lee [19] adopted PCM to measure the profitability in his study on Determinants of Cyclical Properties of PCM in Manufacturing Industries of the US. Yean [18] too used PCM to measure profitability in his empirical study on the Market Structure and Performance in Korean Manufacturing Industries. Berndt and Morrison [12] and Amato and Amato [13] used PCM to measure the impact of High-Tech Investments on competetiveness.

Factors Affecting PCM in the Manufacturing Industry

Growth Rate of Output (GRO): It is logical to expect that output growth can lead to higher factor productivity growth because it affords the "economies of scale" advantage. As output grows, capacity utilisation is bound to increase leading to a fall in the average cost. Therefore, it is rational to postulate a positive relationship between output growth and PCM. Yean [18], Lee [19], Round [21], Amato and Amato [13], Go, Kamerschen and Delorme [17] and Shiraz [16] considered GRO as an explanatory variable of PCM. GRO was measured as the increase in output between two consecutive years divided by the output of the previous year. The value of output for each MSIC five-digit industry was obtained from ASMI and deflated using GDPD.

Industry Concentration (CR4): Oligopoly theory explains that the higher the level of concentration, the more likely it is that the dominant firms will be able to collude, tacitly or expressly, to raise prices above the long run average costs [16]. Therefore, it is reasonable to include this variable in the PCM models as it can affect the profitability in a given industry. Industry concentration is widely expressed in terms of four-firm concentration ratio (CR4) i.e. sales of the four largest firms divided by the total sales in an industry. Lee [19], Yeaoon [18] and Amato and Amato [13], Shiraz [16], Go, Kamerschen and Delorme [17], Round [21] consider CR4 as an explanatory variable of PCM. In this study, CR4 was measured as the percentage of industry sales contributed by the four largest firms in each MSIC five digit industry for each year. The figures for CR4 are not published annually in Malaysia, so these figures were worked out using the data obtained from the Economic census data maintained by the DOS.

Capital Intensity (CAPIN): It is rational to expect that capital-intensive industries offer more scope for technological progress and learning by doing and thereby influencing their profitability. However, the efficiency of capital intensity is more likely to depend on the availability of efficient infrastructure. Therefore, the direction of the relationship is contentious. The researchers Amato and Amato [13], Lee [19], Go, Kamerschen and Delorme [17], Round [20] and Shiraz [16] considered CAPIN as an explanatory variable of PCM. A significant variability could be observed among the researchers in their approach to quantifying CAPIN. Sharma [3] defined it as Fixed Capital divided by Total Employment. Goldar and Kumari [22] used the ratio of Investment to Capital. Leung [23], McGuckin and Streitwieser [24] and Mahadevan [25] used the capital labour ratio, measured as the capital per employee. Amato and Amato [13] defined capital intensity as the value of shipment divided by capital stock. In the present study, CAPIN was measured by the value of assets divided by the total number of employees. The data required for this computation were obtained from ASMI.

Advertisement Intensity (ADV): Advertising helps to make a product and its characteristics known to the public and as a result industry sales are bound to increase. Yoon [18] in his PCM study in Korea has considered ADV as an explanatory variable having a positive relationship with PCM. In this study ADV was measured by the total expenditure on advertisement divided by the total sales. However, ADV is not published annually and so is computed using the industry-wise data obtained from the Economic Census data maintained by the DOS.

Export Intensity (EXP): Export Intensity can lead to higher productivity due to a number of reasons namely, opportunity for greater capacity utilisation, particularly in industries in which the minimum efficient size of plant is large relative to the domestic market; greater horizontal specialisation as each firm concentrates on a narrow range of products; and increasing familiarity with and absorption of new technologies [3]. Moreover, with foreign exchange earned from export growth, firms would have better access to imported inputs and new technology the effects of which can evidently enhance profitability. Yean [18] considered EXP as an explanatory variable of PCM. In the present study, EXP was measured as the ratio of the value of manufactured exports to the total sales. The "value of export" data were obtained from the unpublished Economic Census data maintained by the DOS.

The eight types of FMT considered in this study are given below:

- Computer Numerical Control Machine Tools (CNC): Measured as the percentage of firms in each MSIC five-digit industry using microprocessor based numerical control technologies, referred to as computer numerical control machine tools.

- Numerical Controlled Machine Tools (NC): Measured as the percentage of firms in each MSIC fivedigit industry using numerical controlled machine tools.

- Robotics (ROB): Measured as the percentage of firms in each MSIC five-digit industry using robotics.

- Programmable Logic Controllers (PLC): Measured as the percentage of firms in each MSIC five-digit industry using programmable logic controllers.

- Automated Inspections (INS): Measured as the percentage of firms in each MSIC five-digit industry using automated sensor-based inspection, either during the production process or final product.

- Automated Storage and Retrieval Systems (ASR): Measured as the percentage of firms in each MSIC five-digit industry using automated storage and retrieval systems.

- Computer Aided Design (CAD): Measured as the percentage of firms in each MSIC fivedigit industry using computer aided design to control manufacturing machinery.

- Local Area Networks (LAN): Measured as the percentage of firms in each MSIC five-digit industry using local area networks.

Industry Fixed Effects Dummy Variables (IND_j): The study involved 50 five-digit industries included in 16 three-digit industries. It is logical to assume that industry characteristics among these 16 three-digit industries can be diverse and need to be captured by a variable. Therefore, 16 dummy variables (IND_j) were incorporated into the PCM model to capture industry fixed effects. However, most contemporary researchers have not considered industry fixed effects in the PCM model. While Mahadevan [26] and Amato & Amato [13] have incorporated industry fixed effect dummy variables, Goldar & Kumari [21], Yean [27], Leung [23] and Sharma [3] in their similar studies, have not made any reference to industry fixed effects, let alone considering them in their models.

The model representing the relationship among PCM, explanatory variables and FMT variables can be specified and shown below in Equation (2).

Price Cost Margin Model

PCM =

$\beta_0 + \beta_1 GRO + \beta_2 CR4 + \beta_3 CAPIN + \beta_4 ADV$

$\quad + \beta_5 EXP + \beta_6 CNC + \beta_7 NC + \beta_8 ROB + \beta_9 PLC$

$\quad + \beta_{10} INS + \beta_{11} ASR + \beta_{12} CAD + \beta_{13} LAN$

$\quad + \sum \beta_{13+j} IND_j + \mu$
(2)

DATA AND ESTIMATION

Inclusion Criteria

According to the Malaysian Standard Industrial Classification 2000 (MSIC 2000), there are 53 three-digit industries [15]. In order to obtain a rational outcome, the study needs to be conducted only within industries in which FMT is intensively adopted. On account of this, inclusion criteria were formulated in an effort to select FMT intensively adopted MSIC three-digit industries for the sample, which is shown below:

Industries with high "capital/labour" ratio;

Industries in which product features need to be varied as a marketing strategy;

Industries in which demand for products are susceptible to fluctuation.

The "capital/labour" ratios for all 53 three-digit Industries were tabulated after computing them as the ratio of total value of assets to the total number of employees. The questionnaire carried two separate questions: one question sought the response of each firm regarding the degree to which product features need to be varied as a marketing strategy; another sought the response of each firm regarding the degree to which demand for products are susceptible to fluctuation. Using these responses that were indicated on a Likert scale, two separate values were computed for either criterion for each industry. After careful evaluation of the values obtained for the three criteria, a sample of 16 MSIC three-digit industries which together comprised 50 five-digit industries was selected.

Data

Primary Data

The data that indicate the degree of adoption of FMT is not published by any organisation in Malaysia. Hence, a questionnaire survey was conducted to gather information necessary to compute the percentage of establishments adopting each specific type of FMT in a given year, within a given MSIC five-digit industry. The questionnaires were sent to all the establishments listed under the 50 MSIC five-digit industries, in the directory of Federation of Malaysian Manufacturers (FMM) [28].

Secondary Data

In order to compute PCM, industry-wise data is required for output, intermediate input, capital input and labour input. The closest indicators for these values were obtained from Table 3 of the ASMI published for the years 2000 through 2005 by the DOS of Malaysia. The variables GRO and CAPIN were computed using the data obtained from Table 3. EXP, CR4 and ADV were computed using the data obtained from the Economic Censes conducted by the DOS Malaysia.

EMPIRICAL RESULTS

Multicollinearity of FMT

Since only FMT intensively used industries were included in the sample, naturally some similarity in the sequence and characteristics of the production processes could be expected even amongst different five-digit industries. Hence, there could be a tendency for a similarity in the technology adopted amongst these industries. Due to the similarities in technologies, a high prevalence of multicollinearity among the eight types of FMT could be anticipated. In this study, bivariate Pearson productmoment correlation analysis has been conducted using SPSS to test for multicollinearity amongst FMT. The output that reveals potential multicollinearity among FMT variables is displayed in Table 1. According to Coakes, Steed and Price [29] and Field [30], when a considerable number of correlations are exceeding 0.3, the matrix is suitable for Principal Component Analysis (PCA).

PCA was adopted using SPSS in order to obtain underlying dimensions (Principal Components) of FMT as a remedy for multicollinearity. As per both standard methods (i.e. screen test and eigen values greater than one) of extracting the optimal number of components, three Principal Components (PCs) were extracted that account for 67 percent of the variation in the FMT. According to Table 2, the loadings of variables onto three PCs obtained from both types of rotations (Orthogonal and Oblique) are quite similar. Hence, due to simplicity, PCs obtained from orthogonal rotation was used in the rest of the analysis.

Once the most appropriate type of rotation and the resultant PCs were decided, the variables loading onto each of these PCs were examined as the next step. An examination of the component loadings depicted in Table 2 indicates that LAN, CAD, PLC and CNC load onto PC1; ASR, INS and ROB load onto PC2 while only NC loads onto PC3. Usually it is difficult to give clear cut themes or names to PCs that only relate to or encompass particular variables that are loading onto it. Hence, only the best possible names have been assigned to the PCs extracted from this analysis. The technologies LAN, CAD, PLC and CNC are used in the manufacturing set up as process control technologies. Since these load onto PC1, they can be named as "process control" technologies.

Table 1: Correlations among FMT

		CNC	NC	ROB	PLC	INS	ASR	CAD	LAN
CNC	Pearson Correlation	1.000	0.160**	0.351**	0.634**	0.307**	0.237**	0.248**	0.322**
	Sig. (2-tailed)		0.005	0.000	0.000	0.000	0.000	0.000	0.000
	N	300	300	300	300	300	300	300	300
NC	Pearson Correlation	0.160**	1.000	0.012	0.164**	0.177**	0.126*	0.141*	0.171**
	Sig. (2-tailed)	0.005		0.836	0.005	0.002	0.030	0.014	0.003
ROB	Pearson Correlation	0.351**	0.012	1.000	0.368**	0.250**	0.427**	0.391**	0.236**
	Sig. (2-tailed)	0.000	0.836		0.000	0.000	0.000	0.000	0.000
PLC	Pearson Correlation	0.634**	0.164**	0.368**	1.000	0.302**	0.257**	0.394**	0.380**
	Sig. (2-tailed)	0.000	0.005	0.000		0.000	0.000	0.000	0.000
INS	Pearson Correlation	0.307**	0.177**	0.250**	0.302**	1.000	0.564**	0.115*	0.186**
	Sig. (2-tailed)	0.000	0.002	0.000	0.000		0.000	0.046	0.001
ASR	Pearson Correlation	0.237**	0.126*	0.427**	0.257**	0.564**	1.000	0.308**	0.129*
	Sig. (2-tailed)	0.000	0.030	0.000	0.000	0.000		0.000	0.025
CAD	Pearson Correlation	0.248**	0.141*	0.391**	0.394**	0.115*	0.308**	1.000	0.609**
	Sig. (2-tailed)	0.000	0.014	0.000	0.000	0.046	0.000		0.000
LAN	Pearson Correlation	0.322**	0.171**	0.236**	0.380**	0.186**	0.129*	0.609**	1.000
	Sig. (2-tailed)	0.000	0.003	0.000	0.000	0.001	0.025	0.000	

**Correlation is significant at the 0.01 level (2-tailed) *Correlation is significant at the 0.05 level (2-tailed)

Table 2: Comparison of components obtained from two types of rotations

	Component One			Component Two			Component Three		
	Orthogonal	Oblique		Orthogonal	Oblique		Orthogonal	Oblique	
		Pattern	Structure		Pattern	Structure		Pattern	Structure
LAN	0.816	0.861	0.811						
CAD	0.816	0.841	0.801						
PLC	0.666	0.640	0.722			0.445			
CNC	0.555	0.517	0.621			0.467			
ASR				0.845	0.858	0.851			
INS				0.816	0.844	0.826			
ROB	0.477	0.412	0.542	0.526	0.460	0.573			
NC							0.883	0.871	0.883

The technologies ASR, INS and ROB load onto PC2, so can be named as "production and quality control" technologies. PC3 has only one variable i.e. NC, loading onto it so can be called the "general control" technology.

As the next step, the eight FMT variables were substituted with the three PCs namely, PC1, PC2 and PC3. Therefore, the PCM model was reformulated and presented in Equation (3) below (the changes in PCs are considered here to be consistent with PCM which too is measured as a change):

PCM =

$\beta_0 + \beta_1 GRO + \beta_2 CR4 + \beta_3 CAPIN + \beta_4 ADV$

$+ \beta_5 EXP + \beta_6 PC1 + \beta_7 PC2 + \beta_8 PC3$

$+ \sum \beta_{8+j} IND_j + \mu$

(3)

Multiple Regression Analysis of PCM

As described, the model contains a set of 16 industry fixed dummy variables (IND_j) to account for the differences of technological opportunity among industries. Although it is theoretically desirable to include IND_j, the consequent impact of adding these 16 extra variables needs to be examined by comparing and contrasting the results obtained without considering the IND_j in the model. A separate regression was performed for this scenario and the tables of Model Summary, ANOVA and Coefficients were obtained. In order to facilitate easy comparison of the results, the tables of output obtained from regression analysis for the two situations, one with the IND_j included and the other without the IND_j have been combined into one.

The tables of Model Summary, ANOVA and Coefficients contained in the SPSS output have been reproduced in Tables 3-5 respectively. According to the Model Summary, Adjusted R square is 0.367. This indicates that explanatory variables in the model have the ability to explain 36.7 percent of the variance in PCM.

According to ANOVA, the F statistics (8.534) is much larger than the critical values of the F distribution, obtainned from the F distribution calculator for $\alpha = 0.05$ level of significance for the situations of Ind_j excluded (1.97) and Ind_j excluded (1.57).

H_0: $\beta_1 + \beta_2 + \beta_3 = \ldots \beta_k = 0$ H_1: Not all the β_i (i= 1,2,…,24) are zero.

As the F statistic is in the rejection region, H_0 was rejected and H_1 was accepted. Since "p < 0.000" it can be concluded that there is strong evidence of PCM having a linear regression relationship with any of the explanatory variables in the model with a probability of less than 0.1 percent of making an error in this conclusion.

IND_j Included

According to Table 5, CAPIN (0.000) is very highly significant at "p < 0.001". This implies that the chances of making an error by assuming that this variable correlates with PCM is less than 0.1 percent. While GRO (0.002) is highly significant at "0.001 < p < 0.01", EXP (0.038) is significant at "0.01 < p < 0.05". All variables CAPIN, GRO and EXP are positively correlated. While ADV (0.072) is moderately significant at "0.05 < p < 0.1" and positively correlated, CR4 (0.13) is marginally significant at "0.1 < p < 0.15" and negatively correlated. Since the main focus of the study is to test the significance of the correlation of FMT with PCM, an examination of the correlation of the three PCs with PCM becomes necessary. Both PC1 (0.069) and PC2 (0.098) are moderately significant at "0.1 < p < 0.05" and display positive correlation with PCM. However, PC3 is highly insignificant and shows a negative relationship.

IND$_j$ Excluded

According to the output of the model that excluded the IND$_j$, the significance of explanatory variables, GRO, CR4, CAPIN, ADV, EXP, PC1, PC2 and PC3 are, 0.001, 0.001, 0.632, 0.000, 0.007, 0.125, 0.032, 1.000 and 0.671 respectively.

Table 3: Model summary[a]

	R		R Square		Adjusted R Square		Std. Error of the Estimate	
	IND$_j$ Included	IND$_j$ Excluded	IND$_j$ Included	IND$_j$ Excluded	IND$_j$ Included	IND$_j$ Excluded	IND$_j$ Included	IND$_j$ Excluded
	0.645	0.418	0.416	0.175	0.367	0.152	0.062700	0.072548

[a]Dependent Variable PCM

Table 4: Anova[a]

	Sum of Squares		df		Mean Square		F		Sig (p-value)	
	Ind$_j$ Included	Ind$_j$ Excluded	Ind$_j$ Included	Ind$_j$ Excluded	Ind$_j$ Included	Ind$_j$ Excluded	Ind$_j$ Included	Ind$_j$ Excluded	Ind$_j$ Included	Ind$_j$ Excluded
Regression	0.772	0.325	23	8	0.034	0.041	8.534	7.722	0.000[b]	0.000
Residual	1.085	1.532	276	291	0.004					
Total	1.857	1.857	299	299						

[a]Dependent Variable PCM

Table 5: Coefficients[b]

Variable	IND$_j$ Included		IND$_j$ Excluded	
	B	Sig (p-value)	B	Sig (p-value)
(Constant)	0.159	0.000	0.196	0.000
GRO	0.010	0.002	0.012	0.001
CR4	−0.030	0.130	−0.008	0.632
CAPIN	3.415E−7	0.000	3.947E−7	0.000
ADV	1.239	0.072	−0.047	0.007
EXP	0.036	0.038	1.170	0.125
PC1	0.007	0.069	0.009	0.032
PC2	0.008	0.098	3.031E−7	1.000
PC3	−0.001	0.739	−0.002	0.671

[b]Dependent Variable PCM

The differences in this model are: CR4 is highly significant (marginally significant in the first PCM model), ADV is moderately significant (marginally significant in the first PCM model). While PC1 is significant (moderately significant in the first PCM model), PC2 is very highly significant (moderately significant in the first PCM model). Similar to the first PCM model, PC3 is negative and very highly insignificant. Hence, it is inferred that the reliability of PCM model which included IND_j variable is much higher.

The variables, PC1 and PC2 of the FMT which represent two important themes (dimensions), namely "process control technologies" and "production and quality control technologies" which together account for 53 percent of the variance in FMT is significant. PC3 which represents "general control technology" is insignificant and it only accounts for 12 percent of the variance in FMT. According to the first PCM model, the null hypotheses that PC1 and PC2 have no partial correlation with PCM (i.e. $\beta_6 = 0$ and $\beta_7 = 0$) can be rejected. However, according to the second PCM, model only the null hypotheses that PC1 has no partial correlation with PCM (i.e. $\beta_6 = 0$) can be rejected. Therefore, the alternative hypothesis can be accepted which means that CP1 has a significant correlation with PCM which is positive (since λ_6 is positive). However, in the second PCM model, the null hypothesis that PC2 has no partial correlation with PCM has to be accepted (i.e. $\beta_7 = 0$). However, as explained above, according to the first model, PC2 has a positive partial correlation with PCM. Since, as was shown, according to all indicators the predictability of the first PCM model is greater, it can be concluded that PC2 has a positive partial correlation with PCM. This leads to the acceptance of research Hypothesis; a high degree of FMT adoption enhances PCM of the Manufacturing Industry of Malaysia.

For all the cases, Mahalanobis distance and Cooks distance which indicate the impacts of outliers had been saved in the SPSS data editor. The critical chi-square value of 51.179 at $\alpha = 0.001$ level of significance was taken as the critical value for the Mahalanobis distance. There were 12 cases which exceeded the critical value indicating that there were 12 multivariate outliers among the 300 cases. The critical value considered for the Cooks distance was one and only in two cases the critical value was exceeded.

CONCLUSIONS

The main purpose of this study was to evaluate the impact of the degree of FMT adoption on the profitability of the manufacturing industry of Malaysia. The types of FMT considered are namely, Computer Numerical Control Machine Tools (CNC), Numerical Controlled Machine Tools (NC), Robotics (ROB), Programmable Logic Controllers (PLC), Automated Inspections

(INS), Automated Storage and Retrieval Systems (ASR), Computer Aided Design (CAD) and Local Area Networks (LAN). On account of the potential multicollinearity among the eight types of FMT, three PCs were extracted to substitute the individual FMT variables. The FMT variables load onto PCs as follows: LAN, CAD, PLC and CNC load onto PC1; ASR, INS and ROB load onto PC2 and NC only loads onto PC3. The three PCs were labelled so that they best describe the respective constituents; PC1- "process control" technologies, PC2-"production and quality control" technologies and the PC3- "general control" technology.

Separate PCM models were solved for the two scenarios: one included the IND_j; and the other excluded the IND_j. One of the important contributions of the present study is that it reveals regarding the models specified to study the impact of FMT, that by including an industry fixed dummy variable to account for the differences in technological opportunity among different industries, the credibility of the models can be increased considerably.

The most significant finding of the study is that both PC1 and PC2 show significant and positive correlation with PCM. In contrast, PC3 shows a highly insignificant and negative relationship with PCM. This indicates that the degree of adoption of process control technologies and production and quality control technologies have a positive influence on PCM of the FMT intensively adopted sub sector of the manufacturing industry. Since both PC1 and PC2 together account for (53 percent) greater variation and PC3 account for (12 percent) relatively smaller variation among the eight FMT and it can be concluded that a high degree of FMT adoption enhances PCM of the Manufacturing Industry of Malaysia. This is consistent with the a priori expectations regarding the investments in FMT.

The present study has been made different from previous studies by incorporating most of the aspects overlooked by other studies. As a result of this, the findings of the present study afford a more realistic picture of the relationship between PCM and the degree of FMT adoption in the manufacturing industry of Malaysia. According to Mohamed, Mohamed, Abdullah and Jalil [31], findings of the empirical estimation appear to lend some support to the idea that in the Malaysian manufacturing industry, production flexibility is one of the forces that explains the lasting presence of small firms alongside their larger counterparts in the market.

In its customary call for future research, the authors recommend studies that investigate the relationship of investments in FMT rather than the degree of adoption of FMT have with the PCM of the manufacturing industry. It can be safely admitted that the accuracy of findings can be increased further by taking precautionary measures to elicit information regarding the value of investments. Hence, it is proposed that future studies need be undertaken in collaboration with the industry monitoring institutes of the state sector that makes establishments obligatory to divulge investments made in FMT to evaluate the impact of investments in FMT on PCM.

REFERENCES

1. P. Kumar and S. G. Deshmukh, "A Model for Flexible Supply Chain through Flexible Manufacturing," Global Journal of Flexible Systems Management, Vol. 7, No. 3, 2006, pp. 17-24. doi:10.5465/AMR.1988.4307510

2. P. L. Nemetz and L. W. Fry, "Flexible Manufacturing Organizations: Implications for Strategy Formulation and Organization Design," Academy of Management Review, Vol. 13, No. 4, 1988, pp. 627-638.

3. O. P. Sharma, "Managing Flexibility in Manufacturing and Operations," Journal of Management Research, Vol. 2, No.3, 2002, pp. 147-163.

4. H. F. Gale, T. R. Wojan and J. C. Olmsted, "Skills, Flexible Manufacturing Technology and Work Organization," Industrial Relations, Vol. 41, No. 1, 2002, pp. 48-79.doi:10.1111/1468-232X.00235

5. L. H. Roller and M. M. Tombak, "Competition and Investment in Flexible Technologies," Management Science, Vol. 39, No. 1, 1993, pp. 107-114. doi:10.1287/mnsc.39.1.107

6. R. Slagmulder and W. Bruggeman, "Investment Justification of Flexible Manufacturing Technologies: Inferences from Field Research," International Journal of Operations & Production Management, Vol. 12, No. 7-8, 1992, pp. 168-186.doi:10.1108/EUM0000000001310

7. C. H. Fine and R. M. Freund, "Optimal Investment in Product-Flexible Manufacturing," Management Science, Vol. 36, No. 4, 1990, pp. 449-466. doi:10.1287/mnsc.36.4.449

8. R. Sinha and C. H. Noble, "The Adoption of Radical Manufacturing Technologies and Firm Survival," Strategic Management Journal, Vol. 29, No. 9, 2008, pp. 943-962.doi:10.1002/smj.687

9. Malaysian Industrial Development Authority (MIDA), "Malaysia-Investment in the Manufacturing Sector," MIDA, 2007.

10. J. Wind and A. Rangaswamy, "Customisation: The Next Revolution in Mass Customization," Journal of Interactive Marketing, Vol. 15, No. 1, 2001, pp. 13-31.doi:10.1002/1520-6653(200124)15:1<13::AID-DIR1001>3.0.CO;2-#

11. G. Da Silveria and F. S. Fogliatto, "Effects of Technology Adoption on

Mass Customisation Ability of Broad and Narrow Market Firms," Gestao & Producao, Vol. 12, No. 3, 2005, pp. 347-359. doi:10.1590/S0104-530X2005000300006

12. E. R. Brendt and C. J. Morrsison, "High Tech Capital Formation and Economic Performance in USA Manufacturing Industries: An Exploratory Analysis," Journal of Econometrics, Vol. 65, 1995, pp. 9-43. doi:10.1016/0304-4076(94)01596-R

13. L. H. Amato and C. H. Amato, "The Impact of High Tech Production Techniques on Productivity and Profitability in Selected U.S. Manufacturing Industries," Review of Industrial Organization, Vol. 16, No. 4, 2000, pp. 327- 342.doi:10.1023/A:1007800121100

14. Y. Okada, "Competition and Productivity in Japanese Manufacturing Industries," Journal of Japanese and International Economies, Vol. 19, No. 4, 2005, pp. 586-616.doi:10.1016/j.jjie.2005.10.003

15. Department of Statistics Malaysia, DOS, "Annual Survey of Manufacturing Industries," DOS, Malaysia, 2000- 2005.

16. J. K. Shiraz, "Market Structure and Price Cost Margins in United Kingdom Manufacturing Industries," The Review of Economics and Statistics, Vol. 56, No. 1, 1973, pp. 67- 76.

17. G. L. Go, D. R. Kamerschen and C. D. Delorme, "Market Structure and Price-Cost Margins in Philippine Manufacturing Industries," Applied Economics, Vol. 31, No. 7, 1999, pp. 857-864. doi:10.1080/000368499323814

18. T. S. Yean, "Determinants of Productivity Growth in the Malaysian Manufacturing Sector," ASEAN Economic Bulletin, Vol. 13, No. 3, 1997, pp. 333-343. doi:10.1355/AE13-3D

19. I. K. Lee, "Determinants of Cyclical Properties of the Price Cost Margin in US Manufacturing Industries," International Economic Journal, Vol. 18, No. 3, 2004, pp. 353-364.

20. N. H. Collins and L. E. Preston, "Concentration and Price Cost Margins in Manufacturing Industries," University of California Press, Berkeley, 1968.

21. D. K Round, "Price Cost Margins in Australian Manufacturing Industries," University of Adelaide Press, Adelaide, 1971-1972.

22. B. Goldar and A. Kumari, "Import Liberalisation and Productivity Growth in Indian Manufacturing Industries in the 1990s," The Developing Economies, Vol. 41, No. 4, 2003, pp. 436-460.

23. H. M. Leung, "Total Factor Productivity Growth in Singapore's

Manufacturing Industries," Applied Economics Letters, Vol. 4, No. 8, 1997, pp. 525-528. doi:10.1080/758536639

24.   R. H. McGuckin and M. L. Streitwieser, "The Effect of Technology Use on Productivity Growth," Economics of Innovation and New Technology, Vol. 7, No. 1, 1998, pp. 1-26.

25.   R. Mahadevan, "Assessing the Output and Productivity Growth of Malaysia's Manufacturing Sector," Journal of Asian Economics, Vol. 12, No. 4, 2001, pp. 587-597.doi:10.1016/S1049-0078(01)00104-X

26.   R. Mahadevan, "A Frontier Approach to Measuring Total Factor Productivity Growth in Singapore's Services Sector," Journal of Economic Studies, Vol. 29, No. 1, 2002, pp. 48-58. doi:10.1108/01443580210414111

27.   S. Yoon, "A Note on the Market Structure and Performance in Korean Manufacturing Industries," Journal of Policy Modeling, Vol. 26, No. 6, 2004, pp. 733-746.doi:10.1016/j.jpolmod.2004.03.005

28.   Directory of Federation of Malaysian Manufacturers 2007, Federation of Malaysian Manufacturers, Malaysia.

29.   S. J. Coakes, L. Steed and J. Price, "SPSS 15.0, Analysis without Anguish," John Wiley & Sons, Hoboken, 2008.

30.   A. Field, "Discovering Statistics Using SPSS," SAGE Publications Ltd, London, 2005.

31.   N. Mohamed, N. G Mohamed, A. Z Abdullah and S. A Jalil, "Flexibility and Small Firm's Survival: Further Evidence from Malaysian Manufacturing," Applied Economics Letters, Vol. 14, No. 12, 2007, pp. 931-934. doi:10.1080/13504850600706065

Chapter 3

FLEXIBLE MANUFACTURING SYSTEM SELECTION USING PREFERENCE RANKING METHODS : A COMPARATIVE STUDY

Prasenjit Chatterjee[a], Shankar Chakraborty[b]

[a]Mechanical Engineering Department, MCKV Institute of Engineering, Howrah - 711 204, West Bengal, India

[b]Department of Production Engineering, Jadavpur University, Kolkata - 700 032, INDIA

ABSTRACT

Flexible manufacturing systems (FMSs) offer opportunities for the manufacturers to improve their technology, competitiveness and profitability through a highly efficient and focused approach to manufacturing effectiveness. Justification, evaluation and selection of FMSs have now been receiving significant attention in the manufacturing environment. Evaluating alternative FMSs in the presence of multiple conflicting criteria and performance measures is often a difficult task for the decision maker. Preference ranking tools are special types of multi-criteria decision-making methods in which the decision maker's preferences on criteria are aggregated together to arrive at the final evaluation and selection of the alternatives. This paper deals with the application of six most potential preference ranking methods for selecting the best FMS for a given manufacturing organization. It is observed that although the performances of these six methods are almost similar, ORESTE (Organization, Rangement Et Synthese De Donnes Relationnelles) method slightly outperforms the others. These methods use some preference function or utility value or Besson ranking of criteria and alternatives, to indicate how much an alternative is preferred to the others. Most of these methods need quantification of criteria weights or different preference parameters, but ORESTE method, being an ordinal outranking approach, only requires ordinal data and attribute rankings according to their importance. Therefore, it is particularly applicable to those situations where the decision maker is unable to provide crisp evaluation data and attribute weights.

INTRODUCTION

In today's global competitive environment, manufacturing organizations need to be more flexible, adaptive and responsive to changes to produce a variety of products in a short time at the minimum cost. High competition, technological advancements and continuous change in customers' demands have made the manufacturing organizations realize the importance of manufacturing flexibility, which can only be achieved through the adoption and augmentation of flexible manufacturing systems (FMSs) (O'Grady and Menon, 1986). The FMS has been a focal theme in the manufacturing related research since the early 1970s. A high level of flexibility can enable the manufacturing organizations to provide faster response to market changes, while maintaining increased product quality standards. The FMS can present opportunities for the manufacturing organizations to improve their technology, competitiveness and profitability through a highly efficient and focused approach to manufacturing effectiveness. However, implementation of FMS is extremely capital-intensive. Prior to its implementation, a careful analysis regarding its feasibility and performance is needed, in which the impact of various long- and medium-term managerial, social and economic factors associated with FMS adoption can be assessed.

An FMS consists of computerized numerical control machines and/or robots, physically linked by a conveyance network to move parts and/or tools, and an overall effective computer control to create an integrated system. The reason the FMS is called 'flexible' is that it is capable of processing a variety of different part styles simultaneously at various workstations, and the mix of part styles and production quantities can be easily adjusted in response to changing demand patterns. Potential benefits of an FMS implementation include reduced inventory levels, manufacturing lead times, floor space, and setup and labor costs, in addition to higher flexibility, quality, speed of response and a longer useful life of the equipment over successive generations of products. An FMS can manufacture a wide range of products in batch sizes from one to thousands. As an FMS implementation involves a huge capital investment, the selection of the most appropriate FMS from a set of candidate configurations requires extensive analysis and evaluation. Thus, the selection of an FMS requires trading off among various performance attributes of FMS alternatives so as to achieve the maximum possible benefits from its implementation. Among these attributes, some are quite difficult to quantify, some are conflicting in nature and some are to be balanced against each other while taking into account the preferences of the decision maker in the manufacturing organizations. High quality management is not enough for dealing with this type of complex and ill-structured decision-making problem.

Hence, there is a need for simple mathematical tools to help the decision maker to select the most suitable FMS for a given industrial application. As the evaluation and selection of the most appropriate FMS for an industrial application involves different conflicting criteria, it is a unique example where multi-criteria decision-making (MCDM) methods can be successfully applied. These FMS selection criteria can be categorized as objective and subjective attributes or beneficial and non-beneficial attributes. Objective attributes can be numerically defined, e.g. capital and maintenance cost of an FMS, floor space required for an FMS, reduction in work-in-process (WIP) etc. On the other hand, subjective attributes are qualitative in nature, like increase in market response, improvement in quality etc. Beneficial attributes are those whose higher values are desirable (reduction in WIP, improvement in quality etc.) and non-beneficial attributes are those whose lower values are always preferable (capital and maintenance cost of an FMS, floor space required for an FMS etc.).

The past researchers have already successfully applied different mathematical techniques, like analytic hierarchy process (Wabalackis, 1988; Chan et al., 2000; Bayazit, 2005), digraph and matrix approach (Rao, 2006; Rao and Parnichkun, 2009), compromise ranking method (Rao, 2009), artificial neural network (Bhattacharya et al., 2007), data envelopment analysis (Shang & Sueyoshi, 1995; Sarkis, 1997; Talluri et al., 2000; Karsak, 2008; Liu, 2008), technique for order preference by similarity to ideal solution (TOPSIS) (Karsak, 2002; Rao, 2008), Euclidean distance-based integrated approach (Rao & Singh, 2011), axiomatic design method (Kulak & Kahraman, 2005), multi-objective programming method (Lotfi, 1995; Karsak & Kuzgunkaya, 2002), fuzzy decision algorithm (Karsak, 2002; Karsak & Tolga, 2001; Mehrabad & Anvari, 2010), multi-attribute value function (Borenstein, 1998) etc. to solve the FMS selection problems. However, there is still a need to search for some more efficient and accurate methods that can give more precise ranking of FMS alternatives. Preference ranking-based methods are observed to have immense potential to deal with complex decision-making problems in conflicting situations. This paper mainly focuses on the applications of six preference ranking methods for effectively solving the FMS selection problems. The ranking performance of these six methods is also compared.

Section 2 of this paper deals with the detailed mathematical formulations of the six considered preference ranking methods. In Section 3, a real time FMS selection problem is solved using these preference ranking methods. A comparative performance study between these methods is shown in Section 4. The results and discussions are presented in Section 5. The applicability of these methods is provided in Section 6 and Section 7 concludes the paper.

PREFERENCE RANKING METHODS

Preference ranking methods are special types of MCDM approaches in which the decision maker's preferences and preferences on criteria are aggregated together to reach the final evaluation and decision about the alternatives considering all the selection attributes. These methods require information on the preferences among the instances of an attribute and the preferences across the existing attributes. The decision maker may express or define a ranking for the attributes as importance/weights. In classical preference ranking methods, the decision maker judges two alternatives based on the notion that one alternative is preferred to another or the two alternatives are indifferent or the decision maker is unable to compare them. According to these cases, three binary relations are defined, i.e. a) the strict preference relation (P), b) the indifference relation (I), and c) the incomparability relation (R). Thus, a preference structure on a set of alternatives (X) is defined as a triplet (P, I, J) of binary relation (Ovchinnikov & Roubens, 1992).

In these methods, preferences are usually incorporated in the decision-making process by assigning a preference function. The starting point is the decision matrix, which presents the performance of each alternative with respect to each criterion. Using the data of the decision matrix, the alternatives are pairwise compared with respect to every single criterion, and the concordance and discordance indices are determined. The results are expressed by the preference functions, which are calculated for each pair of alternatives and can range from 0 to 1. A 0 value signifies that there is no difference between the pair of alternatives, whereas, 1 denotes a big difference. Then an outranking degree is estimated by multiplying the preferences by the criteria's weights and adding the single values, and subsequently the global preferences are calculated. Although there are several preference ranking methods, this paper mainly deals with the following six methods which have the potential to be popular, widely acceptable and accurate for giving more precise ranking of the candidate alternatives.

- Evaluation of mixed data (EVAMIX) method,
- Complex proportional assessment (COPRAS) method,
- Extended PROMETHEE II (EXPROM2) method,
- ORESTE (Organization, Rangement Et Synthese De Donnes Relationnelles) method,
- Operational competitiveness rating analysis (OCRA) method, and
- Additive ratio assessment (ARAS) method.

The computational details of these above-mentioned methods are presented as below.

EVAMIX Method

The EVAMIX method was mainly established by Voogd in 1983, and later advocated by Martel and Matarazzo (Martel & Matarazzo, 2005). It is based on the determination of the dominance score of an alternative on criterion-by-criterion basis. This method is especially designed to deal with the mixed (quantitative and qualitative) data. The main difference between EVAMIX and other MCDM methods is that it can treat the qualitative (ordinal) criteria and quantitative (cardinal) criteria separately. Both the ordinal and cardinal data are separately normalized in the range of 0 to 1 using a linear normalization procedure. In this method, the degree of pair-wise dominance for each pair of alternatives is calculated as the difference in score received by the higher performing alternative compared to the poorer performing alternative. The weighted sum of the dominance scores is then assigned to each alternative. The outcome of this aggregation procedure is similar to the outcome of the weighted sum method.

The EVAMIX method consists of the following procedural steps enlisted as below (Martel & Matarazzo, 2005; Hajkowicz & Higgins, 2008): Step 1: In the decision matrix, at first, differentiate between the ordinal and cardinal criteria. Step 2: For beneficial attributes (where higher values are desired), normalize the decision matrix using the following equation:

$$r_{ij} = [x_{ij} - min(x_{ij})]/[max(x_{ij}) - min(x_{ij})] \quad (i = 1,2,...,m; \ j = 1,2,...,n) \quad (1)$$

where x^{ij} is the performance measure of ith alternative with respect to j^{th} criterion and rij is the normalized value of x^{ij}.

For non-beneficial attributes (where lower values are preferable), Eq. (1) can be rewritten as follows:

$$r_{ij} = [max(x_{ij}) - x_{ij}]/[max(x_{ij}) - min(x_{ij})] \quad (2)$$

Step 3: Calculate the evaluative differences of ith alternative on each ordinal and cardinal criteria with respect to other alternatives. This step involves the calculation of differences in criteria values between different alternatives pair-wise.

Step 4: Compute the dominance scores of each alternative pair, (i,i') for all the ordinal and cardinal criteria using the following equations:

$$\alpha_{ii'} = \left[\sum_{j \in O} \{w_j \, sgn(r_{ij} - r_{i'j})\}^k \right]^{1/c} ,$$

$$\text{where } sgn(r_{ij} - r_{i'j}) = \begin{cases} +1 \text{ if } r_{ij} > r_{i'j} \\ 0 \text{ if } r_{ij} = r_{i'j} \\ -1 \text{ if } r_{ij} < r_{i'j} \end{cases}$$

(3)

$$\gamma_{ii'} = \left[\sum_{j \in C} \{w_j \, sgn(r_{ij} - r_{i'j})\}^k \right]^{1/c} .$$

(4)

where the symbol c is a scaling parameter, for which any arbitrary positive odd number, like 1,3,5,... may be chosen, O and C are the sets of ordinal and cardinal criteria respectively, a $_{ii}$ ¢ and ¢ γ_{ii} are the dominance scores for alternative pair, (i, i') with respect to ordinal and cardinal criteria respectively, and w_j is the weight (relative importance) of jth criterion.

Step 5: Calculate the standardized dominance scores. Martel and Matarazzo (2005) proposed an additive interval method to derive the standardized ordinal dominance score)(d $_{ii}$ ¢ and cardinal dominance score (d $_{ii \, \varrho}$) for the alternative pair, (i, i') as follows:

Standardized ordinal dominance score $(\delta_{ii'}) = \dfrac{(\alpha_{ii'} - \alpha^-)}{(\alpha^+ - \alpha^-)}$,

where $\alpha^+ (\alpha^-)$ is the highest (lowest) ordinal dominance score for the alternative pair, (i, i').

Standardized cardinal dominance score $(d_{ii'}) = \dfrac{(\gamma_{ii'} - \gamma^-)}{(\gamma^+ - \gamma^-)}$,

where $\gamma^+ (\gamma^-)$ is the highest (lowest) cardinal dominance score for the alternative pair, (i, i')

Step 6: Determine the overall dominance score. The overall dominance score, (D $_{ii'}$) ¢ for each pair of alternatives, (i, i') is calculated to measure the degree by which alternative i dominates alternative i'

$$D_{ii'} = w_O \delta_{ii'} + w_C d_{ii'} ,$$

where w_O is the sum of the weights for the ordinal criteria $(w_o = \sum_{j \in O} w_j)$ and w_C is the sum of the weights

for the cardinal criteria $(w_c = \sum_{j \in C} w_j)$.

Step 7: Calculate the appraisal score.

Appraisal score (Si) $= \sum_{i'} \left(\dfrac{D_{ii'}}{D_{i'i}} \right)^{-1}$

The appraisal score for i^{th} alternative (Si) is computed which gives the final preference of the alternatives. Higher the appraisal score, better is the performance of the alternative.

COPRAS Method

The COPRAS method assumes direct and proportional dependences of the significance and utility degree of the available alternatives under the presence of mutually conflicting criteria (Kaklauskas et al., 2006; Kaklauskas et al., 2007; Zavadskas et al., 2008). It takes into account the performance of the alternatives with respect to different criteria and also the corresponding criteria weights. This method selects the best decision considering both the ideal and the ideal-worst solutions. The COPRAS method which is used here for evaluating and selecting the alternative FMSs adopts a stepwise ranking and evaluating procedure of the alternatives in terms of their significance and utility degree. The steps of COPRAS method are presented as below: Step 1: Normalize the decision matrix using linear normalization procedure (Kaklauskas et al., 2006). Step 2: Determine the weighted normalized decision matrix, D.

$$D = [y_{ij}]_{mxn} = r_{ij} \times w_j \quad (i = 1,2,\ldots,m; j = 1,2,\ldots,n)$$

The sum of dimensionless weighted normalized values of each criterion is always equal to the weight for that criterion.

$$\sum_{i=1}^{m} y_{ij} = w_j$$

Thus, it can be said that the weight, w_j of jth criterion is proportionally distributed among all the alternatives according to their weighted normalized value, y_{ij}

Step 3: The sums of weighted normalized values are calculated for both beneficial and non-beneficial attributes using the following equations:

$$S_{+i} = \sum_{j=1}^{n} y_{+ij}$$

$$S_{-i} = \sum_{j=1}^{n} y_{-ij}$$

where y_{+ij} and y_{-ij} are the weighted normalized values for the beneficial and non-beneficial attributes respectively.

The greater the value of S_{+i}, the better is the alternative; and the lower the value of S_{-i}, the better is the alternative. The S_{+i} and S-i values express the degree of goals attained by each alternative. In any case, the sums of 'pluses'

S_{+i} and 'minuses' S_{-i} of the alternatives are always respectively equal to the sums of weights for the beneficial and non-beneficial attributes as expressed by the following equations:

$$S_+ = \sum_{i=1}^{m} S_{+i} = \sum_{i=1}^{m} \sum_{j=1}^{n} y_{+ij}$$

$$S_- = \sum_{i=1}^{m} S_{-i} = \sum_{i=1}^{m} \sum_{j=1}^{n} y_{-ij}$$

Step 4: Determine the significances of the alternatives on the basis of defining the positive alternatives S_{+i} and negative alternatives S_{-i} characteristics.

Step 5: Determine the relative significances or priorities (Q_i) of the alternatives.

$$Q_i = S_{+i} + \frac{S_{-min} \sum_{i=1}^{m} S_{-i}}{S_{-i} \sum_{i=1}^{m} (S_{-min}/S_{-i})} \quad (i = 1,2,\ldots,m)$$

where S_{-min} is the minimum value of S_{-i}. The greater the value of Q_i, the higher is the priority of the alternative. The relative significance value of an alternative shows the degree of satisfaction attained by that alternative. The alternative with the highest relative significance value (Q_{max}) is the best choice among the candidate alternatives.

Step 6: Calculate the quantitative utility (U_i) for ith alternative. The degree of an alternative's utility which leads to a complete ranking of the candidate alternatives, is determined by comparing the priorities of all the alternatives with the most efficient one and can be denoted as below:

$$U_i = \left[\frac{Q_i}{Q_{max}} \right] \times 100\%,$$

where Q_{max} is the maximum relative significance value. These utility values of the alternatives range from 0% to 100%.

EXPROM2 Method

The extended PROMETHEE II (EXPROM2) is basically a modified version of PROMETHEE II (preference ranking organization method for enrichment evaluation) method. In this method, the relative performance of one alternative over the other is defined by two preference indices. The first one is weak preference index based on the aggregated preference function considering the criteria weights as determined in PROMETHEE II method. The second one is strict preference index based on the notion of ideal and anti-ideal solutions. The ideal and anti-ideal values are directly derived from the decision matrix, and they reflect the extreme limits for a particular criterion. A total preference index is also computed by adding the strict and the weak preference indices, which gives an accurate measure of intensity of preference of one alternative over the other considering all the criteria.

The procedural steps of EXPROM2 method are given as below (Raju & Kumar, 1999; Doumpos & Zopounidis, 2004):

Step 1: Normalize the decision matrix.

Step 2: Calculate the evaluative differences of ith alternative with respect to other alternatives. This step involves the calculation of differences in criteria values (d_{jj} between different alternatives pair-wise.

Step 3: Calculate the preference function, $P_j(i,i')$.

$$P_j(i,i')=0 \text{ if } r_{ij} \leq r_{i'j} \tag{17}$$

$$P_j(i,i')=(r_{ij} - r_{i'j}) \text{ if } r_{ij} > r_{i'j} \tag{18}$$

Step 4: Calculate the weak preference index taking into account the criteria weights.

$$WP(i,i') = \left[\sum_{j=1}^{n} w_j \times P_j(i,i')\right] / \sum_{j=1}^{n} w_j \tag{19}$$

Step 5: Define the strict preference function, $SP_j(i,i')$.

The strict preference function is based on the comparison of the difference values (dmj) with the range of values as defined by the evaluation of the whole set of alternatives for a criterion.

$$SP_j(i,i') = [max(0,d_j - L_j)]/[dm_j - L_j]. \tag{20}$$

where L_j = limit of preference (0 for usual criterion preference function, and indifference values for other five preference functions) and d_{mj} = difference between the ideal and anti-ideal values of jth criterion.

Step 6: Compute the strict preference index.

$$SP(i,i') = \left[\sum_{j=1}^{n} w_j \times SP_j(i,i')\right] / \sum_{j=1}^{n} w_j \tag{21}$$

Step 7: Calculate the value of total preference index.

$$TP(i,i') = Min[1, WP(i,i') + SP(i,i')] \tag{22}$$

Step 8: Determine the leaving and the entering outranking flows using the following equations:

Leaving   (positive)   flow   for i^{th} alternative,

$$\varphi^+(i) = \frac{1}{m-1} \sum_{i'=1}^{m} TP(i,i') \quad (i \neq i') \tag{23}$$

Entering (negative) flow for i^{th} alternative $\varphi^-(i) = \dfrac{1}{m-1}\sum_{i=1}^{m} TP(i'.i)$ $(i \neq i')$ (24)

where m is the number of alternatives.

The leaving flow expresses how much an alternative dominates the other alternatives, while the entering flow denotes how much an alternative is dominated by the other alternatives. Based on these flow values, EXPROM2 method can give the complete preorder of the candidate alternatives by using a net flow.

Step 6: Calculate the net outranking flow for each alternative.

$\varphi(i) = \varphi^+(i) - \varphi^-(i)$

(26)

Step 7: Determine the ranking of all the considered alternatives depending on the values of $\varphi(i)$. The higher the value of $\varphi(i)$, the better is the alternative.

ORESTE Method

The ORESTE method is a compensatory preference ranking approach, as it uses the differences between the ranks of pairs of actions based on their evaluations (Roubens, 1982; Pastijn & Leysen, 1989). In many real time decision-making situations having several quantitative as well as qualitative criteria, this method is particularly appropriate to support the conflicting decisions in absence of crisp numerical values and weights of the attributes.

The ORESTE method deals with the situation where an alternative ai (i = 1,2,…,m) is ranked according to criterion cj (j = 1,2,…,n), and the main objective is to find a global preference structure on a set of alternatives, which reflects the evaluation of alternatives on each criterion and the preference among the criteria (Pastijn & Leysen, 1989). Since it only takes into account the ranking of alternatives and criteria, it is mainly suited to problems with ordinal data, but it can also be used for problems with cardinal or mixed data. The main advantage of this method is that as it uses only the ordinal ranking of criteria which avoids the occurrence of lengthy discussions among the decision makers to set weight importance of the attributes, it speeds up the decision-making process.

The steps of ORESTE method are presented as below (Teghem et al., 1989):

Step 1: From the decision matrix, determine the weak order of the criteria indicating their relative importance as follows:

c_1 P c_2 I c_3 P c_4........c_n

This means c_1 is the most important and preferred criterion, while c_2 and c_3 are tied as the intermediate important criteria and c_4 is the least important criterion.

Step 2: Determine for each criterion a weak order of the alternatives similar to step (1).

c_1: a_1 P a_2 P a_3........a_m
c_2: a_1 P a_2 I a_3........a_m
c_3: a_1 P a_2 I a_3........a_m

. . . .

c_n: a_1 I a_2 R a_3........a_m

Step 3: Obtain Besson rankings of the alternatives and criteria. In ORESTE method, each alternative is given a Besson rank based on its weak order among the other alternatives with respect to each criterion and also each criterion is given a Besson rank according to its position in the weak order among all the criteria. The Besson rank of an alternative a_1 with respect to a criterion j is denoted by $r_j(a_1)$ and the Besson rank of criterion j is denoted by r_{cj}.

Step 4: Calculate the projection distances. The projection distances correspond to the relative positions of the alternatives with respect to an arbitrary origin O and are defined by $d(O,a_1)$.

Pastijn and Leysen (1989) discussed about different types of projections. In this paper, the linear orthogonal projection is adopted as expressed by the following equation

$$d_j(O,a_i) = (1/2)[rc_j + r_j(a_i)].$$

(26)

Projection distances are such calculated that if an alternative a1 is preferred to another alternative a2 (a1 P a_2) for criterion j, then $d_j(a_1) < d_j(a_2)$, i.e. the smaller projection distance, the better is the position of the alternative.

Step 5: Rank the projections to obtain the global ranks. A mean global Besson rank $r_j(a_i)$ is assigned to all the projection distances from the lowest to the highest ones. Smaller $r_j(a_i)$ indicates better position of the particular alternative.

Step 6: Calculate the mean ranks. For each alternative, a mean rank is computed by the summation of their global Besson ranks over the entire set of criteria using the following expression.

$$r(a_i) = \sum_{j=1}^{n} r_j(a_i)$$

(27)

These mean ranks are simply sorted increasingly to determine the global weak order of the alternatives.

OCRA Method

The OCRA method was developed to measure the relative performance of a set of production units, where resources are consumed to create value-added outputs. In this method, in the first step, the preference ratings with respect to non-beneficial or input criteria are determined; in the second step, the preference ratings of the output criteria are determined, and in the last step, the overall preference ratings of the available alternatives are evaluated where both the cardinal and ordinal data are used. OCRA uses an intuitive method for incorporating the decision maker's preferences about the relative importance of the criteria (Parkan & Wu, 1997; Parkan & Wu, 1997). The preference ratings of the alternatives in OCRA method reflect the decision maker's preferences for the criteria. Besides this, the main advantage of OCRA method is that it can deal with those MCDM situations when the relative weights of the criteria are dependent on the alternatives and different weight distributions are assigned to the criteria for different alternatives as well as some of the criteria are not applicable to all the alternatives.

The general OCRA procedure is described as below (Parkan & Wu, 2000):

Step 1: Compute the preference ratings with respect to the non-beneficial criteria. In this step, OCRA method is only concerned with the scores that various alternatives receive for the input criteria without considering the scores received for the beneficial criteria. The lower values of 324 non-beneficial or input criteria are more preferable. The aggregate performance of ith alternative with respect to all the input criteria is calculated using the following equation:

$$\bar{I}_i = \sum_{j=1}^{n} w_j \frac{\max(x_j^m) - x_j^i}{\min(x_j^m)} \quad (i = 1,2,\ldots,m; \, j = 1,2,\ldots,n; \, i \neq m) \tag{28}$$

where $\bar{I}_i$ is the measure of the relative performance of ith alternative and x_j^i is the performance score of ith alternative with respect to j^{th} input criterion. If i^{th} alternative is preferred to m^{th} alternative with respect to j^{th} criterion, then $x_j^i < x_j^m$. The term $\frac{\max(x_j^m) - x_j^i}{\min(x_j^m)}$ indicates the difference in performance scores for criterion j, between i_{th} alternative and the alternative whose score for criterion j is the highest among all the alternatives considered. The calibration constant w_j (relative importance of j_{th} criterion) is used to increase or reduce the impact of this difference on the rating $\bar{I}_i$ with respect to j_{th} criterion.

Step 2: Calculate the linear preference rating for the input criteria.

$$\bar{\bar{I}} = \bar{I} - \min(\bar{I})$$

(29)

This linear scaling is done to assign a zero rating to the least preferable alternative. $\bar{I}$ represents the aggregate preference rating for i[th] alternative with respect to the input criteria. Step 3: Compute the preference ratings with respect to the beneficial criteria. The aggregate performance for i[th] alternative on all the beneficial or output criteria is measured using the following expression:

$$\bar{O}_i = \sum_{h=1}^{H} w_h \frac{x_h^i - \min(x_h^m)}{\min(x_h^m)},$$

(30)

where h = 1,2,...,H indicates the number of beneficial criteria and wh is the calibration constant (weight importance) of hth output criteria. The higher an alternative's score for an output criterion, the higher is the preference for that alternative.

It can be mentioned that $\sum_{j=1}^{n} w_j + \sum_{h=1}^{H} w_h = 1$

Step 4: Calculate the linear preference rating for the output criteria using the following equation:

$$\bar{\bar{O}}_i = \bar{O}_i - \min(\bar{O}_i)$$

(31)

Step 5: Compute the overall preference ratings.

The overall preference rating for each alternative is calculated by scaling the sum $(\bar{\bar{I}}_i + \bar{\bar{O}}_i)$ so that the least preferable alternative receives a rating of zero. The overall preference rating (P$_i$) is calculated as follows:

$$P_i = (\bar{\bar{I}}_i + \bar{\bar{O}}_i) - \min(\bar{\bar{I}}_m + \bar{\bar{O}}_m)$$

(32)

The alternatives are ranked according to the values of the overall preference rating. The alternative with the highest overall performance rating receives the first rank.

ARAS Method

The ARAS method is based on quantitative measurements and utility theory. In this method, a utility function value determines the relative efficiency of an alternative over other alternatives. This utility function is directly proportional to the relative effect of the criteria values and weight importance of the considered criteria. The utility value of an alternative is determined by a comparison of variant with the ideally best alternative. The steps of ARAS method are as follows (Turskis & Zavadskas, 2010; Zavadskas & Turskis, 2010):

Step 1: For beneficial attributes, determine the normalized decision matrix, using linear normalization procedure (Zavadskas & Turskis, 2010). For non-beneficial attributes, the normalization procedure follows two steps. At first, the reciprocal of each criterion with respect to all the alternatives is taken as follows:

$$x_{ij}' = \frac{1}{x_{ij}}$$

(33)

In the second step, the normalized values are calculated as follows:

$$R = [r_{ij}]_{m \times n} = \frac{x_{ij}'}{\sum_{i=1}^{m} x_{ij}'}$$

(34)

Step 2: Determine the weighted normalized decision matrix, D, using Eq. (9).

Step 3: Determine the optimality function (S_i) for i^{th} alternative.

$$S_i = \sum_{j=1}^{n} y_{ij}$$

(35)

Higher the S_i value, the better is the alternative. The optimality function Si has a direct and proportional relationship with values in the decision matrix and criteria weights.

Step 4: Calculate the degree of the utility (U_i) for each alternative. It is determined by a comparison of the variant with the most efficient one (S_0). The equation used for calculating the value of U_i is given as below:

$$U_i = \frac{S_i}{S_0}$$

(36)

The utility values of the alternatives range from 0% to 100%. The alternative with the highest utility value is the best choice among the candidate alternatives. In some preference ranking methods, pair-wise comparison of the alternatives is performed to compute a preference function for each criterion and based on this preference function, a preference index is determined to show the preference of an alternative over the other. This preference index is the measure to support the hypothesis that there is some preference of an alternative over the other. In some other preference ranking methods, a utility or priority function value determines the relative efficiency of an alternative over the other. In all these approaches, the ranking index is nothing but an aggregation of the normalized criteria values, the relative importance of the criteria, and a balance between total and individual satisfaction or preference either by defining a preference function or an utility value. So basically, all the preference ranking methods whether it uses preference function or weighted sum utility value, indicate how much an alternative is preferred to the other.

Industrial Example

Today's global competition has compelled the manufacturing organizations to improve their product quality in a cost effective manner. Use of proper manufacturing technologies, like FMS, offers great potentials for improving manufacturing performance and helps to attain the organizational objectives in an efficient way. A wrong alternative selection may result in loss of productivity and profitability. Thecomplexity of decision-making makes multi-criteria analysis an invaluable tool in the engineering design and selection process. Thus, the main objective of this paper is set to reveal the computational easiness of the six preference ranking methods in dealing with FMS selection problems, involving both ordinal and cardinal attribute data. It mainly focuses on introducing these multi-beneficial MCDM methods that can make FMS selection easier and compatible with most of the situations. These preference ranking methods are applied to an existing problem, dealing with the selection of the best FMS alternative for a given manufacturing environment. In these methods, the decision makers' preferences and preferences on alternatives' performances are aggregated together to reach the final evaluation and selection decision. The past researchers have adopted different mathematical tools for evaluating, justifying and selecting FMS technologies, but all those methods are either very complicated or require lengthy computations. For decision-making problems with large number of attributes and small number of alternatives, those approaches may occasionally provide poor results. This paper takes the opportunity to explore the application viability and potentiality of six popular preference ranking methods to provide more precise and accurate ranking of the feasible FMS alternatives. According to the best knowledge of the authors, there have been very few applications of these preference ranking methods in manufacturing environment. Few successful implementations of these methods can be found in construction engineering, financial analysis and waste water management. Even till date, very less effort has been devoted to study the relative performance of these methods as employed in discrete manufacturing environment. Furthermore, no attempt has been made to map/match any FMS selection problem to these methods. All these methods are successfully applied and the results are compared for better visualization. Four performance analysis tests are also executed to assess the degree of agreement between the ranking orders as obtained by these methods, while keeping the performance measures in the evaluation matrix of the considered example constant. These six preference ranking methods are also qualitatively compared in terms of their suitability for solving different FMS selection problems, operational similarities and other model characteristics, like information type and criteria requirement, methodological aspect, operational approach, compensatory character and nature of the obtained results.

Thus, in order to apply these six preference ranking methods and compare their relative ranking performance, the FMS selection problem as considered by Karsak (2002) is cited here. Karsak (2002) considered eight alternative FMSs and eight attributes affecting the FMS selection decision for a given industrial application. Based on several literatures on evaluation and justification of FMS investments, Karsak (Karsak, 2002) considered capital and operating costs (COS), required floor space (RFS), work in progress (WIP), product flexibility (PF), volume flexibility (VF), expansion flexibility (EF), lead time reduction (LTR), and quality improvement (QI) as the main selection criteria for the considered problem. Product flexibility was defined as the ability to start producing a new set of part types quickly and economically. Volume flexibility was the ability to operate an FMS profitably at varying production levels, whereas, expansion flexibility was defined as the ability to easily add capability and capacity to an existing system. Among these, three attributes are quantitative in nature and the remaining five are judged in an interval of 0-1 scale. Karsak (2002) solved that FMS selection problem using a distance-based fuzzy TOPSIS method and obtained a comparative ranking of FMS alternatives as 6-2-1-7-5-8-4-3, which indicates that the first FMS alternative attains a rank of 6, the second one has a rank of 2, the third one achieves a rank of 1, and so on. More precisely, the ranking order of FMS alternatives as obtained by Karsak (2002) is FMS3 > FMS2 > FMS8 > FMS7 > FMS5 > FMS1 > FMS4 > FMS6. Table 1 represents the performance characteristics of the considered FMS alternatives with respect to all the criteria. Here, COS, RFS and WIP are non-beneficial attributes where lower values are always desirable. The weight values or relative importance of the eight criteria are determined using entropy method, as given in Table 2.

Table 1: Quantitative data for the FMS selection problem (Karsak, 2002)

FMS	Capital and operating costs ($ millions)	Required floor space (m²)	Work in progress (units)	Product flexibility	Volume flexibility	Expansion flexibility	Lead time reduction	Quality improvement
1	3.8	630	42	0.7	0.5	0.7	0.5	0.2
2	3.1	620	37	0.7	0.7	0.5	0.9	0.5
3	5	425	32	0.9	0.7	0.7	0.7	0.7
4	6.4	500	54	0.7	0.5	0.5	0.7	0.9
5	3.6	600	44	0.2	0.7	0.7	0.2	0.7
6	6.7	780	59	0.9	0.9	0.5	0.5	0.5
7	3.4	740	37	0.5	0.2	0.2	0.7	0.9
8	3.7	550	36	0.5	0.9	0.9	0.5	0.5

Table 2: Criteria weights determined using entropy method

Criteria	COS	RFS	WIP	PF	VF	EF	LTR	QI
Weight	0.1297	0.1420	0.1400	0.1173	0.1173	0.1194	0.1194	0.1150

EVAMIX Method

Now, this FMS selection problem is solved using EVAMIX method. In the original decision matrix, as shown in Table 1, the ordinal and cardinal criteria are at first separated (PF, VF, EF, LTR and QI are the qualitative attributes) and then, the decision matrix is normalized using Eq. (1) and Eq. (2) respectively for beneficial and non-beneficial attributes. This normalized decision matrix is shown in Table 3. From the normalized decision matrix, the evaluative differences of ith alternative for each ordinal and cardinal criteria with respect to all other alternatives are calculated. Now, the dominance scores of each pair of FMS alternatives for all the ordinal and cardinal criteria are estimated applying Eq. (3) and Eq. (4) respectively, and are given in Table 4. While calculating the dominance scores, the value of c is taken as 1. Table 4 also exhibits the standardized dominance scores for all the pairs of FMS alternatives, as computed employing Eq. (5) and Eq. (6) respectively for the ordinal and cardinal criteria. The overall dominance score for each FMS alternative pair, (i,i') is calculated using Eq. (7) which shows the degree by which FMS i dominates FMS i'. These overall dominance scores for all the pairs of FMS alternatives are given in Table 5. Now, using Eq. (8), the appraisal score for each FMS alternative is calculated, as shown in Table 6 and based on the descending values of this appraisal score, the final ranking of FMS alternatives is obtained as 8-3-1-5-4-6-7-2. The best choice is FMS 3. FMS 8 is the second choice and the last choice is FMS 1.

Table 3: Normalized decision matrix for the FMS selection problem

FMS	COS	RFS	WIP	PF	VF	EF	LTR	QI
1	0.8056	0.4225	0.6296	0.7143	0.4286	0.7143	0.4286	0
2	1	0.4507	0.8148	0.7143	0.7143	0.4286	1	0.4286
3	0.4722	1	1	1	0.7143	0.7143	0.7143	0.7143
4	0.0833	0.7887	0.1852	0.7143	0.4286	0.4286	0.7143	1
5	0.8611	0.5070	0.5556	0	0.7143	0.7143	0	0.7143
6	0	0	0	1	1	0.4286	0.4286	0.4286
7	0.9167	0.1127	0.8148	0.4286	0	0	0.7143	1
8	0.8333	0.6479	0.8519	0.4286	1	1	0.4286	0.4286

Table 4: Dominance and standardized dominance scores of each alternative FMS pair

FMS pair	$\alpha_{ii'}$	$\gamma_{ii'}$	$\delta_{ii'}$	$d_{ii'}$	FMS pair	$\alpha_{ii'}$	$\gamma_{ii'}$	$\delta_{ii'}$	$d_{ii'}$
(1,2)	-0.2323	-0.4116	0.1611	0	(5,1)	-0.0043	0.1317	0.3163	0.3849
(1,3)	-0.4690	-0.1522	0	0.1838	(5,2)	-0.0022	-0.1276	0.3177	0.2012
(1,4)	-0.1150	0.1276	0.2409	0.3821	(5,3)	-0.2366	-0.1522	0.1581	0.1838
(1,5)	0.0043	-0.1317	0.3222	0.1983	(5,4)	-0.1150	0.1276	0.2409	0.3821
(1,6)	-0.2302	0.4116	0.1625	0.5832	(5,6)	-0.1195	0.4116	0.2379	0.5832
(1,7)	0.1195	-0.1276	0.4006	0.2012	(5,7)	-0.1150	-0.1276	0.2409	0.2012
(1,8)	1	1	1	1	(5,8)	1	1	1	1
(2,1)	0.2323	0.4116	0.4774	0.5832	(6,1)	0.2302	-0.4116	0.4759	0
(2,3)	-0.2323	-0.1522	0.1611	0.1838	(6,2)	0.1151	-0.4116	0.3976	0
(2,4)	0.1216	0.1276	0.4020	0.3821	(6,3)	-0.2366	-0.4116	0.1582	0
(2,5)	0.0022	0.1276	0.3208	0.3821	(6,4)	0.0001	-0.4116	0.3193	0
(2,6)	-0.1151	0.4116	0.2409	0.5832	(6,5)	0.1195	-0.4116	0.4006	0
(2,7)	0.3582	0.2717	0.5631	0.4841	(6,7)	0.1195	-0.4116	0.4006	0
(2,8)	0	-0.1522	0.3192	0.1838	(6,8)	1	1	1	1
(3,1)	0.4690	0.1522	0.6385	0.3995	(7,1)	-0.1195	0.1276	0.2379	0.3821
(3,2)	0.2323	0.1522	0.4774	0.3995	(7,2)	-0.3582	-0.2717	0.0754	0.0991
(3,4)	0.2389	0.4116	0.4818	0.5832	(7,3)	-0.2389	-0.1522	0.1566	0.1838
(3,5)	0.2366	0.1522	0.4803	0.3995	(7,4)	-0.3539	0.1276	0.0783	0.3821
(3,6)	0.2366	0.4116	0.4803	0.5832	(7,5)	0.1150	0.1276	0.3976	0.3821
(3,7)	0.2389	0.1522	0.4818	0.3995	(7,6)	-0.1195	0.4116	0.2379	0.5832
(3,8)	1	1	1	1	(7,8)	-0.0022	-0.1522	0.3177	0.1838
(4,1)	0.1150	(4,1)	0.3976	0.2012	(8,1)	0.2344	0.4116	0.4788	0.5832
(4,2)	-0.1216	(4,2)	0.2365	0.2012	(8,2)	0	0.1522	0.3192	0.3995
(4,3)	-0.2389	(4,3)	0.1566	0	(8,3)	-0.1150	-0.1522	0.2409	0.1838
(4,5)	0.1150	(4,5)	0.3976	0.2012	(8,4)	-0.1150	0.1276	0.2409	0.3821
(4,6)	-0.0001	(4,6)	0.3192	0.5832	(8,5)	0.3582	0.1522	0.5631	0.3995
(4,7)	0.3539	(4,7)	0.5602	0.2012	(8,6)	0.0021	0.4116	0.3207	0.5832
(4,8)	1	(4,8)	1	1	(8,7)	0.0022	0.1522	0.3208	0.3995

Table 5: Overall dominance scores for EVAMIX method

FMS pair	$D_{ii'}$	FMS pair	$D_{ii'}$	FMS pair	$D_{ii'}$	FMS pair	$D_{ii'}$
(1,2)	0.0948	(3,1)	0.5401	(5,1)	0.3445	(7,1)	0.2973
(1,3)	0.0757	(3,2)	0.4453	(5,2)	0.2697	(7,2)	0.0852
(1,4)	0.2990	(3,4)	0.5236	(5,3)	0.1687	(7,3)	0.1678
(1,5)	0.2712	(3,5)	0.4470	(5,4)	0.2990	(7,4)	0.2034
(1,6)	0.3357	(3,6)	0.5227	(5,6)	0.3801	(7,5)	0.3912
(1,7)	0.3185	(3,7)	0.4479	(5,7)	0.2246	(7,6)	0.3801
(1,8)	1	(3,8)	1	(5,8)	1	(7,8)	0.2626
(2,1)	0.5210	(4,1)	0.3167	(6,1)	0.2800	(8,1)	0.5218
(2,3)	0.1704	(4,2)	0.2219	(6,2)	0.2339	(8,2)	0.3523
(2,4)	0.3938	(4,3)	0.0922	(6,3)	0.0931	(8,3)	0.2174
(2,5)	0.3460	(4,5)	0.3167	(6,4)	0.1879	(8,4)	0.2990
(2,6)	0.3818	(4,6)	0.4279	(6,5)	0.2357	(8,5)	0.4957
(2,7)	0.5306	(4,7)	0.4124	(6,7)	0.2357	(8,6)	0.4288
(2,8)	0.2635	(4,8)	1	(6,8)	1	(8,7)	0.3532

Table 6: Appraisal score and rank of each alternative FMS in EVAMIX method

FMS	1	2	3	4	5	6	7	8
S_i	0.0323	0.0689	0.1565	0.0489	0.0589	0.0379	0.0357	0.0776
Rank	8	3	1	5	4	6	7	2

COPRAS Method

In this method, at first, the weak order of the criteria is determined indicating their relative importance. Based on the criteria values, as shown in Table 2, the weak order of the criteria is C_2 P C_3 P C_1 P C_6 I C_7 P C_4 I C_5 P C_8 which indicates that required floor space (C_2) is the most preferred criterion and quality improvement (C_8) is the least important criterion. Then for each criterion, a weak order of the alternatives is determined, as shown in Table 12. Now, based on step (3) of sub-section 2.4, the Besson ranking of all the FMS selection criteria and also the Besson ranking of the considered FMS alternatives are determined, as given in Tables 13 and 14 respectively. Now, applying Eq. (26), the corresponding projection distances are computed, as shown in Table 15. From this table, the rankings of the projections or global ranks are obtained, as given in Table 16. In the last step of this method, the mean ranks of FMS alternatives are obtained from which the final ranking of the candidate alternatives is derived, as shown in Table 17. From this table, the ranking of FMS alternatives is observed as 7-2-1- 4-6-8-5-3. FMS 3 is the best choice and FMS 6 is the worst chosen alternative.

Table 12: Weak order of the alternatives for each criterion in ORESTE method

Criteria	Weak order of alternatives
C_1	2 P 7 P 5 P 8 P 1 P 3 P 4 P 6
C_2	3 P 4 P 8 P 5 P 2 P 1 P 7 P 6
C_3	3 P 8 P 2 I 7 P 1 P 5 P 4 P 6
C_4	3 I 6 P 1 I 2 I 4 P 7 I 8 P 5
C_5	6 I 8 P 2 I 3 I 5 P 1 I 4 P 7
C_6	8 P 1 I 3 I 5 P 2 I 4 I 6 P 7
C_7	2 P 3 I 4 I 7 P 1 I 6 I 8 P 5
C_8	4 I 7 P 3 I 5 2 I 6 I 8 P 1

Table 13: Besson ranking of criteria in ORESTE method

Criteria	C_1	C_2	C_3	C_4	C_5	C_6	C_7	C_8
Besson ranking	3	1	2	6.5	6.5	4.5	4.5	8

Table 14: Besson ranking of the alternatives in ORESTE method

Criteria	FMS1	FMS2	FMS3	FMS4	FMS5	FMS6	FMS7	FMS8
C_1	5	1	6	7	3	8	2	4
C_2	6	5	1	2	4	8	7	3
C_3	5	3.5	1	7	6	8	3.5	2
C_4	4	4	1.5	4	8	1.5	6.5	6.5
C_5	6.5	4	4	6.5	4	1.5	8	1.5
C_6	3	6	3	6	3	6	8	1
C_7	6	1	3	3	8	6	3	6
C_8	8	5	5	1.5	5	5	1.5	5

Table 15: Projection distances of FMS alternatives in ORESTE method

Criteria	FMS1	FMS2	FMS3	FMS4	FMS5	FMS6	FMS7	FMS8
C_1	4	2	4.5	5	3	5.5	2.5	3.5
C_2	3.5	3	1	1.5	2.5	4.5	4	2
C_3	3.5	1.5	1.5	4.5	4	5	2.75	2
C_4	5.25	5.25	4	5.25	7.25	4	6.5	6.5
C_5	6.5	5.25	5.25	6.5	5.25	4	7.25	4
C_6	3.75	5.25	3.75	5.25	3.75	5.25	6.25	2.75
C_7	5.25	2.75	3.75	3.75	6.25	5.25	3.75	5.25
C_8	8	6.5	6.5	4.75	6.5	6.5	4.75	6.5

Table 16: Mean global Besson rankings of the projections in ORESTE method

Criteria	FMS1	FMS2	FMS3	FMS4	FMS5	FMS6	FMS7	FMS8
C_1	27	6	32	36.5	13.5	50	8.5	16
C_2	16	13.5	1	3	8.5	32	27	6
C_3	16	3	3	32	27	36.5	11	6
C_4	43.5	43.5	27	43.5	62.5	27	57	57
C_5	57	43.5	43.5	57	43.5	27	62.5	27
C_6	20.5	43.5	20.5	43.5	20.5	43.5	51.5	11
C_7	43.5	11	20.5	20.5	51.5	43.5	20.5	43.5
C_8	64	57	57	34.5	57	57	34.5	57

Table 17: Mean and overall ranks in ORESTE method

FMS	1	2	3	4	5	6	7	8
Mean rank	287.5	221	204.5	270.5	284	316.5	272.5	223.5
Rank	7	2	1	4	6	8	5	3

OCRA Method

At first, using Eq. (28), the aggregate performance of the alternatives with respect to all the input criteria is calculated. Then based on these values, the linear preference ratings for the input criteria are computed. Applying Eq. (30), the aggregate performance of the alternatives on all the beneficial or output criteria are then determined and subsequently, the linear preference ratings for the output criteria are calculated. Lastly, using Eq. (32), the overall preference rating for each of the FMS alternative is determined. The detailed computations of this method are illustrated in Table 18. In this method, the ranking of FMS alternatives is obtained as 6-3-1-4-7-5-8-2, which suggests that FMS 3 attains the top rank. FMS 8 is the second best choice and FMS 7 has the last rank.

Table 18: Computation details for OCRA method

FMS	I_i	$\overline{I}_i$	O_i	$\overline{O}_i$	P_i	Rank
1	0.2458	0.2458	0.9466	0.0696	0.0677	6
2	0.3003	0.3003	1.3558	0.4788	0.5314	3
3	0.3078	0.3078	1.5881	0.7111	0.7712	1
4	0.1280	0.1280	1.3492	0.4722	0.3525	4
5	0.2555	0.2555	0.8792	0.0022	0.0100	7
6	0	0	1.3515	0.4745	0.2269	5
7	0.2477	0.2477	0.8770	0.0000	0	8
8	0.3030	0.3030	1.3558	0.4788	0.5341	2

ARAS Method

From the weighted normalized decision matrix, as given in Table 19, and using Eq. (35), the optimality function (S_i) for each of the FMS alternative is calculated. Then the corresponding values of the utility degree (U_i) are determined for all the alternatives. The values of Si and U_i, and the ranking achieved by the FMS alternatives are exhibited in Table 20. It is revealed from this table that FMS 3 is the best chosen alternative and FMS 5 obtains the last rank. FMS 2 has the second rank.

Table 19: Weighted normalized decision matrix for ARAS method

FMS	COS	RFS	WIP	PF	VF	EF	LTR	QI
1	0.0177	0.0165	0.0171	0.0161	0.0115	0.0178	0.0127	0.0047
2	0.0216	0.0168	0.0194	0.0161	0.0161	0.0127	0.0229	0.0117
3	0.0134	0.0244	0.0224	0.0207	0.0161	0.0178	0.0178	0.0164
4	0.0105	0.0208	0.0133	0.0161	0.0115	0.0127	0.0178	0.0211
5	0.0186	0.0173	0.0163	0.0046	0.0161	0.0178	0.0051	0.0164
6	0.0100	0.0133	0.0122	0.0207	0.0207	0.0127	0.0127	0.0117
7	0.0197	0.0140	0.0194	0.0115	0.0046	0.0051	0.0178	0.0211
8	0.0181	0.0189	0.0199	0.0115	0.0207	0.0229	0.0127	0.0117

Table 20: Si and Ui values in ARAS method

FMS	S_i	U_i	Rank
1	0.1140	0.7647	6
2	0.1373	0.9209	2
3	0.1491	1.0000	1
4	0.1237	0.8302	4
5	0.1122	0.7530	8
6	0.1140	0.7649	5
7	0.1132	0.7597	7
8	0.1364	0.9153	3

Comparative Analzysis

In order to validate the applicability and suitability of the six considered preference ranking methods to solve this FMS selection problem, their relative ranking performance is compared using the following measures:

- Spearman's rank correlation coefficient,
- Kendall's coefficient of concordance,
- agreement between the top three ranked alternatives, and
- number of ranks matched, as the percentage of the number of considered alternatives.

Using Spearman's rank correlation coefficient (r_s) value, the similarity between two sets of rankings can be measured. Usually, its value lies between −1 and +1, where the value of +1 denotes a perfect match between two rank orderings. Table 21 shows the Spearman's rank correlation coefficients when the rankings of FMS alternatives as obtained using all the six preference ranking methods are compared between themselves and also with respect to the rank ordering as derived by Karsak (Karsak, 2002). It is observed that the r_s value ranges between 0.5238 and 0.9761. The performance of ORESTE method is satisfactory with respect to r_s value. Other methods, except COPRAS and ARAS, also perform well. It is also observed that CORRAS is almost similar to OCRA, and OCRA to ARAS.

The relative performance of these methods with respect to the ranking of FMS alternatives as obtained by Karsak (2002) is well visualized using the value of Z, which can be expressed as below:

$$Z = r_s \sqrt{m-1}$$ (37)

Fig. 1 plots the Z values for all the considered preference ranking methods and it reveals that ORESTE is the best method. On the other hand, the performance of OCRA and ARAS methods are not so much satisfactory

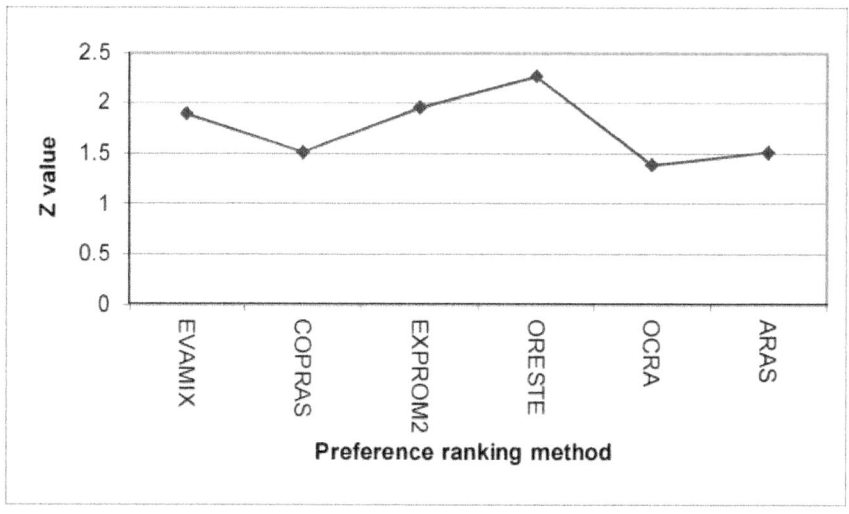

Figure 1: Z values for the six preference ranking methods

The similarity of rankings obtained by these methods is also measured using Kendall's coefficient of concordance (z). Its value lies between 0 and 1, where a value of 1 results in a perfect match. In this case, the value of z is computed as 0.8705, which suggests that there is an almost perfect agreement between the considered methods. When the ranking of FMS alternatives as derived by Karsak (2002) is taken into consideration, the z value is observed to be 0.8318, which is also quite high suggesting a similarity of the rank orderings between those obtained by the six preference ranking methods and that of Karsak (2002). A high z value signifies the suitability of these methods to solve the considered FMS selection problem. Sometimes, the decision maker may be interested to select the best FMS as the single choice. So, another test is performed based on the agreement between the top three ranked FMS alternatives. Here, a result of (1,2,3) means the first, second and third ranks match; (1,2,#) means the first and second ranks match; (1,#,#) means only the first ranks match; and (#,#,#) means no match. Table 21 shows the results of this test, which indicates that ARAS method has the maximum number of mismatches with respect to the ranking of the top three FMS alternatives. It is also quite interesting to note that for all the methods, the top ranked alternative is FMS 3.

Table 21: A comparative study on ranking performance for six preference ranking methods

Method	EVAMIX	COPRAS	EXPROM2	ORESTE	OCRA	ARAS
Karsak	0.7142, (1,#,#), 12.5	0.5714,(1,#,#), 12.5	0.7380, (1,#,#), 25	0.8571, (1,2,3), 50	0.5238, (1,#,#), 25	0.5714, (1,2,3), 50
EVAMIX		0.7619, (1,2,3), 37.5	0.7857,(1,2,3), 37.5	0.8095, (1,#,#), 12.5	0.8095, (1,2,3), 37.5	0.7142, (1,#,#), 25
COPRAS			0.9285, (1,2,3), 62.5	0.7619, (1,#,#), 25	0.9761, (1,2,3), 75	0.9285, (1,#,#), 25
EXPROM2				0.8809, (1,#,#), 50	0.8571, (1,2,3), 50	0.8095, (1,#,#) 37.5
ORESTE					0.7380, (1,#,#), 25	0.7857, (1,2,3), 50
OCRA						0.9523, (1,#,#), 50

Table 22: Characteristics of the six preference ranking methods

Method	Criteria	Information of the criteria	Approach	Operational approach	Compensatory character	Characteristics of result
EVAMIX	Mixed	Cardinal information on criteria weights	Rank-problem statement	Preference dominance	Yes	Total pre-order
COPRAS	Mixed	Preference weight	Rank-problem statement	Preference priority	Yes	Total pre-order
EXPROM2	Mixed	Preference function	Rank-problem statement	Preference outranking method	Yes	Total pre-order
ORESTE	Mixed	Besson ranking	Rank-problem statement	Preference outranking method	Yes	Total pre-order
OCRA	Mixed	Preference rating	Rank-problem statement	Preference rating analysis	No	Total pre-order
ARAS	Mixed	Utility function and weight importance	Rank-problem statement	Preference utility	No	Total pre-order

The last test is performed with respect to the number of ranks matched, expressed as the percentage of the number of alternatives considered. These results are also shown in Table 21. In this test, it is observed that ORESTE emerges out as the best preference ranking method. Table 22 exhibits various characteristics of the six considered preference ranking methods (Teghem et al., 1989). It is observed that all these methods are quite capable to deal with both the cardinal and ordinal data, and can provide the total ranking of the considered alternatives, although they have different mathematical treatments and operational approaches.

RESULTS AND DISCUSSIONS

In order to validate the superiority of ORESTE method over the other considered preference ranking methods, two more FMS selection problems are studied. The first problem (Rao and Parnichkun, 2009) considers selection of an FMS from a set of eight alternatives which are evaluated based on seven criteria, i.e. reduction in labour cost (%), reduction in WIP (%), reduction in set up cost (%), increase in market response, increase in quality, capital and maintenance cost ($000), and floor space used (sq. ft.). While solving this problem using digraph and matrix approach, Rao and Parnichkun (2009) derived a ranking of FMS alternatives as 3-4-7-2-5-6-1-8. FMS 7 was the best alternative and FMS 8 was the worst choice. The same FMS selection problem is now solved employing the six preference ranking methods and their relative ranking performance is shown in Table 23. From this table, it is found that there exists a perfect match between the rank ordering as obtained by ORESTE method and that derived by Rao and Parnichkun (2009). The Kendal's coefficient of concordance value is also estimated to be quite high (z = 0.9553) which indicates the existence of a near to perfect agreement between the considered methods.

Table 23: Performance test results of six preference ranking methods for example 2

Method	EVAMIX	COPRAS	EXPROM2	ORESTE	OCRA	ARAS
Rao and Parnichkun	0.8810, (#,#,#),12.5	0.9524, (#,#,3),50	0.9048, (1,2,#),37.5	1.0000, (1,2,3),100	0.9762, (#,#,3),62.5	0.9524, (#,#,3),50
EVAMIX		0.9524, (1,#,#),50	0.8810, (#,#,#),50	0.8810, (#,#,#),12.5	0.9286, (1,#,#),25	0.9524, (1,#,#),50
COPRAS			0.9286, (#,#,#),25	0.9524, (#,#,3),50	0.9762, (1,2,3),75	1.0000,(1,2,3),100
EXPROM2				0.9048, (1,2,#),37.5	0.8810, (#,#,#),12.5	0.9286, (#,#,#),25
ORESTE					0.9762, (#,#,3),75	0.9524, (#,#,3),50
OCRA						0.9762, (1,2,3),75

Rao and Parnichkun (2009) also cited another FMS selection problem which consists of four FMS alternatives and six evaluation criteria, and solved that problem using digraph and matrix approach to obtain a comparative ranking of FMS alternatives as 2-1-3-4. The six considered criteria are annual depreciation and maintenance costs, quality of results, ease of use, competitiveness, adaptability and expandability. When this FMS selection problem is solved using the six preference ranking methods, their comparative ranking performance is derived as shown in Table 24. A very high z value (z = 0.9778) assures the applicability of these preference ranking methods to solve this FMS selection problem too. From Tables 23 and 24, it becomes quite clear that ORESTE method provides the exact rank orderings of FMS alternatives when compared to those obtained by Rao and Parnichkun (2009). It can be concluded that ORESTE method outperforms the other methods with respect to their ranking performance.

Table 24: Performance test results of six preference ranking methods for example 3

Method	EVAMIX	COPRAS	EXPROM2	ORESTE	OCRA	ARAS
Rao and Parnichkun	0.4090, (1,#,#),25	0.4000, (1,#,#),25	1.0000, (1,2,3),100	1.0000, (1,2,3),100	0.2000, (1,#,#),50	1.0000, (1,2,3),100
EVAMIX		1.0000, (1,2,3),100	0.4000, (1,#,#),25	0.4000, (1,#,#),25	0.8000, (1,2,#),50	0.4000, (1,#,#),25
COPRAS			0.4000, (1,#,#),25	0.4000, (1,#,#),25	0.8000, (1,2,#),50	0.4000, (1,#,#),25
EXPROM2				1.0000, (1,2,3),100	0.2000, (1,#,#),25	1.0000, (1,2,3),100
ORESTE					0.2000, (1,#,#),25	1.0000, (1,2,3),100
OCRA						0.2000, (1,#,#),25

MAPPING TO INDUSTRIAL PROBLEMS

The main aim of this paper is to investigate the efficacy of different state-of-the-art preference ranking methods for arriving at the best FMS selection decision. The performance of six preference ranking methods is compared with respect to their suitability to solve MCDM problems, simplicity of use, consistency of choice and degree of decision maker's involvement. As discussed earlier, most of these methods need the definition of criteria weights, which express the relative importance of one criterion over the other. This is one of the most decisive phases for executing any preference ranking-based MCDM method. Application of EVAMIX method starts with the initial decision matrix. A linear normalization technique is adopted to convert all the criteria values into dimensionless numbers ranging from 0 to 1. This is required to compare all the non-commensurable units in the decision matrix. After normalization, the dominance scores for each pair of alternatives are calculated on a criterion-bycriterion comparison basis and an additive interval model is then adopted followed by a weighted summation method. For COPRAS and ARAS methods, a simple weighted summation technique is separately adopted for the normalized beneficial and non-beneficial attributes, leading to the calculation of an overall significance or utility for each of the alternatives. The main difference between the operational procedures of COPRAS and ARAS methods lies in how they normalize the original decision matrix. In COPRAS method, a straightforward linear normalization procedure is used, whereas in ARAS method, a two-step linear normalization technique is adopted. EXPROM2 method is based on the comparison of alternatives considering the deviations that the alternatives show for each criterion. It allows direct computation on the normalized variables on the basis of a 'usual criterion', which is the simplest preference function, requiring no intervention from the decision maker. It also allows the involvement of different preference models to each criterion, each one being characterized by a certain degree of intricacy and a given involvement of some preference parameters to be set by the decision maker. Application of ORESTE method starts with defining a weak order of criteria indicating their relative importance. Then, the alternatives are compared on the basis of Besson rankings and finally, global mean ranks for the alternatives are obtained. This procedure is just like an aggregation technique as followed in most of the popular preference ranking methods. Lastly, application of OCRA method starts with the calculation of preference ratings for each alternative with respect to all the beneficial and non-beneficial criteria. This same concept is also used for COPRAS and ARAS methods. Finally, an aggregation technique is adopted to determine the overall preference rating of each alternative showing the ranking of the alternatives. Thus, it can be said that although the mathematical and operational procedures of the considered preference ranking methods substantially differ from each other, but there are similarities in the concepts they use to reach the final evaluation and ranking of the alternatives in terms of overall utility or significance or preference rating. Hence, mapping an industrial example (like FMS selection) to different preference ranking methods will not be quite troublesome.

CONCLUSIONS

Although different MCDM methods have already been proposed by the past researchers to address the issue of FMS evaluation and selection, it is still not clear which MCDM method is the best for a given FMS selection problem. This paper considers six potential preference ranking-based methods and compares their ranking performance while selecting an FMS for a specific industrial application. Four performance tests are conducted for this ranking performance comparison and also for measuring the degree of agreement between the rankings derived by the considered methods. It is found that although ORESTE performs well, any preference ranking method can be successfully applied for FMS selection problems as the change of the method does not produce any difference in the top ranked FMS alternative. The main reason behind it is that although these methods are based on different mathematical models, they either consider formulation of dominance scores of alternatives on criterion by-criterion basis (as in EVAMIX method) or defining relative priorities of alternatives based on weighted normalized decision matrix (as in COPRAS method) or defining some preference functions (as in EXPROM2 method) or calculating preference ratings of the alternatives considering all the selection criteria (as in OCRA method) or determining optimality functions (as in ARAS method), ultimately leading to the development of a ranking score, which is nothing but an aggregation of all the weighted normalized criteria and a balance between total and individual satisfaction or preference. So basically, all these methods whether they adopt preference function or weighted sum utility value, indicate how much an alternative is preferred to other alternatives. The minor discrepancy that appears between the intermediate rankings obtained by different methods can be attributed to the difference in their mathematical and operational approaches to select the best alternative, the way of dealing with criteria weights in their calculations and introduction of additional parameters affecting the final ranking of the alternatives. In few cases where strong disagreement between these methods occurs, it is due to presence of mixed ordinal-cardinal data in the decision matrix. Thus, the focus would lie not on the selection of the most appropriate preference ranking method to be adopted, but on proper structuring of the decision problem considering relevant criteria and decision alternatives. Future scope may include the application of other preference ranking methods, like COPRAS-G, PSI (preference selection index) etc. to solve complex decision-making problems with crisp, grey and fuzzy criteria.

REFERENCES

1. Selection of low e-windows in retrofit of public buildings by applying multiple criteria method COPRAS: A Lithuanian case. Energy and Buildings, 38(5), 454-462.

2. Kaklauskas, A., Zavadskas, E.K., & Trinkunas, V. (2007). A multiple criteria decision support on-line system for construction. Engineering Applications of Artificial Intelligence, 20(2), 163-175.

3. Karsak, E.E., & Tolga, E. (2001). Fuzzy multi-criteria decision-making procedure for evaluating advanced manufacturing system investments. International Journal of Production Economics, 69(1), 49-64.

4. Karsak, E.E., & Kuzgunkaya, O. (2002). A fuzzy multiple objective programming approach for the selection of a flexible manufacturing system. International Journal of Production Economics, 79(2), 101-111.

5. Karsak, E.E. (2002). Distance-based fuzzy MCDM approach for evaluating flexible manufacturing system alternatives. International Journal of Production Research, 40(13), 3167-3181.

6. Karsak, E.E. (2008). Using data envelopment analysis for evaluating flexible manufacturing systems in the presence of imprecise data. International Journal of Advanced Manufacturing Technology, 35(9-10), 867-874.

7. Kulak, O., & Kahraman, C. (2005). Multi-attribute comparison of advanced manufacturing systems using fuzzy vs. crisp axiomatic design approach. International Journal of Production Economics, 95(3), 415-424.

8. Liu, S-T. (2008). A fuzzy DEA/AR approach to the selection of flexible manufacturing systems. Computers & Industrial Engineering, 54(1), 66-76.

9. Lotfi, V. (1995). Implementing flexible automation: A multiple criteria decision making approach. International Journal of Production Economics, 38(2-3), 255-268.

10. Martel, J.M., & Matarazzo, B. (2005). Other outranking approaches. In: Figueira J, Salvatore G, Ehrgott M. (Eds.) Multiple Criteria Decision Analysis: State of the Art Surveys. Springer: New York.

11.   Mehrabad, M.S., & Anvari, M. (2010). Provident decision making by considering dynamic and fuzzy environment for FMS evaluation. International Journal of Production Research, 48(15), 4555-4584.

12.   O'Grady, P.J., & Menon, U. (1986). A concise review of flexible manufacturing systems and FMS literature. Computers in Industry, 7(2), 155-167.

13.   Ovchinnikov, S., & Roubens, M. (1992). On fuzzy strict preference, indifference, and incomparability relations. Fuzzy Sets and Systems, 49(1), 15-20.

14.   Pastijn, H., & Leysen, J. (1989). Constructing an outranking relation with ORESTE. Mathematical and Computer Modelling, 12(10-11), 1255-1268.

15.   Parkan, C., & Wu, M-L. (1997). On the equivalence of operational performance measurement and multiple attribute decision making. International Journal of Production Research, 35(11), 2963-2988.

16.   Parkan, C., & Wu, M-L. (2000). Comparison of three modern multicriteria decision-making tools. International Journal of Systems Science, 31(4), 497-517.

17.   Raju, K.S., & Kumar, D.N. (1999). Multicriterion decision making in irrigation planning. Agricultural Systems, 62(2), 117-129.

18.   Rao, R.V. (2006). A decision-making framework model for evaluating flexible manufacturing systems using digraph and matrix methods. International Journal of Advanced Manufacturing Technology, 30(11-12), 1101-1110.

19.   Rao, R.V. (2008). Evaluating flexible manufacturing systems using a combined multiple attribute decision making method. International Journal of Production Research, 46(7), 1975-1989.

20.   Rao, R.V., & Parnichkun, M. (2009). Flexible manufacturing system selection using a combinatorial mathematics-based decision-making method. International Journal of Production Research, 47(24), 6981-6998.

21.   Rao, R.V. (2009). Flexible manufacturing system selection using an improved compromise ranking method. International Journal of Industrial and System Engineering, 4(2), 198-215.

22.   Rao, R.V., & Singh, D. (2011). Evaluating flexible manufacturing systems using Euclidean distance-based integrated approach. International Journal of Decision Sciences, Risk and Management, 3(1-2), 32-53.

23.   Roubens, M. (1982). Preference relations on actions and criteria in

multicriteria decision making. European Journal of Operational Research, 10(1), 51-55.

24. Sarkis, J. (1997). Evaluating flexible manufacturing systems alternatives using data envelopment analysis. The Engineering Economist, 43(1), 25-47.

25. Shang, J., & Sueyoshi, T. (1995). A unified framework for the selection of a flexible manufacturing system. European Journal of Operational Research, 85(2), 297-315.

26. Talluri, S., Whiteside, M.M., & Seipel, S.J. (2000). A nonparametric stochastic procedure for FMS evaluation. European Journal of Operational Research, 124(3), 529-538.

27. Teghem, J., Delhaye, C., & Kunsch, P.L. (1989). An interactive decision support system (IDSS) for multicriteria decision aid. Mathematical and Computer Modelling, 12(10-11), 1311-1320

28. Turskis, Z., & Zavadskas, E.K. (2010). A novel method for multiple criteria analysis: Grey additive ratio assessment (ARAS-G) method. Informatica, 21(4), 597-610.

29. Wabalackis, R.N. (1988). Justification of FMS with the analytic hierarchy process. Journal of Manufacturing Systems, 7(3), 175-182.

30. Zavadskas, E.K., Kaklauskas, A., Turskis, Z., & Tamo?aitien?, J. (2008). Selection of the effective dwelling house walls by applying attributes values determined at intervals. Journal of Civil Engineering & Management, 14(2), 85-93.

31. Zavadskas, E.K., & Turskis, Z. (2010). A new additive ratio assessment (ARAS) method in multicriteria decision-making. Technological and Economic Development of Economy, 16(2), 159-172.

Chapter 4

QUALITY EVALUATION IN FLEXIBLE MANUFACTURING SYSTEMS: A MARKOVIAN APPROACH

Jingshan Li[1] and Ningjian Huang[2]

[1]Department of Electrical and Computer Engineering and Center for Manufacturing, University of [1]Kentucky, Lexington, KY 40506, USA

[2]Manufacturing Systems Research Lab, General Motors Research and Development Center, General Motors Corporation, Warren, MI 48090-9055, USA

ABSTRACT

The flexible manufacturing system (FMS) has attracted substantial amount of research effort during the last twenty years. Most of the studies address the issues of flexibility, productivity, cost, and so forth. The impact of flexible lines on product quality is less studied. This paper intends to address this issue by applying a Markov model to evaluate quality performance of a flexible manufacturing system. Closed expressions to calculate good part probability are derived and discussions to maintain high product quality are carried out. An example of flexible fixture in machining system is provided to illustrate the applicability of the method. The results of this study suggest a possible approach to investigate the impact of flexibility on product quality and, finally, with extensions and enrichment of the model, may lead to provide production engineers and managers a better understanding of the quality implications and to summarize some general guidelines of operation management in flexible manufacturing systems.

INTRODUCTION

Manufacturing system design and product quality have been studied extensively during the last 50 years. However, most of the studies address the problems independently. In other words, the majority of the publications on quality research seek to maintain and improve product quality while ignoring the production system concerns. Similarly, the majority of the production system research seeks to maintain the desired productivity while neglecting the question of quality. Little research attention has been paid to investigate the coupling or interaction between production system design and product quality. However, it has been shown in [1] that production system design and product quality are tightly coupled, that is, production system design has a significant impact on product quality as well as other factors. The analysis in this area, which is important but largely unexplored, will open a new direction of research in production systems engineering. To stimulate research in this area, [1] presents several research opportunities from the automotive industry perspective, and flexibility is one of them. To satisfy the rapidly changing markets and varying customer demands, manufacturing systems are becoming more and more flexible. For example, in automotive industry, flexible manufacturing is "becoming even more critical" [2]. Substantial amount of research effort and practices have been devoted to flexible manufacturing systems (FMSs), and it has taken an explicit role in production system design. Much of the work related to flexibility addresses the issues of investment cost, flexibility measurement, and the tradeoffs between productivity and flexibility. However, interactions not only exist between flexibility and productivity, but also between flexibility and quality (as suggested by [1]). The latter one is much less studied.

For example, in many flexible machining systems, a flexible fixture restricts and is the core enabler to flexibility of the whole system, and the cost of designing and fabricating fixtures can amount to 10%–20% of the total manufacturing system cost [3, 4]. A flexible fixture often is a programmable fixture designed to support multiple distinguished parts being manufactured (assembled or machined) on the same line. With the flexible fixture, system flexibility can be achieved with little or no loss of production. In automotive industry, a flexible fixture might be clamps/locators held by robots or other "smart" mobile apparatuses. The challenge, however, with the flexible fixture is the accuracy of the locator measured by the variance. Whenever there is a product change, the fixture needs to adapt itself to the desired corresponding location. As we know, the quality of the manufacturing operation heavily depends on the fixture. The discrepancy of the fixture location from its "ideal" one, in many cases, dominates the quality of the products. For instance, consider a production line producing two products, A and B. Assuming that the fixture is located in a "good" position, that is, within the nominal tolerance, for product A, then if the subsequent parts belong to product A, it is more likely that good quality parts can be produced. Analogously, if the fixture is in a "bad" location, then more defective parts can be produced. However, when the subsequent part is switched to product B, then the fixture needs to readjust its location and either good quality or defective parts may be produced (more detailed description is introduced in Section 4). Therefore, the quality characteristic of the current part is dependent on the part type and quality of the previous one. A study to evaluate that the quality performance in flexible machining environment is valuable, however, has been missing in current literature.

An automotive paint shop is typically capable of painting different models with desired colors. However, the number of available paint colors can significantly impact product quality [2]. Whenever a color change happens, previous paints and solvent need to be purged and spray guns need to be cleaned to remove any residue. The paint quality may temporarily decline after the switch [5]. Thus, the previous vehicle's color may affect next vehicle's quality, as well as other factors (e.g., paint mixing, vehicle cleaning, dirty air, and equipment, etc.). Therefore, vehicles with the same colors are usually grouped into a batch before entering the painting booths without sacrificing much on vehicle delivery. In addition, it is typical to sequence the light color vehicles before the darker ones [6]. Through this, the change-over time (or paint purging time) and the cost of paint purging are reduced. More importantly, the paint quality can be improved by reducing the possibility of incomplete cleaning during purging [7]. However, no analytical study has been found to investigate how flexibility (in terms of number of colors) impacts paint quality, and what would the appropriate batch size and batch sequence be to obtain good paint quality and to satisfy throughput and order delivery requirements as well

Additional examples can be found in welding, assembly operations, and so forth, as well. These examples suggest that flexibility and quality are tightly coupled and much more work is needed to fully understand this coupling. Such an issue is very important but almost neglected. We believe that quality should be integrated into the considerations when designing production systems as well as objectives of productivity and flexibility. The goal of this study is to investigate the coupling between flexibility and product quality, and to provide production engineers and managers a better understanding of the quality implications in flexible manufacturing systems and to offer some general guidelines for management of flexible operations. To start such a study, a simple Markovian model to analyze the quality performance of a flexible manufacturing system is developed. Specifically, a closed-form expression is derived to evaluate the system quality in terms of good part probability and some discussions are carried out based on the analysis. Although inventory, flow control, scheduling, and so forth are also important parts of FMS studies, we limit our work in this paper to quality performance only. Enrichment of the model by integrating quality with other performance measures (e.g., throughput, inventory, cost, etc.) will be a topic for future work. The rest of the paper is structured as follows. Section 2 reviews the related literature. Models and analysis are developed and carried out in Section 3. Using the method developed, an example of quality performance evaluation in a flexible machining system is introduced in Section 4. Finally, Section 5 concludes the paper. All proofs are presented in the appendix.

LITERATURE REVIEW

Although significant research effort has been devoted separately to manufacturing system design and product quality, the coupling or interaction between them has not been studied intensively. Paper [1] reviews the related literature and suggests that this is an open area with promising research opportunities. Limited work addressing this coupling can be found in [8–14]. Specifically, [8] studies the perturbation in the average steady state production rate by quality inspection machines for an asymptotically reliable twomachine one-buffer line. The tradeoffs between productivity and product quality as well as their impact on optimal buffer designs are investigated in [9]. Paper [10] delineates the tradeoff between throughput and quality for a robot whose repeatability deteriorates with speed. Paper [11] uses stochastic search techniques (generic algorithms and simulated annealing) to investigate the impact of inspection allocation in manufacturing systems (serial and nonserial) from the cost perspective. The competing effects of large or small batch sizes are studied in [12] and a model for the interaction between batch size and quality is developed. In addition, [13] uses quantitative measures to deduce that Ushaped lines produce better quality products. A new line balancing approach is proposed in [14] to improve quality by reducing work overload. The recent advances in this area are contained in [15–19]. In [15], a multistage variation propagation model is presented. Paper [16] studies a transfer production line with Andon. It is shown that to produce more good quality parts, Andon is preferable only when average repair time is short and the line should be stopped to repair all the defects. The impact of repair capacity and first time quality on the quality buy rate of an automotive paint is analyzed in [17, 18]. Paper [19] introduces an integrated model of a two-machine one-buffer line with inspection and information feedback to study both quality and quantity performances in terms of good production rate.

Flexibility has attracted a significant amount of research in the last two decades. Most of the work related to flexibility focus on the definition, meaning, and measurement of manufacturing flexibility, and performance modeling of flexible manufacturing systems, and so forth (see, e.g., monographs [20–23], and review papers [24–31]). However, as pointed out in [3], most of the flexibility studies assume that quality-related issues, such as rejects, rework, have minimal impact and that only products of acceptable quality are produced. The production of high quality parts in an FMS requires significant effort and investments. Only a few publications are found discussing the impact of manufacturing flexibility on product quality [32–35]. Specifically, a measure of productivity, quality, and flexibility for production systems is presented in [32]. Paper [33] studies the issues of flexibility, productivity, and quality from an extensive search and analysis of empirical studies. In [34], a method is developed to model the fuzzy flexibility elements such as quality level, efficiency, versatility, and availability. In addition, paper [35] surveys the existing literature related to mass customization. In particular, it points out that quality control issues should be taken into account and current literature lacks in-depth study on how to assure quality in mass-customized products.

In spite of the above effort, the current literature does not provide a quantitative model which enables us to investigate the correlation between quality and number of products and to predict the quality performance of a flexible manufacturing system. We still need to fully understand the coupling or interactions between flexible manufacturing system design and product quality. An in-depth analytical study of the impact of flexibility on quality is necessary and important. This paper is intended to contribute to this end.

MODELS AND ANALYSIS

One Product Type

Consider a flexible manufacturing system producing one product type and let g and d denote the states that the system is producing a good quality part or a defective part in steady states, respectively. Note that here we only study the working or production period of the system. In other words, machine breakdowns are not considered. When the system is in state g, it has a transition probability λ to produce a defective part in the next cycle, and probability $1 - \lambda$ to continue producing a good part. Similarly, when the system is in state d, it can produce a good part with probability μ and a defective part with probability $1 - \mu$ in the next cycle (see Figure 3.1). λ and μ can be viewed as quality failure and repair probabilities, respectively. Similar to throughput analysis, constant transition probabilities are assumed to simplify the analysis for steady state operations.

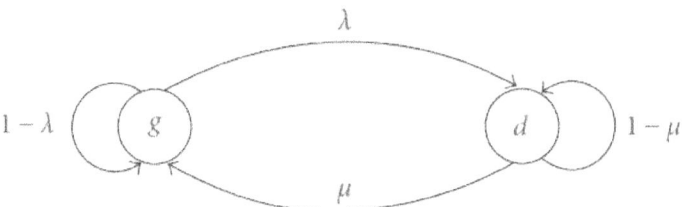

Figure 3.1: State transition diagram in one-product-type case.

Let P(g,t) and P(d,t) denote the probabilities that the system is in states g or d at cycle t, respectively. Clearly, states g and d are similar to the up- and down-states in throughput analysis. Therefore, by extending the method used in throughput analysis to study quality performance, we obtain

$P(g,t+1) = P(\text{produce a good part at } t+1 \mid \text{produce a good part at } t)P(g,t)$

$\qquad + P(\text{produce a good part at } t+1 \mid \text{produce a defective part at } t)P(d,t)$

$\qquad = P(g,t+1 \mid g,t)P(g,t) + P(g,t+1 \mid d,t)P(d,t)$

$\qquad = (1-\lambda)P(g,t) + \mu P(d,t).$

(3.1)

In terms of the steady states, P(g) and P(d) are used to denote the probabilities to produce a good or a defective part during a cycle, respectively, that is,

$$\lim_{t \to \infty} P(g,t) := P(g), \qquad \lim_{t \to \infty} P(d,t) := P(d).$$

(3.2)

It follows that

$P(g) = (1-\lambda)P(g) + \mu P(d),$

(3.3)

which implies that

$$P(d) = \frac{\lambda}{\mu} P(g).$$

(3.4)

From the fact that total probability equals 1,

$P(g) + P(d) = 1,$

(3.5)

it follows that the system good product ratio is

$$P(g) = \frac{\mu}{\lambda + \mu}.$$

(3.6)

Clearly, as expected, (3.6) has a similar form as machine efficiency in throughput analysis. Below, we will extend this study to multiple-product-types case.

TWO PRODUCT TYPES

Now we consider a flexible system producing two types of products, types 1 and 2. Introduce $P(g_i)$ and $P(d_i)$ as the probabilities to produce a good part type i, i = 1,2, or defective part type i, i = 1,2, during a cycle, respectively. Again $P(g)$ and $P(d)$ are used to represent the good or defective part probability (of both products). Then we obtain

$$P(g_1) + P(g_2) = P(g), \qquad P(d_1) + P(d_2) = P(d). \tag{3.7}$$

In addition, introduce the following assumptions.

(i)   A flexible system has four states: producing good part type 1, type 2, and producing defective part type 1 and type 2, denoted as g_1, g_2, d_1, and d_2, respectively.

(ii)   The transition probabilities from good states g_i, i = 1,2, to defective states d_j, j = 1,2, are determined by λ_{ij}. The system has probabilities vij to stay in good states g_j, j = 1,2. Similarly, when the system is in defective states d_i, i = 1,2, it has probabilities μ_{ij} to transit to good states g_j, j = 1,2, and probabilities η_{ij} to stay in defective states d_j, j = 1,2.

Remark 3.1. Similar to one-product-type case, λ_{ii} and μ_{ii}, i = 1,2, can be viewed as nonswitching quality failure and repair probabilities, respectively (i.e., product types are not switched). Analogously, λ_{ij} and μ_{ij}, i, j = 1,2, i ≠ j, can be viewed as switching quality failure and repair rates, respectively

(iii)   When incoming parts are in random order without correlations (nonsequenced), the part flow is identically and uniformly distributed with probabilities $P(1)$ and $P(2)$ for part types 1 and 2, respectively. In other words, every cycle the system has probability $P(1)$ or $P(2)$ to work on part types 1 and 2, respectively.

Remark 3.2. Assumptions (ii) and (iii) imply that probabilities $P(1)$ and $P(2)$ are embedded in the transition probabilities λ_{ij}, μ_{ij}, vj, and η_{ij}, i, j = 1,2. For example, λ_{ij} defines the transition probability that the incoming part is type j and the system produces a defective part at cycle t + 1 given that it produces a good type i part at cycle t.

Based on the above assumptions, we can describe the system using a discrete Markov chain illustrated in Figure 3.2. In addition, since total probabilities equal 1, we have

$$P(1) + P(2) = 1, \qquad P(g_1) + P(d_1) = P(1), \qquad P(g_2) + P(d_2) = P(2),$$
$$\lambda_{11} + \lambda_{12} + v_{11} + v_{12} = 1, \qquad \lambda_{22} + \lambda_{21} + v_{22} + v_{21} = 1,$$
$$\mu_{11} + \mu_{12} + \eta_{11} + \eta_{12} = 1, \qquad \mu_{22} + \mu_{21} + \eta_{22} + \eta_{21} = 1. \tag{3.8}$$

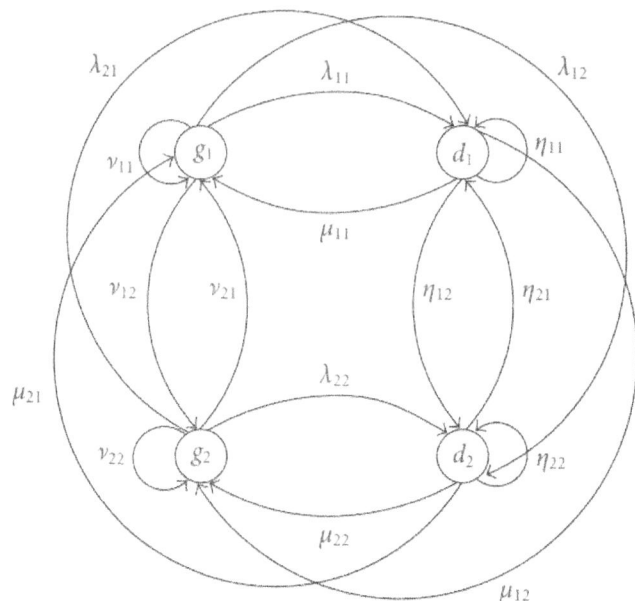

Figure 3.2: State transition diagram in two-product-type case.

Analogously to Section 3.1, the transitions to state g_1 can be described as

$$P(g_1, t+1) = P(g_1, t+1 \mid g_1, t) P(g_1, t) + P(g_1, t+1 \mid d_1, t) P(d_1, t)$$

$$+ P(g_1, t+1 \mid g_2, t) P(g_2, t) + P(g_1, t+1 \mid d_2, t) P(d_2, t)$$

$$= v_{11} P(g_1, t) + v_{21} P(g_2, t) + \mu_{11} P(d_1, t) + \mu_{21} P(d_2, t).$$

(3.9)

Considering the steady state probability $P(g_1)$, we have

$$P(g_1) = v_{11} P(g_1) + v_{21} P(g_2) + \mu_{11} P(d_1) + \mu_{21} P(d_2). \tag{3.10}$$

Similarly,

$$P(g_2) = v_{12} P(g_1) + v_{22} P(g_2) + \mu_{12} P(d_1) + \mu_{22} P(d_2), \tag{3.11}$$

$$P(d_1) = \lambda_{11} P(g_1) + \lambda_{21} P(g_2) + \eta_{11} P(d_1) + \eta_{21} P(d_2), \tag{3.12}$$

$$P(d_2) = \lambda_{12} P(g_1) + \lambda_{22} P(g_2) + \eta_{12} P(d_1) + \eta_{22} P(d_2). \tag{3.13}$$

Solving the above equations, we obtain a closed formula to calculate the probability of good quality part P(g).

Theorem 3.3. Under assumptions (i)–(iii), the good part probability P(g) can be calculated as

$$P(g) = \frac{\mathcal{F}}{\mathcal{F} + \mathcal{G}},$$

(3.14)

where

$$\mathcal{F} = (\lambda_{11} - \lambda_{21})(\mu_{12}\mu_{21} - \mu_{11}\mu_{22}) + (1 - \nu_{22} + \nu_{12})[(1 - \eta_{11})\mu_{21} + \eta_{21}\mu_{11}]$$
$$+ (1 - \nu_{11} + \nu_{21})[(1 - \eta_{11})\mu_{22} + \eta_{21}\mu_{12}].$$
$$\mathcal{G} = (\mu_{21} - \mu_{11})[(1 - \nu_{22})\lambda_{11} + \nu_{12}\lambda_{21}] - (\mu_{12} - \mu_{22})[(1 - \nu_{11})\lambda_{21} + \lambda_{11}\nu_{21}]$$
$$+ [(1 - \nu_{11})(1 - \nu_{22}) - \nu_{12}\nu_{21}](1 - \eta_{11} + \eta_{21}).$$

(3.15)

For the proof, see the appendix.

Multiple (n > 2) Product Types.

Now consider a flexible manufacturing system producing more than two types of product. The same assumptions and notations in Section 3.2 will be used with the exception that now $i = 1,...,n$, denoting n product types. Therefore, we have

$$\sum_{i=1}^{n} P(i) = 1, \qquad \sum_{i=1}^{n} P(g_i) = P(g), \qquad \sum_{i=1}^{n} P(d_i) = P(d),$$

$$P(g_i) + P(d_i) = P(i), \quad i = 1,...,n, \qquad \sum_{i=1}^{n} P(g_i) + \sum_{i=1}^{n} P(d_i) = 1,$$

$$\sum_{j=1}^{n} (\lambda_{ij} + \nu_{ij}) = 1, \quad i = 1,...,n, \qquad \sum_{j=1}^{n} (\mu_{ij} + \eta_{ij}) = 1, \quad i = 1,...,n.$$

(3.16)

Analogously to Section 3.2, we obtain the following transition equations:

$$P(g_j) = \sum_{i=1}^{n} \nu_{ij} P(g_i) + \sum_{i=1}^{n} \mu_{ij} P(d_i), \quad j = 1,...,n,$$

$$P(d_j) = \sum_{i=1}^{n} \lambda_{ij} P(g_i) + \sum_{i=1}^{n} \eta_{ij} P(d_i), \quad j = 1,...,n-1,$$

$$1 = \sum_{i=1}^{n} P(g_i) + \sum_{i=1}^{n} P(d_i).$$

(3.17)

Rearranging them and writing into a matrix form, we have

$$AX = B,$$

(3.18)

where

$$A = \begin{pmatrix} \nu_{11}-1 & \nu_{21} & \cdots & \nu_{n1} & \mu_{11} & \mu_{21} & \cdots & \mu_{n-1,1} & \mu_{n1} \\ \nu_{12} & \nu_{22}-1 & \cdots & \nu_{n2} & \mu_{12} & \mu_{22} & \cdots & \mu_{n-1,2} & \mu_{n2} \\ & \vdots & & \vdots & & \vdots & & \vdots & \\ \nu_{1n} & \nu_{2n} & \cdots & \nu_{nn}-1 & \mu_{1n} & \mu_{2n} & \cdots & \mu_{n-1,n} & \mu_{nn} \\ \lambda_{11} & \lambda_{21} & \cdots & \lambda_{n1} & \eta_{11}-1 & \eta_{21} & \cdots & \eta_{n-1,1} & \eta_{n1} \\ \lambda_{12} & \lambda_{22} & \cdots & \lambda_{n2} & \eta_{12} & \eta_{22}-1 & \cdots & \eta_{n-1,2} & \eta_{n2} \\ & \vdots & & \vdots & & \vdots & & \vdots & \\ \lambda_{1,n-1} & \lambda_{2,n-1} & \cdots & \lambda_{n,n-1} & \eta_{1,n-1} & \eta_{2,n-1} & \cdots & \eta_{n-1,n-1}-1 & \eta_{n,n-1} \\ 1 & 1 & \cdots & 1 & 1 & 1 & \cdots & & 1 \end{pmatrix}$$

$$(3.19)$$

$$X = \left(P(g_1), P(g_2), \ldots, P(g_n), P(d_1), P(d_2), \ldots, P(d_n) \right)^T, \tag{3.20}$$

$$B = (0,0,\ldots,1)^T. \tag{3.21}$$

Therefore, we obtain the following.

Theorem 3.4. Under assumptions (i)–(iii), the good part probability P(g) can be calculated from

$$P(g) = \sum_{i=1}^{n} P(g_i) = \sum_{i=1}^{n} x_i, \tag{3.22}$$

where $x_1 = P(g_1)$, i = 1,...,n, are the elements in X and can be solved from

$$X = A^{-1}B, \tag{3.23}$$

and A, B are defined in (3.19) and (3.21), respectively.

Note that the inverse of matrix A exists due to the fact that an irreducible Markov chain with finite number of states has a unique stationary distribution [36].

In the case of "equal product types," that is, n product types are equally composed (1/n each) and have identical transition probabilities, we have

$$\mu_{11} = \mu_{ii}, \quad \nu_{11} = \nu_{ii}, \quad \lambda_{11} = \lambda_{ii}, \quad \eta_{11} = \eta_{ii}, \quad i = 1,\ldots,n,$$
$$\mu_{12} = \mu_{ij}, \quad \nu_{12} = \nu_{ij}, \quad \lambda_{12} = \lambda_{ij}, \quad \eta_{12} = \eta_{ij}, \quad i,j = 1,\ldots,n, \; i \neq j,$$

$$(3.24)$$

which implies that the transitions from one product type to another are reversible (or equivalent) in terms of quality. Then we obtain the following.

Corollary 3.5. Under assumptions (i)–(iii), the good part probability P(g) for n equal product types is described by

$$P(g) = \frac{\mu_{11} + (n-1)\mu_{12}}{\lambda_{11} + \mu_{11} + (n-1)(\lambda_{12} + \mu_{12})}. \tag{3.25}$$

In addition, P(g) is monotonically increasing and decreasing with respect to μ_{1i} and λ_{1i}, i = 1,2, respectively for the proof, see the appendix.

In order to avoid messy notations, the following discussions are limited to equal product types only.

Discussions

Single versus multiple product types. Similar to throughput analysis (e.g., [20, 21]), let

$$e_{1i} = \frac{\mu_{1i}}{\lambda_{1i} + \mu_{1i}}, \quad i = 1, 2,$$

(3.26)

where e_{12} and e_{11} denote the "switching and nonswitching quality efficiencies," respectively. In other words, $e1_i$ represents the efficiency to produce a good quality part if product type is kept constant (i = 1) or changed (i = 2). By comparing the results with the results of one product case, the following is derived.

Corollary 3.6. Under assumptions (i)–(iii), the following statements hold for the equal product-type case:

a

$$P(g) = \frac{\mu_{11}}{\lambda_{11} + \mu_{11}} \quad \text{if } e_{11} = e_{12},$$

(3.27)

b

$$P(g) < \frac{\mu_{11}}{\lambda_{11} + \mu_{11}} \quad \text{if } e_{11} > e_{12},$$

(3.28)

c

$$P(g) > \frac{\mu_{11}}{\lambda_{11} + \mu_{11}} \quad \text{if } e_{11} < e_{12}.$$

(3.29)

From (3.26), we have $\lambda_{1i} + \mu_{1i} = \mu_{1i}/e_{1i}, i = 1,2.$ Then expression (3.25) can be rewritten into

$$P(g) = \frac{\mu_{11} + (n-1)\mu_{12}}{\mu_{11}/e_{11} + (n-1)(\mu_{12}/e_{12})} = e_{11}\left[\frac{\mu_{11} + (n-1)\mu_{12}}{\mu_{11} + (n-1)\mu_{12} \cdot (e_{11}/e_{12})}\right].$$

(3.30)

The statements follow immediately by replacing e_{12} with e= in the denominator

Corollary 3.6 implies that when $e_{11} = e_{12}$, that is, quality efficiency does not change whether the product types are changed or not, we can obtain P(g) with the same method as in one product case. In other words, if introducing a new product does not change the quality failure or repair probabilities and the product mix does not affect the quality efficiency, then the same quality performance can be achieved, which agrees with our intuition. However, if $e_{11} > e_{12}$, that is, switching quality efficiency is decreased compared to nonswitching, then introducing an additional product will lead to a decrease in system quality performance. Finally, a flexible system can perform better on different products in terms of quality only when the switching quality efficiency is improved with the additional products, that is, $e_{12} > e_{11}$.

Since in many cases much more effort may be needed to keep e_{12} the same as or larger than e_{11}, this result indicates that frequently changing product types may lead to quality degradation in a multiple-product environment. Therefore, using batch operation to reduce product transitions may be an alternative solution to keep both product flexibility and high quality performance

Less Versus More Product Types

Now we consider how the number of product types may affect quality. This is based on the investigation of the monotonic property of $P(g)$ as a function of number of product types n.

Corollary 3.7. Under assumptions (i)–(iii), the good part probability $P(g)$ is monotonically decreasing or increasing with respect to the number of product types n if $e_{11} > e_{12}$ or $e_{11} < e_{12}$, respectively.

Corollary 3.7 suggests that when the switching quality efficiency is not as good as nonswitching efficiency, introducing more products may be harmful for overall quality performance of the system. Therefore, to ensure maintaining desired quality performance, every effort has to be made to achieve $e_{12} \geq e_{11}$.

Random Versus Sequenced Part Flows

To further investigate this phenomenon, consider the following two systems, A and B, both producing n equal part types. System A adopts a sequencing policy with part types 1 to n being mixed randomly with uniform distribution (as described in assumption (iii)), while system B keeps strict alternative sequences 1,2,...,n,1,2,...,n,1,2,..., that is, product type changes at the end of every cycle. Clearly, from (3.25),

$$P(g)^A = \frac{\mu_{11} + (n-1)\mu_{12}}{\lambda_{11} + \mu_{11} + (n-1)(\lambda_{12} + \mu_{12})},$$

$$(3.31)$$

where $P(g)^A$ defines the good job probability of system A. For system B, product type is changed at every cycle, therefore,

$$P(g)^B = \frac{\mu_{12}}{\lambda_{12} + \mu_{12}}.$$

$$(3.32)$$

mparing $P(g)^A$ and $P(g)^B$, we have

$$P(g)^A - P(g)^B = \frac{\mu_{11} + (n-1)\mu_{12}}{\lambda_{11} + \mu_{11} + (n-1)(\lambda_{12} + \mu_{12})} - \frac{\mu_{12}}{\lambda_{12} + \mu_{12}}$$

$$= \frac{\lambda_{12}\mu_{11} - \lambda_{11}\mu_{12}}{[\lambda_{11} + \mu_{11} + (n-1)(\lambda_{12} + \mu_{12})](\lambda_{12} + \mu_{12})}$$

$$= \frac{\mu_{11}\mu_{12}(e_{11} - e_{12})}{e_{11}e_{12}[\lambda_{11} + \mu_{11} + (n-1)(\lambda_{12} + \mu_{12})](\lambda_{12} + \mu_{12})}. \tag{3.33}$$

Therefore, if $e_{11} > e_{12}$, we obtain $P(g)^A > P(g)^B$. It implies that when quality efficiency is decreased for changing products, using randomly mixed sequence has better quality performance than using strictly alternating sequence policy, since the former one has less transitions among products.

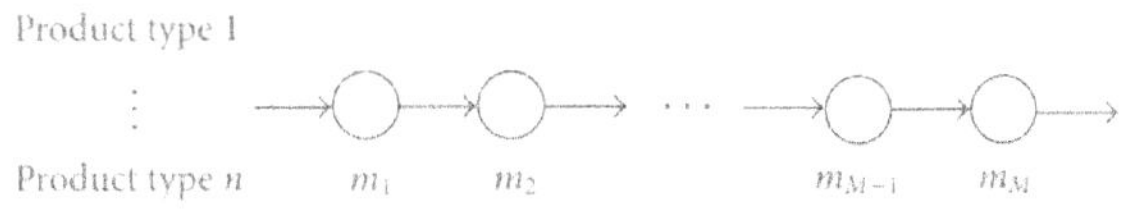

Figure 3.3: M-machine line.

Again, it indicates that using batch processing may lead to a better quality performance than the sequencing policy. A thorough investigation of batch production is important and is a topic in future work.

Extensions to Multistage Flexible Systems

Now we consider a flexible system consisting of multistages as shown in Figure 3.3, where the circles represent each stage. Introduce the following additional assumption.

(iv) Each stage of flexible system, m_i, only performs its own function, and therefore each stage is independent. In other words, downstream stages could not correct the defects introduced by upstream stages.

Let $P(g(i))$, $i = 1,...,M$, be the probability of producing a good part at stage i, then the overall probability to produce a good part for an M-stage flexible line would be

$$P(G) = \prod_{i=1}^{M} P(g(i)). \tag{3.34}$$

Introduce $\lambda_{i,k\,j}$ and $\mu_{i,k\,j}$, $i = 1,...,M$, $k, j = 1,...,n$, to be the transition probabilities from state g_k to state d_j, or from d_k to g_j for machine i. Then for the case of n equal part types, we obtain

$$P(G) = \prod_{i=1}^{M} \frac{\mu_{i,11} + (n-1)\mu_{i,12}}{\lambda_{i,11} + \mu_{i,11} + (n-1)(\lambda_{i,12} + \mu_{i,12})}. \tag{3.35}$$

In the case where all stages are identical, the first subscripts in $\lambda_{i,kj}$ and $\mu_{i,kj}$ can be omitted, we have

$$P(G) = [P(g(i))]^M = \left[\frac{\mu_{11} + (n-1)\mu_{12}}{\lambda_{11} + \mu_{11} + (n-1)(\lambda_{12} + \mu_{12})}\right]^M. \qquad (3.36)$$

Similar insights can be obtained when we compare the results with the single-stage multiple-product-type case (where quality performance is $[\mu_{11}/(\lambda_{11} + \mu_1 1)]M$). In other words, when switching quality efficiency is kept the same as nonswitching in mixed products environment, that is, $e_{11} = e_{12}$, the same quality performance as single product case can be achieved. However, if quality efficiency is decreased for changing products, $e_{11} > e_{12}$, then additional product type can decrease the system quality performance. Only when $e_{12} > e_{11}$, multiple-product system has better quality performance. Therefore, to ensure a flexible manufacturing system having high quality performance, the quality effi- ciency for changing products must be equivalent to or better than that for single product.

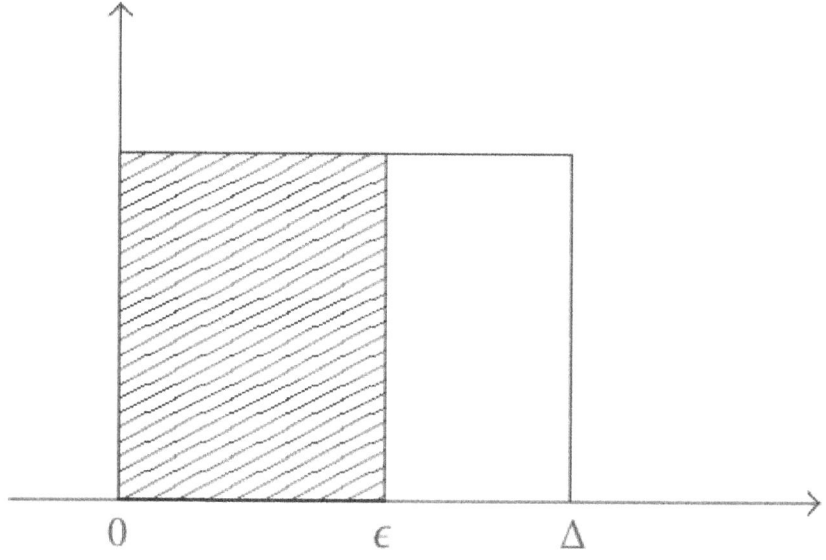

Figure 4.1: Locator discrepancy and tolerance range.

Remark 3.8. Note that in the above multistage model of flexible systems, only quality performance is addressed and the issues of buffers and inventory are not investigated. However, even if buffers are considered, since the current formulation does not include machine breakdown, all parts, no matter good or defective, will flow into and out of the buffer without interruptions. Moreover, even when productivity (e.g., machine breakdowns) is taken into consideration, a separation principle can be applied, that is, as long as there are no actions (e.g., scrap, rework, etc.) taking at each stage, we can simply separate the analysis of quality and productivity (similar to the separation principle in control theory) by evaluating the good part probability and production volume independently. Only when we reach the stage where some actions are taken, integrated analysis is needed. Such integrated study will be a topic of future work.

AN EXAMPLE IN FLEXIBLE MACHINING SYSTEM

Consider a drilling operation in a flexible machining system that drills a hole on part type A and part type B. The system has a flexible fixture. When a job comes in, the fixture can adapt itself to predesigned locations (referred to as L_a and L_b for part types A and B, resp.) in order to hold the part, then the drill will take place. Now assuming incoming parts are in a random order mixed with types A and B (assumption (iii)), then the fixture may move to location La when part type A is coming, then to L_b when B is coming, and may return to L_a after some time to process A again. Since the fixture is not perfect, the Las (correspondingly, L_{bs}) may not be the same as the designed L_a (correspondingly, L_b). One way of evaluating it is to measure the distance between the real L_a (correspondingly, L_b) and the ideal location. Figure 4.1 shows discrepancy of a locator from its nominal position, assuming the locator can be anywhere between the "ideal" location 0 and distance Δ_a or Δ_b with uniform distribution for parts A and B, respectively. It is clear that when the locator (e.g., La) is too far from the designed (ideal) location, the hole will be drilled on a wrong place, which will cause a quality defect. On the other hand, when the locator is within the designed tolerance (shown in Figure 4.1 as), it will not hurt the hole drilling.

Now we assume that the flexible fixture is the only factor that causes quality defects. (It is common that the locating error is much larger than the tooling error.) Then the probability of a part with good quality is $/\Delta_a$ for part type A (correspondingly, $/\Delta_b$ for part type B), denoted as δ_a (correspondingly, δ_b), indicating the probability that the locator moves to a satisfactory location.

Assuming δ_a and δ_b are independent of the locator's starting location, then the transition matrix of the states of this problem (making part A and part B) becomes

$$P_{transition} = \begin{pmatrix} \nu_{11} & \lambda_{11} & \nu_{12} & \lambda_{12} \\ \mu_{11} & \eta_{11} & \mu_{12} & \eta_{12} \\ \nu_{21} & \lambda_{21} & \nu_{22} & \lambda_{22} \\ \mu_{21} & \eta_{21} & \mu_{22} & \eta_{22} \end{pmatrix}$$

(4.1)

$$= \begin{pmatrix} P(\overline{move}, good)P_a & P(\overline{move}, bad)P_a & P(move, good)P_b & P(move, bad)P_b \\ P(\overline{move}, good)P_a & P(\overline{move}, bad)P_a & P(move, good)P_b & P(move, bad)P_b \\ P(move, good)P_a & P(move, bad)P_a & P(\overline{move}, good)P_b & P(\overline{move}, bad)P_b \\ P(move, good)P_a & P(move, bad)P_a & P(\overline{move}, good)P_b & P(\overline{move}, bad)P_b \end{pmatrix},$$

(4.2)

where Pa and P_b are the probabilities that the next job is part A or B, respectively, and $P_a + P_b = 1$. $P(\overline{move}, good)$ and $P(\overline{move}, bad)$ are the probabilities that the locator has moved and is in a "good" or "bad" location, respectively. Similarly, $P(\overline{move}, good)$ and $P(\overline{move}, bad)$ are the probabilities that the locator has not moved and is in a "good" or "bad" location, respectively

This matrix can be simplified. For example, when the locator is in "good" location producing part A, then it does not move if the next job is still part A, and the transition probability of making a good part A (correspondingly, bad part A) will be only determined by P_a (correspondingly, 0). (Note that here we assume that location error is the only source for defects.) This is because when the locator is in a good position and the next job belongs to the same type, the probability of making another good job is 1. Similarly, if it is in the "good" location producing part A, but the next job is part B, the locator will move. The probability of moving to a "good" position (making a good part B) is δ_b. Therefore the transition probability from good A location to good B location is $\delta_b P_b$. Repeat this process, and finally we can obtain a simplified transition matrix:

$$P_{transition} = \begin{pmatrix} \nu_{11} & \lambda_{11} & \nu_{12} & \lambda_{12} \\ \mu_{11} & \eta_{11} & \mu_{12} & \eta_{12} \\ \nu_{21} & \lambda_{21} & \nu_{22} & \lambda_{22} \\ \mu_{21} & \eta_{21} & \mu_{22} & \eta_{22} \end{pmatrix} = \begin{pmatrix} P_a & 0 & \delta_b P_b & (1-\delta_b)P_b \\ 0 & P_a & \delta_b P_b & (1-\delta_b)P_b \\ \delta_a P_a & (1-\delta_a)P_a & P_b & 0 \\ \delta_a P_a & (1-\delta_a)P_a & 0 & P_b \end{pmatrix}.$$

(4.3)

With the above relationship, we obtain values for variables $\lambda_{ij}, \mu_{ij}, \nu_{ij}$ and $\eta_{ij}, i, j = 1, 2.$. Then, using Theorem 3.3, the good part probability is obtained as

$$P(g) = \frac{\mathcal{J}}{\mathcal{J} + \mathcal{G}},$$

(4.4)

where

$$
\begin{aligned}
\mathcal{F} &= -P_a(1-\delta_a)\delta_b P_b \delta_a P_a + (1-P_b+\delta_b P_b)(1-P_a)\delta_a P_a + (1-P_a+\delta_a P_a)(1-\delta_a)P_a\delta_b P_b \\
&= P_a\big[-(1-\delta_a)\delta_a\delta_b P_a P_b + (1-P_b+\delta_b P_b)(1-P_a)\delta_a + (1-P_a+\delta_a P_a)(1-\delta_a)\delta_b P_b\big] \\
&= P_a(1-P_a)\big[(1-\delta_a)\delta_b P_b + \delta_a(1-P_b+\delta_b P_b)\big] \\
&= P_a P_b(\delta_a P_a + \delta_b P_b),
\end{aligned}
$$

$$
\begin{aligned}
\mathcal{G} &= \delta_a P_a \delta_b P_b(1-\delta)P_a - \delta_b P_b(1-P_a)(1-\delta_a)P_a \\
&\quad + \big[(1-P_a)(1-P_b) - \delta_b P_b \delta_a P_a\big]\big[1-P_a+(1-\delta_a)P_a\big] \\
&= \delta_a\delta_b P_a^2 P_b(1-\delta_a) - \delta_b P_a P_b^2(1-\delta_a) + P_a P_b(1-\delta_a\delta_b)\big[P_b+(1-\delta_a)P_a\big] \\
&= P_a P_b\big[P_a\delta_a\delta_b(1-\delta_a) - P_b\delta_b(1-\delta_a) + (1-\delta_a\delta_b)(P_b+(1-\delta_a)P_a)\big] \\
&= P_a P_b\big[(1-\delta_a)(P_a\delta_a - P_b)\delta_b + (1-\delta_a\delta_b)(1-P_a\delta_a)\big].
\end{aligned}
\tag{4.5}
$$

It follows that

$$
\begin{aligned}
\mathcal{F} + \mathcal{G} &= P_a P_b\big[(\delta_a P_a + \delta_b P_b) + (1-\delta_a)(P_a\delta_a - P_b)\delta_b + (1-\delta_a\delta_b)(1-P_a\delta_a)\big] \\
&= P_a P_b(1-\delta_a\delta_b + P_a\delta_a\delta_b + P_b\delta_a\delta_b) = P_a P_b.
\end{aligned}
\tag{4.6}
$$

Therefore, we obtain

$$
P(g) = \frac{P_a(1-P_a)\big[\delta_a(1-P_b)+\delta_b P_b\big]}{P_a P_b} = \delta_a P_a + \delta_b P_b.
\tag{4.7}
$$

Furthermore, it is reasonable to assume that Δ_a and Δ_b would be the same in many cases. Therefore $\delta a = \delta_b = \delta$, and we obtain $P(g) = \delta$, that is, the probability of making a good part depends only on the flexible locators, which is consistent with our intuition.

Applying the same concept to three-product case, we assume that three products A, B, and C are manufactured with the flexible locator. For simplicity, here we only consider the case of $\delta_a = \delta_b = \delta_c = \delta$. We compose the matrix A in (3.19) and simplify it as follows:

$$
A = \begin{pmatrix}
v_{11}-1 & v_{21} & v_{31} & \mu_{11} & \mu_{21} & \mu_{31} \\
v_{12} & v_{22}-1 & v_{32} & \mu_{12} & \mu_{22} & \mu_{32} \\
v_{13} & v_{23} & v_{33}-1 & \mu_{13} & \mu_{23} & \mu_{33} \\
\lambda_{11} & \lambda_{21} & \lambda_{31} & \eta_{11}-1 & \eta_{21} & \eta_{31} \\
\lambda_{12} & \lambda_{22} & \lambda_{32} & \eta_{12} & \eta_{22}-1 & \eta_{32} \\
1 & 1 & 1 & 1 & 1 & 1
\end{pmatrix}
$$

$$
= \begin{pmatrix}
P_a-1 & \delta P_a & \delta P_a & 0 & \delta P_a & \delta P_a \\
\delta P_b & P_b-1 & \delta P_b & \delta P_b & 0 & \delta P_b \\
\delta P_c & \delta P_c & P_c-1 & \delta P_c & \delta P_c & 0 \\
0 & (1-\delta)P_a & (1-\delta)P_a & P_a-1 & (1-\delta)P_a & (1-\delta)P_a \\
(1-\delta)P_b & 0 & (1-\delta)P_b & (1-\delta)P_b & P_b-1 & (1-\delta)P_b \\
1 & 1 & 1 & 1 & 1 & 1
\end{pmatrix}.
\tag{4.7}
$$

After some simplification and rearrangement (see the appendix for details), we can finally reac0068

$$
P(g_a) = \delta P_a, \qquad P(g_b) = \delta P_c, \qquad P(g_c) = \delta P_c,
\tag{4.8}
$$

where g_a, g_b, and g_c denote that the system is in good states producing parts

A, B, and C, respectively. Therefore, the probability of making a good part is

$$P(g) = P(g_a) + P(g_b) + P(g_c) = \delta(P_a + P_b + P_c) = \delta. \quad (4.9)$$

This result again is consistent with the one of two-product case and matches our expectation. It also verifies the analysis presented in Section 3.

For more than three-product case, assume there are n products, and all $\delta i = \delta$, i = 1,...,n. By induction, we can show that $P(g) = \delta$ holds again. The idea of the proof is as follows. We first show that the base case (n = 2) is true (4.7). Next, we assume that the case n = k − 1 is true. Then for case n = k, we can group the first k − 1 products into an aggregated product since they result in good part probability equal to δ. Now we only have two products, the aggregated product and product k. Using the results for n = 2, we prove that the case n = k is also true, which will lead to the good part probability equal to δ for n products as well.

It is not surprising that the probability of making a good part is not dependent on the number of products nor the penetration of each product, since we assume that the quality is only determined by the locators with the same δ. This implies that once we can control the flexible fixture (locator), introducing more products will not hurt product quality. However, when δ's are not identical for different products, then the system quality performance will be dependent on the number of products, their respective δ, and different ratios of product mix.

CONCLUSIONS

Manufacturing system design has a significant impact on product quality as well as other factors. The quality performance of a flexible manufacturing system is less studied and often assumed unchanged compared to dedicated production lines. In this paper, we develop a quantitative model to evaluate the quality performance of a flexible manufacturing system using a discrete Markov chain. We derive closed formulas to calculate good part probability and show that the quality of a flexible system depends on the quality efficiency during transitions of different products. An example in a flexible machining system is presented to illustrate the applicability of the method and verify the results obtained in the paper. The work presented in this paper provides a possible approach for further investigation of the coupling between flexibility and product quality. The future work can be directed to, first, extend the model to multiple-stage flexible lines with correlated quality propagations (e.g., variation stack-up), where the quality performance of a flexible system is also dependent on the condition of incoming parts; second, extend the model to investigate flexible lines with batch or sequenced production to evaluate the impacts of different scheduling and control policies on quality; third, integrate with online and offline inspections, quality repair, and maintenance scheduling, and so forth; fourth, integrate with multiple-product throughput analysis models with quality control devices and study the tradeoffs among productivity, quality, and order delivery; and finally, apply the method to model and analyze different flexible manufacturing systems. The results of such study will provide production engineers and managers a better understanding of the quality implications and to summarize some general guidelines for operation management in flexible manufacturing systems.

APPENDICES

A Proof

Due to page limitation, we provide here only the sketches of the proofs. The complete proof can be found in [37].

Proof of Theorem 3.3. From transition equation (3.10), we have

$$P(d_2) = \frac{1}{\mu_{21}}[(1 - \nu_{11})P(g_1) - \nu_{21}P(g_2) - \mu_{11}P(d_1)].$$

(A.1)

Substituting (A.1) into (3.11), we have

$$(1 - \nu_{22})\mu_{21}P(g_2) = \nu_{12}\mu_{21}P(g_1) + \mu_{12}\mu_{21}P(d_1) + \mu_{22}(1 - \nu_{11})P(g_1) - \mu_{22}\nu_{21}P(g_2) - \mu_{22}\mu_{11}P(d_1),$$

(A.2)

which leads to

$$P(d_1) = \frac{[(1 - \nu_{22})\mu_{21} + \mu_{22}\nu_{21}]P(g_2) - [(1 - \nu_{11})\mu_{22} + \mu_{21}\nu_{12}]P(g_1)}{\mu_{12}\mu_{21} - \mu_{11}\mu_{22}}.$$

(A.3)

Substituting into (A.1), we obtain

$$P(d_2) = \frac{[\mu_{11}\nu_{12} + \mu_{12}(1 - \nu_{11})]P(g_1) - [\mu_{11}(1 - \nu_{22}) + \mu_{12}\nu_{21}]P(g_2)}{\mu_{12}\mu_{21} - \mu_{11}\mu_{22}}.$$

(A.4)

Rewriting (3.12), we have

$$(1 - \eta_{11})P(d_1) = \lambda_{11}P(g_1) + \lambda_{21}P(g_2) + \eta_{21}P(d_2).$$

(A.5)

Again substituting (A.3) and (A.4), we obtain

$$(1 - \eta_{11}) \frac{[(1 - \nu_{22})\mu_{21} + \mu_{22}\nu_{21}]P(g_2) - [(1 - \nu_{11})\mu_{22} + \mu_{21}\nu_{12})P(g_1)}{\mu_{12}\mu_{21} - \mu_{11}\mu_{22}}$$

$$= \lambda_{11}P(g_1) + \lambda_{21}P(g_2)$$

$$+ \eta_{21} \frac{[\mu_{11}\nu_{12} + \mu_{12}(1 - \nu_{11})]P(g_1) - [\mu_{11}(1 - \nu_{22}) + \mu_{12}\nu_{21}]P(g_2)]}{\mu_{12}\mu_{21} - \mu_{11}\mu_{22}}.$$

(A.6)

It follows that

$$[(1 - \eta_{11})(1 - \nu_{22})\mu_{21} + (1 - \eta_{11})\mu_{22}\nu_{21} - \lambda_{21}(\mu_{12}\mu_{21} - \mu_{11}\mu_{22})$$
$$+ \eta_{21}\mu_{11}(1 - \nu_{22}) + \eta_{21}\mu_{12}\nu_{21}]P(g_2)$$
$$= [\lambda_{11}(\mu_{12}\mu_{21} - \mu_{11}\mu_{22}) + \eta_{21}\mu_{11}\nu_{12} + \eta_{21}\mu_{12}(1 - \nu_{11})$$
$$+ (1 - \eta_{11})(1 - \nu_{11})\mu_{22} + (1 - \eta_{11})\mu_{21}\nu_{12}]P(g_1).$$

(A.7)

Therefore,

$$P(g_2)$$
$$= \frac{\lambda_{11}(\mu_{12}\mu_{21} - \mu_{11}\mu_{22}) + \eta_{21}[\mu_{11}\nu_{12} + \mu_{12}(1 - \nu_{11})] + (1 - \eta_{11})[(1 - \nu_{11})\mu_{22} + \mu_{21}\nu_{12}]}{(1 - \eta_{11})[(1 - \nu_{22})\mu_{21} + \mu_{22}\nu_{21}] - \lambda_{21}(\mu_{12}\mu_{21} - \mu_{11}\mu_{22}) + \eta_{21}[\mu_{11}(1 - \nu_{22}) + \mu_{12}\nu_{21}]}P(g_1).$$

(A.8)

From total probabilities equal to 1, that is,

$$P(g_1) + P(g_2) + P(d_1) + P(d_2) = 1,$$ (A.9)

we obtain

$$P(g_1) + P(g_2) + \frac{[(1 - \nu_{22})\mu_{21} + \mu_{22}\nu_{21}]P(g_2) - [(1 - \nu_{11})\mu_{22} + \mu_{21}\nu_{12}]P(g_1)}{\mu_{12}\mu_{21} - \mu_{11}\mu_{22}}$$

$$+ \frac{[\mu_{11}\nu_{12} + \mu_{12}(1 - \nu_{11})]P(g_1) - [\mu_{11}(1 - \nu_{22}) + \mu_{12}\nu_{21}]P(g_2)}{\mu_{12}\mu_{21} - \mu_{11}\mu_{22}} = 1,$$

(A.10)

which implies that

$$\mu_{12}\mu_{21} - \mu_{11}\mu_{22} = P(g_1)[\mu_{12}(1 - \nu_{11} + \mu_{21}) - \mu_{22}(1 - \nu_{11} + \mu_{11}) + (\mu_{11} - \mu_{21})\nu_{12}]$$
$$+ P(g_2)[\mu_{21}(1 - \nu_{22} + \mu_{12}) - \mu_{11}(1 - \nu_{22} + \mu_{22}) + \nu_{21}(\mu_{22} - \mu_{12})].$$

(A.11)

For simplification purpose, introduce the following notations:

$$\mathcal{A} = \mu_{12}\mu_{21} - \mu_{11}\mu_{22},$$
$$\mathcal{B} = (1 - \nu_{11})(\mu_{12} - \mu_{22}) + (\mu_{11} - \mu_{21})\nu_{12} + \mathcal{A},$$
$$\mathcal{C} = (1 - \eta_{11})[(1 - \nu_{22})\mu_{21} + \mu_{22}\nu_{21}] - \lambda_{21}\mathcal{A} + \eta_{21}[\mu_{11}(1 - \nu_{22}) + \mu_{12}\nu_{21}],$$
$$\mathcal{D} = \lambda_{11}\mathcal{A} + \eta_{21}[\mu_{11}\nu_{12} + \mu_{12}(1 - \nu_{11})] + (1 - \eta_{11})[(1 - \nu_{11})\mu_{22} + \mu_{21}\nu_{12}],$$
$$\mathcal{E} = (1 - \nu_{22})(\mu_{21} - \mu_{11}) + (\mu_{22} - \mu_{12})\nu_{21} + \mathcal{A}.$$

(A.12)

Replacing into (A.11), we obtain

$$\mathcal{A} = \mathcal{B}P(g_1) + \mathcal{E}P(g_2).$$

(A.13)

From (A.8), we have

$$P(g_2) = \frac{\mathcal{D}}{\mathcal{C}} P(g_1),$$

(A.14)

then

$$\mathcal{A} = B P(g_1) + \frac{\mathcal{D}}{\mathcal{C}} \mathcal{E} P(g_1).$$

(A.15)

It follows that

$$P(g_1) = \frac{\mathcal{A}}{\mathcal{B} + (\mathcal{D}/\mathcal{C})\mathcal{E}} = \frac{\mathcal{A}\mathcal{C}}{\mathcal{B}\mathcal{C} + \mathcal{D}\mathcal{E}},$$

$$P(g_2) = \frac{\mathcal{D}}{\mathcal{C}} P(g_1) = \frac{\mathcal{A}\mathcal{D}}{\mathcal{B}\mathcal{C} + \mathcal{D}\mathcal{E}}.$$

(A.16)

Therefore,

$$P(g) = P(g_1) + P(g_2) = \frac{\mathcal{A}(\mathcal{C} + \mathcal{D})}{\mathcal{B}\mathcal{C} + \mathcal{D}\mathcal{E}}.$$

(A.17)

To continue simplifying the equations, we obtain

$$\mathcal{C} + \mathcal{D} = (1 - \nu_{22} + \nu_{12})\big[(1 - \eta_{11})\mu_{21} + \eta_{21}\mu_{11}\big] + (\lambda_{11} - \lambda_{21})\mathcal{A}$$
$$+ (1 - \nu_{11} + \nu_{21})\big[(1 - \eta_{11})\mu_{22} + \eta_{21}\mu_{12}\big],$$
$$\mathcal{B}\mathcal{C} + \mathcal{D}\mathcal{E} = \mathcal{A}(\mathcal{C} + \mathcal{D})$$
$$+ \mathcal{A}\big[(\mu_{21} - \mu_{11})\big[(1 - \nu_{22})\lambda_{11} + \nu_{12}\lambda_{21}\big] - (\mu_{12} - \mu_{22})$$
$$\times \big[(1 - \nu_{11})\lambda_{21} + \lambda_{11}\nu_{21}\big] + \big[(1 - \nu_{11})(1 - \nu_{22}) - \nu_{12}\nu_{21}\big](1 - \eta_{11} + \eta_{21})\big].$$

(A.18)

Introduce notations F and G:

$$\mathcal{F} = \mathcal{C} + \mathcal{D},$$

$$\mathcal{G} = (\mu_{21} - \mu_{11})\big[(1 - \nu_{22})\lambda_{11} + \nu_{12}\lambda_{21}\big] - (\mu_{12} - \mu_{22})\big[(1 - \nu_{11})\lambda_{21} + \lambda_{11}\nu_{21}\big]$$
$$+ \big[(1 - \nu_{11})(1 - \nu_{22}) - \nu_{12}\nu_{21}\big](1 - \eta_{11} + \eta_{21}).$$

(A.19)

Then

$$\mathcal{B}\mathcal{C} + \mathcal{D}\mathcal{E} = \mathcal{A}(\mathcal{C} + \mathcal{D}) + \mathcal{A}\mathcal{G}.$$

(A.20)

Finally, we obtain

$$P(g) = \frac{\mathcal{A}\mathcal{F}}{\mathcal{F}\mathcal{A} + \mathcal{G}\mathcal{A}} = \frac{\mathcal{F}}{\mathcal{F} + \mathcal{G}}.$$

(A.21)

Proof of Corollary 3.5. First we aggregate all the good states gi, i = 1,...,n, and all the defective states di, i = 1,...,n, into aggregated good state g_{agg} and defective state d, respectively. Following the logic in Section 3.1, we have

$$P(g_{agg}) = P(g_{agg})(1 - \lambda_{agg}) + P(d_{agg})\mu_{agg},$$

(A.22)

where λ_{agg} and μ_{agg} are the aggregated quality failure and repair probabilities. (The state transition diagram is equivalent to that of Figure 3.1 with subscripts "agg" in all notations.)

Therefore,

$$P(g_{\text{agg}}) = \frac{\mu_{\text{agg}}}{\lambda_{\text{agg}} + \mu_{\text{agg}}}.$$

(A.23)

In addition,

$$
\begin{aligned}
\lambda_{\text{agg}} &= \sum_{i=1}^{n} \lambda_{ii} P(g_i) P(d_i) + \sum_{i=1}^{n} \sum_{j=1, j \neq i}^{n} \lambda_{ij} P(g_i) P(d_j) \\
&= n\lambda_{11} P(g_1) P(d_1) + n(n-1)\lambda_{12} P(g_1) P(d_2), \\
\mu_{\text{agg}} &= \sum_{i=1}^{n} \mu_{ii} P(d_i) P(g_i) + \sum_{i=1}^{n} \sum_{j=1, j \neq i}^{n} \mu_{ij} P(d_i) P(g_j) \\
&= n\mu_{11} P(d_1) P(g_1) + n(n-1)\mu_{12} P(d_1) P(g_2).
\end{aligned}
$$

(A.24)

Since all products are equally distributed, we have

$$P(d_2) = P(d_1), \qquad P(g_2) = P(g_1).$$

(A.25)

Therefore,

$$
\begin{aligned}
\lambda_{\text{agg}} &= n\lambda_{11} P(g_1) P(d_1) + n(n-1)\lambda_{12} P(g_1) P(d_1), \\
\mu_{\text{agg}} &= n\mu_{11} P(d_1) P(g_1) + n(n-1)\mu_{12} P(d_1) P(g_1).
\end{aligned}
$$

(A.26)

Substituting into (A.23), we obtain

$$
\begin{aligned}
P(g_{\text{agg}}) &= \frac{nP(d_1)P(g_1)[\mu_{11} + n(n-1)\mu_{12}]}{nP(g_1)P(d_1)[\lambda_{11} + (n-1)\lambda_{12}] + nP(d_1)P(g_1)[\mu_{11} + (n-1)\mu_{12}]} \\
&= \frac{\mu_{11} + (n-1)\mu_{12}}{\lambda_{11} + (n-1)\lambda_{12} + \mu_{11} + (n-1)\mu_{12}}.
\end{aligned}
$$

(A.27)

Moreover, from

$$
\begin{aligned}
\frac{\partial P(g)}{\partial \mu_{11}} &= \frac{1}{\lambda_{11} + (n-1)\lambda_{12} + \mu_{11} + (n-1)\mu_{12}} - \frac{\mu_{11} + (n-1)\mu_{12}}{[\lambda_{11} + (n-1)\lambda_{12} + \mu_{11} + (n-1)\mu_{12}]^2} \\
&= \frac{\lambda_{11} + (n-1)\lambda_{12}}{[\lambda_{11} + (n-1)\lambda_{12} + \mu_{11} + (n-1)\mu_{12}]^2} > 0, \\
\frac{\partial P(g)}{\partial \mu_{12}} &= \frac{n-1}{\lambda_{11} + (n-1)\lambda_{12} + \mu_{11} + (n-1)\mu_{12}} - \frac{[\mu_{11} + (n-1)\mu_{12}](n-1)}{[\lambda_{11} + (n-1)\lambda_{12} + \mu_{11} + (n-1)\mu_{12}]^2} \\
&= \frac{[\lambda_{11} + (n-1)\lambda_{12}](n-1)}{[\lambda_{11} + (n-1)\lambda_{12} + \mu_{11} + (n-1)\mu_{12}]^2} > 0, \\
\frac{\partial P(g)}{\partial \lambda_{11}} &= -\frac{\mu_{11} + (n-1)\mu_{12}}{[\lambda_{11} + (n-1)\lambda_{12} + \mu_{11} + (n-1)\mu_{12}]^2} < 0, \\
\frac{\partial P(g)}{\partial \lambda_{12}} &= -\frac{[\mu_{11} + (n-1)\mu_{12}](n-1)}{[\lambda_{11} + (n-1)\lambda_{12} + \mu_{11} + (n-1)\mu_{12}]^2} < 0,
\end{aligned}
$$

(A.28)

we obtain the monotonicities of P(g) with respect to μ_{1i} and λ_{1i}, $i = 1,2$.

Proof of Corollary 3.7. From

$$
\frac{\partial P(g)}{\partial n} = \frac{\mu_{12}}{\lambda_{11} + (n-1)\lambda_{12} + \mu_{11} + (n-1)\mu_{12}} - \frac{[\mu_{11} + (n-1)\mu_{12}](\lambda_{12} + \mu_{12})}{[\lambda_{11} + (n-1)\lambda_{12} + \mu_{11} + (n-1)\mu_{12}]^2}
$$

$$
= -\frac{\mu_{11}\mu_{12}(e_{11} - e_{12})}{[\lambda_{11} + (n-1)\lambda_{12} + \mu_{11} + (n-1)\mu_{12}]^2},
$$

(A.29)

we obtain

$$
\frac{\partial P(g)}{\partial n} < 0 \quad \text{if } e_{11} > e_{12}, \qquad \frac{\partial P(g)}{\partial n} > 0 \quad \text{if } e_{11} < e_{12}.
$$

(A.30)

Therefore, P(g) is monotonically decreasing or increasing with respect to n if $e_{11} > e_{12}$ or $e_{11} < e_{12}$, respectively.

B. Solution Procedure for Three-Product Case in SECTION.A

Using matrix A in (4.8), we further simplify (3.18) as follows:

$$
\begin{pmatrix}
\frac{P_a - 1 - \delta P_a}{\delta P_a} & 0 & 0 & -1 & 0 & 0 \\
0 & \frac{P_b - 1 - \delta P_b}{\delta P_b} & 0 & 0 & -1 & 0 \\
0 & 0 & \frac{P_c - 1 - \delta P_c}{\delta P_c} & 0 & 0 & -1 \\
-1 & 0 & 0 & \frac{\delta P_a - 1}{(1-\delta)P_a} & 0 & 0 \\
0 & -1 & 0 & 0 & \frac{\delta P_b - 1}{(1-\delta)P_b} & 0 \\
1 & 1 & 1 & 1 & 1 & 1
\end{pmatrix}
\begin{pmatrix}
P(g_a) \\
P(g_b) \\
P(g_c) \\
P(d_a) \\
P(d_b) \\
P(d_c)
\end{pmatrix}
=
\begin{pmatrix}
-1 \\
-1 \\
-1 \\
-1 \\
-1 \\
1
\end{pmatrix}
$$

(B.1)

Therefore, we obtain

$$
\frac{P_a - 1 - \delta P_a}{\delta P_a} P(g_a) - P(d_a) = -1,
$$

$$
\frac{P_b - 1 - \delta P_b}{\delta P_b} P(g_b) - P(d_b) = -1,
$$

$$
\frac{P_c - 1 - \delta P_c}{\delta P_c} P(g_c) - P(d_c) = -1,
$$

$$
-P(g_a) + \frac{\delta P_a - 1}{(1-\delta)P_a} P(d_a) = -1,
$$

$$
-P(g_b) + \frac{\delta P_b - 1}{(1-\delta)P_b} P(d_b) = -1.
$$

(B.2)

Rearranging the first equation, we have

$$
(P_a - 1 - \delta P_a)P(g_a) - \delta P_a P(d_a) = -\delta P_a.
$$

(B.3)

Using P(g_a) + P(d_a) = P_a, it follows that

$$
(P_a - 1)P(g_a) - \delta P_a^2 = -\delta P_a,
$$

(B.4)

which leads to

$$P(g_a) = \frac{\delta P_a^2 - \delta P_a}{P_a - 1} = \delta P_a.$$

(B.5)

Similarly, we obtain

$$P(g_b) - \delta P_b, \qquad P(g_c) - \delta P_c.$$

(B.6)

Then the probability of making a good part is

$$P(g) = P(g_a) + P(g_b) + P(g_c) = \delta(P_a + P_b + P_c) = \delta.$$

(B.7)

ACKNOWLEDGMENT

The authors thank Dr. Samuel P. Marin of General Motors Research and Development Center for his valuable comments and suggestions.

REFERENCES

1.   R. R. Inman, D. E. Blumenfeld, N. Huang, and J. Li, "Designing production systems for quality: research opportunities from an automotive industry perspective," International Journal of Production Research, vol. 41, no. 9, pp. 1953–1971, 2003.

2.   D. E. Zoia, "Harbour Outlines Who's Winning and Why," September 2005, http:// WardsAuto.com/.

3.   J. Payne and V. Cariapa, "A fixture repeatability and reproducibility measure to predict the quality of machined parts," International Journal of Production Research, vol. 38, no. 18, pp. 4763– 4781, 2000.

4.   Z. M. Bi and W. J. Zhang, "Flexible fixture design and automation: review, issues and future directions," International Journal of Production Research, vol. 39, no. 13, pp. 2867–2894, 2001.

5.   A. Bolat and C. Yano, "Procedures to analyze the tradeoffs between costs of setup and utility work for automobile assembly lines," Tech. Rep. 89-3, Department of Industrial and Operations Engineering, The University of Michigan, Ann Arbor, Mich, USA, 1989.

6.   "Sector Notebook—Profile of the Fabricated Metal Products Industry—Part 2," December 2005, http://www.epa.gov/compliance/resources/ publications/assistance/sectors/notebooks/ fabmetsnpt2.pdf.

7.   "Automobile Production at OPEL, Bochum (Gemany)," http://www. profibus.com/pall/ applications/casestudies/article/3043/.

8.   D. A. Jacobs and S. M. Meerkov, "Asymptotically reliable serial production lines with a quality control system," Computers & Mathematics with Applications, vol. 21, no. 11-12, pp. 85–90, 1991.

9. A. A. Bulgak, "Impact of quality improvement on optimal buffer designs and productivity in automatic assembly systems," Journal of Manufacturing Systems, vol. 11, pp. 124–136, 1992.

10. M. Khouja, G. Rabinowitz, and A. Mehrez, "Optimal robot operation and selection using quality and output trade-off," International Journal of Advanced Manufacturing Technology, vol. 10, no. 5, pp. 342–355, 1995.

11. N. Viswanadham, S. M. Sharma, and M. Taneja, "Inspection allocation in manufacturing systems using stochastic search techniques," IEEE Transactions on Systems, Man, and Cybernetics Part A: Systems and Humans., vol. 26, no. 2, pp. 222–230, 1996.

12. T. L. Urban, "Analysis of production systems when run length influences product quality," International Journal of Production Research, vol. 36, no. 11, pp. 3085–3094, 1998.

13. C. H. Cheng, J. Miltenburg, and J. Motwani, "The effect of straight- and U-shaped lines on quality," IEEE Transactions on Engineering Management, vol. 47, no. 3, pp. 321–334, 2000.

14. S. A. I. Matanachai and C. A. Yano, "Balancing mixed-model assembly lines to reduce work overload," IIE Transactions, vol. 33, no. 1, pp. 29–42, 2001.

15. Y. Ding, J. Jin, D. Ceglarek, and J. Shi, "Process-oriented tolerancing for multi-station assembly systems," IIE Transactions, vol. 37, no. 6, pp. 493–508, 2005.

16. J. Li and D. E. Blumenfeld, "Quantitative analysis of a transfer production line with Andon," IIE Transactions, vol. 38, no. 10, pp. 837–846, 2006.

17. J. Li, D. E. Blumenfeld, and S. P. Marin, "Manufacturing system design to improve quality buy rate: an automotive paint shop application study," IEEE Transactions on Automation Science and Engineering, vol. 4, no. 1, pp. 75–79, 2007.

18. J. Li, D. E. Blumenfeld, and S. P. Marin, "Production system design for quality robustness: theory and application in automotive paint shops," IIE Transactions, vol. 39, 2007.

19. J. Kim and S. B. Gershwin, "Integrated quality and quantity modeling of a production line," OR Spectrum, vol. 27, no. 2-3, pp. 287–314, 2005.

20. N. Viswanadham and Y. Narahari, Performance Modeling of Automated Manufacturing System, Prentice-Hall, Englewood Cliffs, NJ, USA, 1992.

21. J. A. Buzacott and J. G. Shantikumar, Stochastic Models of Manufacturing Systems, Prentice-Hall, Englewood Cliffs, NJ, USA, 1993.

22. H. Tempelmeier and H. Kuhn, Flexible Manufacturing Systems: Decision

Support for Design and Operation, John Wiley & Sons, New York, NY, USA, 1993.

23. M. Zhou and K. Venkatesh, Modeling, Simulation and Control of Flexible Manufacturing Systems: A Petri Net Approach, World Scientific, Singapore, 1999.

24. J. A. Buzacott, "The fundamental principles of flexibility in manufacturing systems," in Proceedings of the 1st International Conference on Flexible Manufacturing Systems, pp. 13–22, Brighton, UK, 1982.

25. J. A. Buzacott and D. D. Yao, "Flexible manufacturing systems: a review of analytical models," Management Science, vol. 32, no. 7, pp. 890–905, 1986.

26. A. K. Sethi and S. P. Sethi, "Flexibility in manufacturing: a survey," International Journal of Flexible Manufacturing Systems, vol. 2, no. 4, pp. 289–328, 1990.

27. N. Viswanadham, Y. Narahari, and T. L. Johnson, "Stochastic modelling of flexible manufacturing systems," Mathematical and Computer Modelling, vol. 16, no. 3, pp. 15–34, 1992.

28. M. Barad and S. Y. Nof, "CIM flexibility measures: a review and a framework for analysis and applicability assessment," International Journal of Computer Integrated Manufacturing, vol. 10, no. 1–4, pp. 296–308, 1997.

29. A. De Toni and S. Tonchia, "Manufacturing flexibility: a literature review," International Journal of Production Research, vol. 36, no. 6, pp. 1587–1617, 1998.

30. R. Beach, A. P. Muhlemann, D. H. R. Price, A. Paterson, and J. A. Sharp, "A review of manufacturing flexibility," European Journal of Operational Research, vol. 122, no. 1, pp. 41–57, 2000.

31. D. Shi and R. L. Daniels, "A survey of manufacturing flexibility: implications for e-business flexibility," IBM Systems Journal, vol. 42, no. 3, pp. 414–427, 2003.

32. Y. K. Son and C. S. Park, "Economic measure of productivity, quality and flexibility in advanced manufacturing systems," Journal of Manufacturing Systems, vol. 6, no. 3, pp. 193–207, 1987.

33. F. F. Chen and E. E. Adam Jr., "The impact of flexible manufacturing systems on productivity and quality," IEEE Transactions on Engineering Management, vol. 38, no. 1, pp. 33–45, 1991.

34. N. Van Hop and K. Ruengsak, "Fuzzy estimation for manufacturing flexibility," International Journal of Production Research, vol. 43, no. 17,

pp. 3605–3617, 2005.

35. G. Da Silveira, D. Borenstein, and F. S. Fogliatto, "Mass customization: literature review and research directions," International Journal of Production Economics, vol. 72, no. 1, pp. 1–13, 2001.

36. P. G. Hoel, S. C. Port, and C. J. Stone, Introduction to Stochastic Processes, Houghton Mifflin, Boston, Mass, USA, 1972.

37. J. Li and N. Huang, "A Markovian model to evaluate quality performance in flexible manufacturing systems," Tech. Rep. R&D-10274, General Motors Research & Development Center, Warren, Mich, USA, 2005.

Chapter 5

META-HIERARCHICAL-HEURISTIC-MATHEMATICAL- MODEL OF LOADING PROBLEMS IN FLEXIBLE MANUFACTURING SYSTEM FOR DEVELOPMENT OF AN INTELLIGENT APPROACH

Ranbir Singh[a] , Rajender Singh[b] and B.K. Khan[c]

[a]Research Scholar, Deptt. of Mech. Engg., DCRUST Murthal, Sonipat (Haryana), 131039, India

[b]Professor, Deptt. of Mech. Engg., DCRUST Murthal, Sonipat (Haryana), 131039, India \

[c]Director, MSIT Jagdishpur, Sonipat (Haryana), India

ABSTRACT

Flexible manufacturing system (FMS) promises a wide range of manufacturing benefits in terms of flexibility and productivity. These benefits are targeted by efficient production planning. Part type selection, machine grouping, deciding production ratio, resource allocation and machine loading are five identified production planning problems. Machine loading is the most identified complex problem solved with aid of computers. System up gradation and newer technology adoption are the primary needs of efficient FMS generating new scopes of research in the field. The literature review is carried and the critical analysis is being executed in the present work. This paper presents the outcomes of the mathematical modelling techniques for loading of machines in FMS's. It was also analysed that the mathematical modelling is necessary for accurate and reliable analysis for practical applications. However, excessive computations need to be avoided and heuristics have to be used for real-world problems. This paper presents the heuristics-mathematical modelling of loading problem with machine processing time as primary input. The aim of the present work is to solve a real-world machine loading problem with an objective of balancing the workload of the FMS with decreased computational time. A Matlab code is developed for the solution and the results are found most accurate and reliable as presented in the paper.

INTRODUCTION

Flexible Manufacturing System in 1960's has evolved with the composition of machines with different capability and capacity constraints. Installation of flexible manufacturing system can be increased through research with physical significance and practical approach & acceptance. In coming decades, the diversity has reduced to negligible amount with technological improvements and advances with the development of advanced CNC's, tool changers, tool transportation systems, automatic material handling system, developments in computer technologies etc. The acceptance and installation of FMS is much lower than expected because of higher installation, running and maintenance cost. FMS is the most accepted manufacturing strategy in the Computer Integrated Manufacturing system. The FMS is composed of a large number of CNC's, with automatic material handling systems, automatic storage and retrieval system, robots, automatic tool changers, tool transporters, which involve a higher installation and running cost. Thus the cost of installation and operating a FMS needs to be initially identified and approved. Production planning is the pilot element in estimating manufacturing cost which is the ever ending research element for any strategy. As the objectives of production planning varies, which requires optimization ideas to implement for different cost reduction manufacturing functions, the need of research arises.

There are a large number of production planning objectives, and different types of manufacturing industries require single or combination of different production planning objectives. Along with the large number of different production planning objectives there are various kinds of objectives. Thus the problem pertaining to multi-production planning objectives coupled with evaluation of multi optimization objectives needs to be investigated. One of the production planning problems is the loading of machines. The elements of loading are the jobs, machines, tools and operations under constraints to achieve some objectives. Manufacturing has different operational requirements; the operations can be performed on different machines using various tools in different times. The same operation can be performed in different times on same machine with various tools, and also in different times on different machines, and there are some capacity and technical constraints and some objectives. Hence as the number of elements, constraints and objectives increases, the complexity of the problem increases too. There are three types of grouping in FMS yielding three various kinds of environments for the loading problem in an FMS, i.e., no grouping, partial grouping, and total grouping (Lee & Kim, 2000).

To increase the acceptability of FMS, the group technology requirement of FMS needs to be modified from no grouping to full grouping as per the requirement of the manufacturing industry for their survival in today's global customer driven market. Also the multi-vendor concept has also evolved in the market, which has changed the concept of FMS from a group of machines to a group of systems. The small scale industries (SSI) and medium scale industries (MSI) these days are striving for their existence. The major factor is the lack of manufacturing strategy in SSI and MSI. Manufacturing strategy is responsible for the life, health and growth of the firm. The stronger is the manufacturing strategy of the firm, the more is its stability in market, the higher the level of its growth. A manufacturing industry survival in the market depends mainly on the manufacturing strategy. The strategy requirement of SSI and MSI is the flexibility requirement of job shop production and productivity of line layout for multi vendor solution for their survival and growth. To optimally utilize the machines and tools the production planning needs to be carried out prior to scheduling, i.e. loading of machines. The present work focuses on the development of Hybrid-Hierarchical-Heuristic-Mathematical-Model of Loading Problems in Flexible Manufacturing System for Throughput Optimization for loading of machines in FMS.

THE LITERATURE REVIEW

A model is a representation of the construction and working of some system of interest which is similar to but simpler than the system it represents. It enables the analyst to predict the effect of changes to the system. The beauty of any model lies in its close approximation to the real system, incorporation of its salient features and minimum complexity. An important issue in modelling is model validity. According to Maria model validation techniques include simulating the model under known input conditions and comparing model output with system output (Maria, 1997). Mathematics, heuristics, queuing theory etc. have been utilized for modelling various types of complex problems of FMS's. Different modelling methods and approaches utilized by earlier researchers for modelling FMS's, particularly the loading problem of FMS's have been identified, analyzed, classified and presented them in tabular form. Table 1 is review of literature on mathematical modelling of loading problem of FMS.

Table 1: Mathematical modelling of loading problem of FMS

Author	Loading objectives	Results
(i) Mixed Integer Programming (MIP)		
Stecke (1981)	Balance assigned machine processing time, maximize number of consecutive operations on each machine and sum of operation priorities	Linearization methods are suggested Results are applicable for a particular range of problems
Stecke (1983b)	Grouping and loading	Need to decrease computational time
Ammons et al. (1985)	General loading problem for discrete optimization	Heuristics improves computational efficiency & effectiveness
Berrada & Stecke (1986)	Minimize machines workload	Heuristics gives efficient solution
Wilson (1992)	Balancing of workload	Used approximate solution technique linearization is necessary
Taboun & Ulger (1992)	Minimize cost	Computational requirements for large size problems are impractical Requirement of real-time FMS control
Stecke & Brian (1995)	Optimize real-time solution of loading problems	Impractical computational time and cost requirements for nonlinear MIPs Optimal solution is cost inefficient in real
Lee et al. (1997)	Minimize subcontracting costs	Iterative algorithms were developed Research on such problems is needed to develop planning software that can be actually implemented in real systems
Lee & Kim (1998)	Minimize earliness, tardiness costs and subcontracting costs	Iterative procedures were developed Computer generated test results
Dobson & Nambimadom (2001)	Minimize scheduling cost	Heuristics provides more optimal solution
Swarnkar & Tiwari (2004)	Minimize system unbalance and maximize throughput	Proposed tabu search and simulated annealing-based hybrid heuristic approach Exhaustive computations were required
Sajono & Lashkari (2007)	Minimize manufacturing cost and maximize compatibility	Validated by numerical example
Jahromi & Tavakkoli (2012)	Minimize production cost	heuristic method is proposed
Kim et al. (2012)	Balancing of workloads	Suggested two-stage heuristics
(ii) Integer Programming (IP)		
Stecke (1983a)	Maximize throughput and machine utilizations	Future need to develop efficient heuristic algorithms for more real life solution
Stecke (1986)	Optimal allocation ratios	Developed queueing network model where information is suppressed
Greene & Sadowski (1986)	Minimize make span, flow time and lateness	Identified variables and constraints necessary to solve real world program
Sarin & Chen (1987)	Minimize machining cost	Lagrangian relaxation is proposed
Ventura et al. (1988)	Minimize make-span	Heuristic algorithms are proposed
Henery et al. (1990)	Balancing of workload and maximize flexibility	Mathematical solution was found impractical
Rajamani & Adil (1996)	Routing flexibility	Routing flexibility is required for rigid loading schedules
Nayak & Acharya (1998)	Minimize number of batches	heuristic has been proposed
Ozdamal & Barbarosoglu (1999)	Minimize the holding cost	GA-SA hybrid heuristics were developed
Lee & Kim (2000)	Minimize maximum workload	Better performance with partial grouping than total grouping, solved by heuristics
Kumar & Shanker (2000)	Genetic algorithms for constrained optimization	GA shows near-optimum performance and need of modern heuristic techniques
Kumar & Shanker (2001)	Balancing of workloads	Results are in agreement with previous findings
Yang & Wu (2002)	Balancing of workloads	Tested for small size test problems only
Gamila & Motavalli (2003)	Minimize total processing time	Used computer generated data for validation
Tadeosi (2004)	Minimize inter-station transfer time	Very high computational effort is required for realistic problems
Chan et al. (2004)	Minimize system unbalance and maximize throughput	Validated only for small set of test problems Require further extension of research
Chen & Ho (2005)	Minimize flow time & tool cost and workload unbalancing	multi-objective genetic algorithm (GA) is proposed
Bilgin & Azizoglu (2006)	Optimization of total processing time	near-optimal solution in reasonable time
Nagarjuna et al. (2006)	Minimize system unbalance	Proposed heuristic yields good results Further extension of work is required
Goswami & Tiwari (2006)	Minimize system unbalance and maximize throughput	Performed extensive computational experiments
Kumar et al. (2006)	Minimize system unbalance and maximize throughput	Proposed constraint-based genetic algorithm comprehensive exploration of research is required
Turkcan et al. (2007)	Minimize tardiness and manufacturing cost	Used sequential and simultaneous approaches for solution
Biswas & Mahapatra (2007)	Minimize system unbalance	need to consider more realistic variables and constraints
Biswas & Mahapatra (2008)	Minimize system unbalance with improved solution quality and reduced computational effort	Proposed particle swarm optimization based meta-heuristic approach Future study to solve the loading problem for multiple-objective framework is required
Ponnambalam & Kiat (2008)	Bi-criterion objective to minimize system unbalance and maximize throughput	Used Particle Swarm Optimization Need of further optimization
Yogeswaran et al. (2009)		GA-SA hybrid algorithms were proposed
Ozpeynirci & Azizoglu (2010)	Maximize total weight of the assigned operations minus total tooling cost	Used Lagrangian relaxation approach for near optimal results
Mandal et al. (2010)	Maximize throughput and minimize system unbalance & make-span	Need to solve the problem in a more realistic environment with more objectives Felt the need of new solution methodology
Yusof et al. (2011)	Balancing of productivity and flexibility	Proposed harmony search algorithm Optimization based methods tend to become impractical with the increase in problem size
Mgwatu (Mgwatu, 2011)	Machining optimisation and part scheduling sub-problems	two-stage sequential methodology was adopted
Yusof et al. (2011)	Minimize system unbalance and increase throughput	Proposed hybrid GA-Harmony Search algorithm Need to solve multi-objective real life large scale machine loading problems
Murat & Erol (2012)	Minimize system unbalance	Proposed hybrid simulated annealing-tabu search algorithm
Yusof et al. (2012)	Minimize system unbalance and maximize throughput	Proposed constraint-chromosome genetic algorithm and identified the need to solve the problem for solve multi-objectives
Kumar et al. (2012)	Minimize system unbalance and maximize throughput simultaneously	Proposed GA-PSO based meta-hybrid heuristic technique
Yaqoub & Abdulghafour (2012)	Meeting delivery dates and reducing manufacturing cost	Need of further research for cost oriented analysis
Abazari et al. (2012)	Maximize profitability and utilization of system	Evaluated unconstrained results by mathematical programming model Felt the need to solve the problem optimally
Mahmudy et al. (2012)	Maximize throughput and balancing of system	Proposed real coded genetic algorithms Stated the requirement of more powerful GA
Kosuoglu & Bilge (2012)	Minimize total distance travelled by parts	Need of research for multi-objective meta-heuristic solution
(iii) Integer constraint		
Kouvelis & Lee (1991)	Minimize operating cost	Need to avoided non-linearity to reduce computational time
(iv) Goal Programming (GP)		
Kumar et al. (1991)	Grouping	Sequential search algorithms were developed Solution obtained by box-complex method
Atmaca & Erol (Atmaca & Erol, 2000)	Maximize throughput, workload balancing and minimize material handling	Tested for small problems

The loading problems of FMS were observed to be modelled with Mathematical Modelling during the period of 1981-2012. Most of the developed mathematical model are not suitable to solve large problems (Nayak & Acharya, 1998). Taboun and Ulger (1992) concluded that computational requirements of mathematical model for large size problems can be impractical (Taboun & Ulger, 1992). Wilson (1992) outlined that linearization is necessary (Wilson, 1992) for near real and optimal results. Further Table 2 outlines the research carried out with modelling loading of machines in FMS with heuristics.

Table 2: Modelling loading of machines in FMS with heuristics

Athor	Title	Conclusion
Stecke (1983a)	Maximize throughput and machine utilizations	Need exists to develop efficient heuristic algorithms for more real life solution
Stecke & Talbot (1983)	Minimize part movements, balancing and unbalancing of workload	None of the developed heuristics was able meet the need of all FMS
Hsu & De-Matta (1997)	Recognize infeasibility of a loading solution	Proposed Lagrangian-based heuristics Need of research to develop better methods
Shankar & Tzen (1985)	Minimize workload & system unbalance and number of late jobs	Developed heuristic methods
Ammons et al. (1985)	General loading problem for discrete optimization	Heuristics improves efficiency and effectiveness
Shankar & Tzen (1985)	Minimize workload & system unbalance and number of late jobs	Proposed heuristic and sequential methods
Shankar & Srinivasulu (1989)	Bi-criterion objective of minimizing workload unbalance and maximizing the throughput	Problem with machine-dependent processing times need to be solved
Mukhopadhyay et al. (1992)	Minimize system unbalance	Heuristic approach was proposed
Kato et al. (1993)	Batch formation to minimize total number of required tools	Heuristic approach was proposed
Roh & Kim (1997)	Minimize total tardiness	Solved with limiting part visit to one machine for entire processing and outlined need of practical research
Farkas et al (1999)	Workload balancing and maximize capacity utilization	Results are demonstrated
Rahimifard & Newman (2000)	Elimination of tardy jobs	Evaluated series of computer based experiments
Tiwari & Vidyarthi (2000)	Minimize system unbalance and maximize throughput	GA-based heuristic were proposed for optimal solution
Tiwari et al. (2007)	Minimize system unbalance and maximize throughput	Genetic algorithm based heuristics were found more efficient than fixed job sequencing rules
Mukhopadhyay et al. (1998)	Minimize system unbalance	Proposed modified insertion scheme Reported higher computational time Hybrid GA was presented
Basnet (Basnet, 2012)	Minimize system unbalance	Stated the need for better heuristics Outlined the need of empirical research

Heuristics was the name of a certain branch of study, not very clearly circumscribed, belonging to logic, or to philosophy or to psychology often outlined, seldom presented in detail. A wide range of heuristics procedures have been developed for different manufacturing strategies. Stecke (1986) stated that for large loading problems, heuristics should be used to find good solutions. The loading problems of FMS excessively depend on efficient heuristics for optimum results. Almost all the researcher during 1983- 2013, felt the need of heuristics development for efficient practically acceptable results because the computational cost and time requirement are very less compared to any other technique (Stecke, 1986). Heuristics has been used by many researchers since 1983 for modelling loading of machines in FMS. Literature review outlines that none of the developed heuristics was able meet the need of all FMS (Stecke, 1983a; Stecke & Talbot, 1983; Hsu & De-Matta, 1997; Basnet, 2012), thus the need to have a better heuristics for realistic solution is major literature gap. The heuristics always showed improved results with realistic and practical nature with reduced computational requirements whenever used to solve the machine loading problem

Major Findings from The Literature Review

Mathematical formulation increases the accuracy of the result on the other hand results in complexity resulting with increased computational requirements. There is a need to develop realistic mathematical model with less computational requirements (Swarnkar & Tiwari, 2004; Tadeusz, 2004). The computational requirements are major identified issues (Stecke, 1983b). literature also reveals that much of the information is usually suppressed in pure mathematical model (Stecke, 1986) may lead to impractical solution (Co et al., 1990). Thus mathematical modelling also needs to be combined with some other techniques to yield practically acceptable realistic results with reasonable computational requirements. There is a need to develop efficient heuristic algorithms for more real life solution (Stecke, 1983a). Requirement of further extension of research was outlined by all researchers (Chan et al., 2004; Nagarjuna et al., 2006; Kumar et al., 2006). A real life solution to machine loading problems of FMS with a new solution methodology is still awaited (Yusof et al., 2012; Biswas & Mahapatra, 2008; Ponnambalam & Kiat, 2008; Mandal et al., 2010; Yusof et al., 2011; Yusof, Budiarto, & Venkat, 2011; Abazari et al., 2012; Petrovic & Akoz, 2008). Researchers also felt the need of real-time FMS control (Stecke & Brian, 1995) and to develop planning software that can be actually implemented in real systems (Lee et al., 1997). Ammons et al. (1985) stated that the use of heuristics in model development improves computational efficiency & effectiveness and provides more optimal solution (Berrada & Stecke, 1986; Ammons et al., 1985; Dobson & Nambimadom, 2001). Heuristic based methods are more robust in practicality (Yusof et al., 2011). Infeasibility of results can be controlled by condition check on heuristics (Hsu & De-Matta, 1997). The major issue for need to further reduce computational requirements was outlined in 1983 (Stecke, 1983b) and is still existing (Mandal et al., 2010; Abazari et al., 2012; Mahmudy et al., 2012; Prakash et al., 2008). Heuristics is found to be most suited. Heuristic reasoning is often based on induction, or on analogy. Heuristics are defined as the set of rules that provides optimal or non-optimal solution to the problem with less computational work (Greene & Sadowski, 1986). With these research gaps and findings to fulfil the research demand the present paper proposes a heuristics--mathematical meta-model for loading of machines in FMS.

Model Presentation

A hybrid hierarchical-heuristic-mathematical modelling and solution methodology has been developed for the optimum utilization of resources in a FMS. The following notations were used for modelling the loading problem.

Variables

J_i	Job number, with i as job index	$i = 1,2,\ldots,I$
M_x	Machine number with x as machine index	$x = 1,2,\ldots,Z$
O_y	Operation number with y as operation index	$y = 1,2,\ldots,Y$
To_z	Tool number with z as tool index	$z = 1,2,\ldots,Z$
t	Time index	$t = 1,2,\ldots,T$

t_{ixyz} Time requirement by job "J_i" on machine "M_x" for operation "O_y" with tool "To_z" (hrs)

t_{ix} Material (Job) handling time for job "J_i" on machine "M_x" (min)

C_{xz} Cost of machining per unit time on machine "M_x" with too "To_z" (in Rs/min)

C_{ix} Handling (Job) cost for job "J_i" on machine "M_x" (in Rs/min)

Av_z Available number of tool type "To_z"

Mc_x Tool Magazine capacity of machine "M_x"

TAl_x Tools allocated to machine "M_x"

To_{ixy} Number of tools required for operation "O_y" on machine "M_x" of job "J_i"

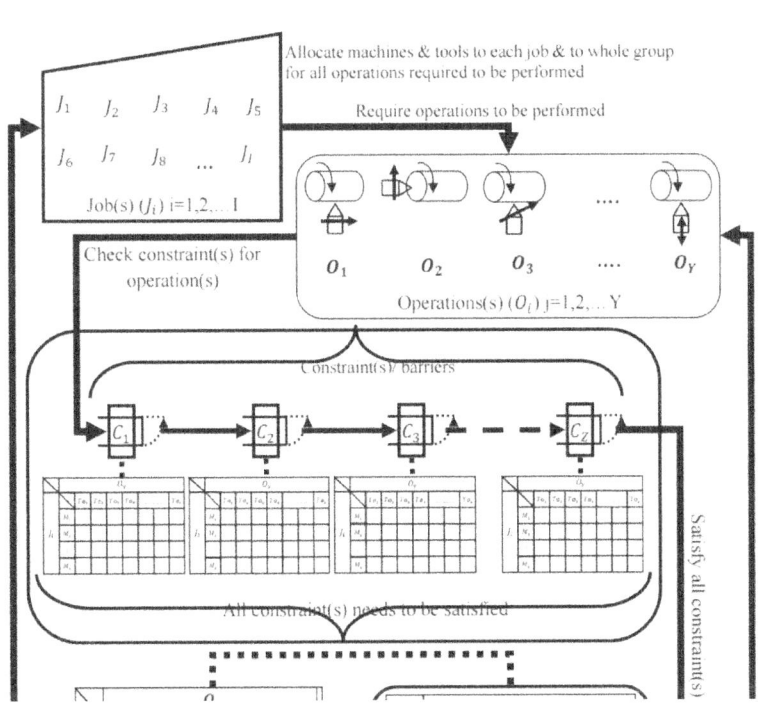

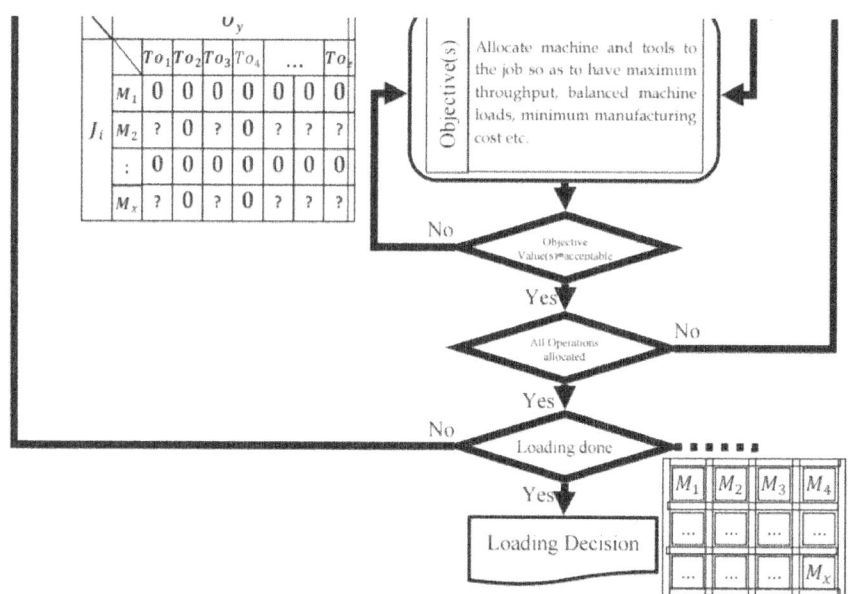

Figure. 1: Diagrammatic representation of the loading problem.

Objective function is $Min\ (FF)$

where,

$$F = Min\left[Min\left(\sum_{x=1}^{X}Min(f_1)\right) - Min(f_2) + Min(f_3)\right]$$

(1)

where,

$$f_1 = Min\left(\sum_{i=1}^{I}\sum_{y=1}^{Y}\sum_{z=1}^{Z}(t_{ixyz})(l_{ixyx})\right)$$

(2)

$$f_2 = Max\left(\frac{I*X}{\sum_{i=1}^{I}\sum_{x=1}^{X}\sum_{y=1}^{Y}\sum_{z=1}^{Z}(t_{ixyz})(l_{ixyx})}\right)$$

(3)

$$f_4 = min\left(\sum_{i=1}^{I}\sum_{y=1}^{Y}\sum_{z=1}^{Z}(l_{ixyz})(t_{ixyz}) - \sum_{i=1}^{I}\sum_{y=1}^{Y}\sum_{z=1}^{Z}(l_{i(x+1)yz})(t_{i(x+1)yz})\right) \approx 0 \quad \forall\ M_y, x$$
$$= 1,2,,(X-1)$$

(4)

Decision variables and constraints

$$\sum_{y=1}^{Y}\sum_{x=1}^{X}\sum_{z=1}^{Z}(l_{ixyz}) = \sum_{y=1}^{Y}\sum_{x=1}^{X}(M_{ixy})(To_{ixy}) \quad \forall\ O_y\ of\ J_i\ and\ i,\ \ i=1,2,,I\ \&\ y = 1,2,,Y$$

(5)

$$\sum_{x=1}^{X}\sum_{z=1}^{Z} l_{ixyz} = \sum_{x=1}^{x}(M_{ixy})(To_{ixy}) \quad \forall \, O_y \& J_i \,, \qquad i = 1,2,,I \, \& \, y = 1,2,,Y \tag{6}$$

$$\sum_{x=1}^{x}(M_{ixy}) = 1 \quad \forall \, O_y \, of \, J_i, \, and \, i, \quad i = 1,2,,I \, \& \, y = 1,2,,Y \tag{7}$$

$$M_{xy} = \begin{cases} 1 \, if \;\; machine \; M_x \; can \; perform \; operation \; O_y \\ 0 \, otherwise \end{cases} \tag{8}$$

$$M_{xy} = \begin{cases} 1 \; if \; \left| \left(\sum_{i=1}^{I}\sum_{y=1}^{Y}\sum_{z=1}^{Z} l_{ixyz} \right) + To_{ixy} \right| \le Mc_x \; \forall \; M_x, \\ 0 \; otherwise \end{cases} \tag{9}$$

$$M_{ixy} = \begin{cases} 1 \, if \, tool \, transportation \, (l = 1) \, is \, allowed \\ 0 \, otherwise \end{cases} \tag{10}$$

$$To_{yz} = \begin{cases} 1 \, if \, machine \, M_x \, can \, perform \, operation \, O_y \, with \, tool \, To_z \\ 0 \, otherwise \end{cases} \tag{11}$$

$$To_{yz} = \begin{cases} 1 \; if \; \left| \left(\sum_{i=1}^{I}\sum_{x=1}^{X}\sum_{y=1}^{Y} l_{ixyz} \right) \right| \le Av_z \; \forall \; To_z \\ 0 \; otherwise \end{cases} \tag{12}$$

$$l = \begin{cases} 1 \, if \, new \, tool \, can \, be \, buyed \\ 0 \, otherwise \end{cases} \tag{13}$$

$$l_{ixyz} = \begin{cases} 1 \, if \, job \, J_i \, is \, loaded \, on \, machine \, M_x \, for \, operation \, O_y \, with \, tool \, To_z \\ \quad 0 \, otherwise \end{cases} \tag{14}$$

$$M_{ixy} = \begin{cases} 1 \, if \;\; machine \, M_x \, is \, allocated \, for \, operation \, O_y \, on \, job \, J_i \\ 0 \, otherwise \end{cases} \tag{15}$$

$$M_{ix} = \begin{cases} 1 \, if \;\; machine \, M_x \, is \, allocated \, to \, job \, J_i \, for \, any \, operation \\ 0 \, otherwise \end{cases} \tag{16}$$

$$To_{iz} = \begin{cases} 1 \, if \;\; tool \, To_z \, is \, allocated \, to \, job \, J_i \\ 0 \, otherwise \end{cases} \tag{17}$$

$$To_{yz} = \begin{cases} 1 \; if \; tool \, To_z \, is \, allocated \, for \, operation \, O_y \\ 0 \, otherwise \end{cases} \tag{18}$$

$$JO_{iy} = \begin{cases} 1 \, if \, job \, J_i \, requires \, operation \, O_y \, to \, be \, performed \\ 0 \, otherwise \end{cases} \tag{19}$$

$$MO_{xy} = \begin{cases} 1 \, if \, Machine \, M_x \, can \, perform \, operation \, O_y \\ 0 \, otherwise \end{cases} \tag{20}$$

$$OTo_{yz} = \begin{cases} 1 \, if \, operation \, O_y \, can \, be \, performed \, by \, tool \, To_z \\ 0 \, otherwise \end{cases} \tag{21}$$

$$e_{iy} = \begin{cases} 1 \, if \, operation \, O_y \, is \, essential \, type \, for \, joi \, J_i \\ 0 \, otherwise \end{cases} \tag{22}$$

$$Tr \quad = \begin{cases} 1 \text{ if tool travel is permitted} \\ 0 \text{ otherwise} \end{cases} \tag{23}$$

$$J_s \quad = \begin{cases} 1 \text{ if job is spiliting} \\ 0 \text{ otherwise} \end{cases} \tag{24}$$

$$J_a \quad = \begin{cases} 1 \text{ if job is allocated} \\ 0 \text{ otherwise} \end{cases} \tag{25}$$

$$TMC_x = \begin{cases} 1 \text{ if tool magazine of machine } M(x) \text{if full} \\ 0 \text{ otherwise} \end{cases} \tag{26}$$

$$ToAv_z = \begin{cases} 1 \text{ if tool of type } To(z) \text{is available} \\ 0 \text{ otherwise} \end{cases} \tag{27}$$

$$To_b \quad = \begin{cases} 1 \text{ if new tool can be buyed} \\ 0 \text{ otherwise} \end{cases} \tag{28}$$

$$0^{ToS_{iyz}} = \begin{cases} 1 \text{ if job } To_z \text{ is specified for operation } O_y \text{ to be performed on Job } J(i) \\ 0 \text{ otherwise} \end{cases} \tag{29}$$

$$y_{iyz} = \begin{cases} 1 \text{ if job } To_z \text{ is specified for operation } O_y \text{ to be performed on Job } J(i) \\ 0 \text{ otherwise} \end{cases} \tag{30}$$

Heusitics procedure

As shown from Fig. 2, the following steps are followed:

Step-1 :Allocate all essential operations

Step-2 :Evaluate the differences $(Max(t_{xz}))- (Min(t_{xz}))$ between maximum $(min(t_{xz}))$ and minimum $(min(t_{xz}))$ time required by the job (J_i) for an operation (O_y) considering all machines and tools in the system, for all jobs $(i= 1,2,3, \dots ,I))$ and operations $(y = 1,2,3, \dots , Y)$ for optional operations only

Step-3 :Select 1st maximum time difference $[Max^{(Max(t_{xz}))- (Min(t_{xz}))}]$, evaluated in step 2

Step-4 :Allocate optional operation corresponding to the selected time on the machine with least processing time

Step-5 :Put the machine out of selection which reaches above the ideal allocation time

Step-6 :Repeat step 3 for next maximum time difference $[Max^{(Max(t_{xz}))- (Min(t_{xz}))}]$, in the order of descending processing times till all operations are allocated

Step-7 :Allocation completed

Step-8 :Analyse the values of objectives

Step-9 :Select the 1st objective need to be modified say cost

Step-10:Provide value of cost that needed to be reduced

Step-11:Evaluate the differences $(Max(c_{xz})) - (Min(c_{xz}))$ between maximum $(Min(c_{xz}))$ and $(min(c_{xz}))$ cost required by the job (J_i) for an operation (O_y) considering all machines and tools in the system, for all jobs $(i = 1,2,3, \ldots ,I))$ and operations $(y = 1,2,3, \ldots , Y)$ for optional operations only

Step-12:Select 1st maximum time cost difference $[Max^{(Max(t_{xz}) - (Min(t_{xz}))}]$, evaluated in step 11

Step-13: Allocate optional operation corresponding to the selected cost on the machine with least cost

Step-14: Put the machine out of selection which reaches above the ideal allocation value

Step-14: Calculate cost reduction achieved by calculating cost differences between previous and current Cost of manufacturing

Step-15: Repeat step 11 for next operation in the order of descending processing times till the desired cost cutting is achieved

Step-16: Allocation completed

Step-17: Analyse the values of objectives

Step-18: Repeat steps 10 to 17 to put any constraint or limitations or any number of objectives

Step-19: when objective values are acceptable, balance the workload on all machines, i.e. all machines when considered to be available for 24×7, the entire machines (×) should run for equal time for optimized loading schedule

Step-20: Select the machine with undesired load, if available shift the tool saving the desired cost/ time

Step-21: Repeat step 20 for all undesired loadings, either saving machining cost at the cost of increased machining time or saving machining time at the cost of increased machining cost.

RESULTS

Test results of a problem with I=5, X=2, Y=3, Z=4 are discussed below for validation of the proposed model and solution methodology. From the table, ideal value of throughput is 0.02, balanced load on the system is 249.5 hours load per machine operating at 100% availability. Here all the operations are considered as optional, with scope of tool travel. The jobs are ordered in sequence of their due dates. All machines are capable of performing all operations, all tools can be loaded on any of the machine. All the jobs can be handled on any of the machines. The results are within known limits of optimal value of optimization, thus are acceptable as per literature guidelines. Thus the developed system is most realistic with least computational requirements best known in the literature.

Table 1: Processing times of various jobs on various machines with different tools for all operations

		O_1				O_2				O_3			
		To_1	To_2	To_3	To_4	To_1	To_2	To_3	To_4	To_1	To_2	To_3	To_4
J_1	M_1	36	6	182	166	137	28	19	103	69	158	22	162
	M_2	81	171	24	21	95	43	63	154	60	59	26	191
J_2	M_1	28	113	200	36	185	181	93	128	151	140	112	139
	M_2	7	173	110	73	22	15	21	19	3	113	99	27
J_3	M_1	190	71	143	12	151	49	202	17	10	81	180	146
	M_2	61	91	202	106	149	11	68	158	135	13	162	23
J_4	M_1	60	11	59	68	114	90	61	183	122	158	149	24
	M_2	68	36	84	36	38	3	13	108	107	69	11	130
J_5	M_1	95	134	94	43	121	182	61	23	148	123	15	67
	M_2	131	67	155	183	61	40	10	167	143	150	18	133

Table 2: Loading of machines in FMS

		O_1				O_2				O_3			
		To_1	To_2	To_3	To_4	To_1	To_2	To_3	To_4	To_1	To_2	To_3	To_4
J_1	M_1	0	0	0	0	0	0	0	0	0	0	0	0
	M_2	0	0	0	1	0	1	0	0	0	0	1	0
J_2	M_1	0	0	0	0	0	0	0	0	0	0	1	0
	M_2	1	0	0	0	0	1	0	0	0	0	0	0
J_3	M_1	0	0	0	0	0	0	0	1	1	0	0	0
	M_2	1	0	0	0	0	0	0	0	0	0	0	0
J_4	M_1	0	1	0	0	0	0	1	0	0	0	0	0
	M_2	0	0	0	0	0	0	0	0	0	0	1	0
J_5	M_1	0	0	0	0	0	0	0	1	0	0	1	0
	M_2	0	1	0	0	0	0	0	0	0	0	0	0

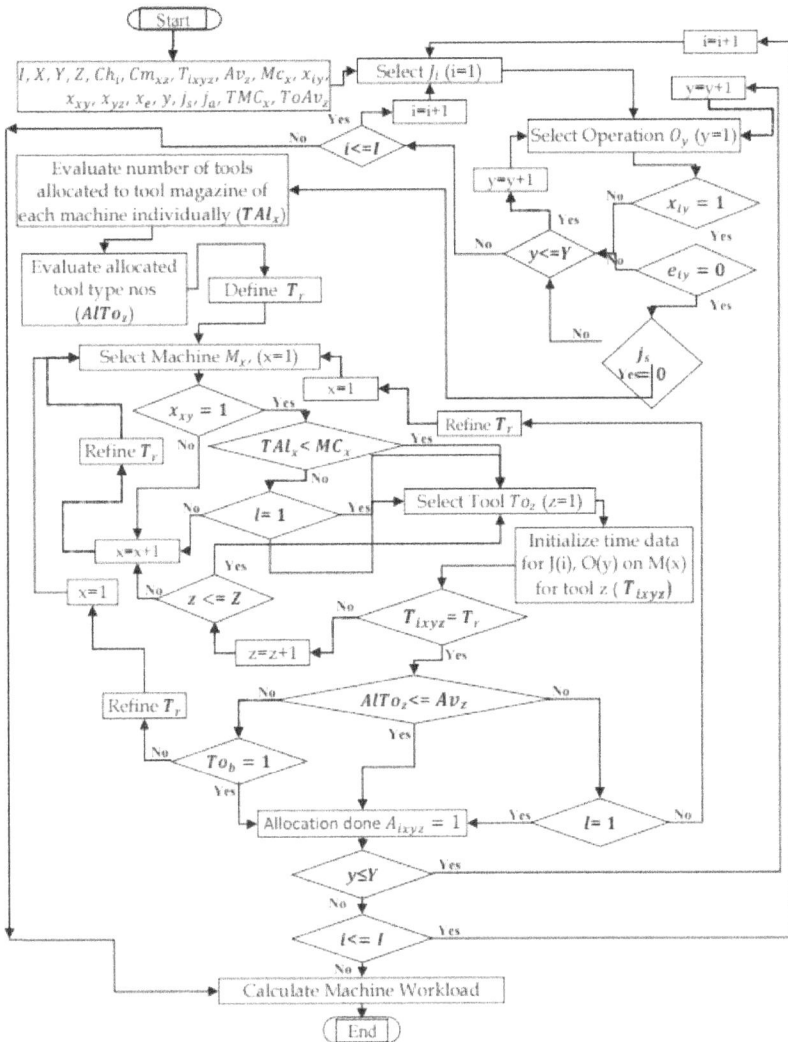

Figure 2: Developed intelligent algorithm LOMFNEO for allocation of machines and tools for non-essential operations in general FMS

Table 2: Results value

	Load/ Machine		Throughput	
Ideal value	249.5		0.0200	
Actual value	249	251	249	251

CONCLUSION

Literature also reveals that validation of the a methodology can be accomplished with computationally randomly generated small set of test problems (Sujono & Lashkari, 2007; Yang & Wu, 2002; Gamila & Motavalli, 2003; Chan et al., 2004; Murat & Erol, 2012; Rahimifard & Newman, 2000; Yeong-Dae & Yano Candace, 1987; Rai et al., 2002). The results of this paper have shown that the on solving the proposed model by developed Matlab codes, yields results very close to the ideal values. For example the ideal and actual throughput through proposed modelling when solved with Matlab codes are nearly similar with negligible percentage difference, which are very real world and acceptable results. Also it is outlined in the literature that the solution are feasible within a known percentage of optimal objective value (Bretthauer & Venkataramanan, 1990).

REFERENCES

1. Abazari, A. M., Solimanpur, M., & Sattari, H. (2012). Optimum loading of machines in a flexible manufacturing system using a mixed-integer linear mathematical programming model and genetic algorithm. Computers & Industrial Engineering, 62(2), 469–478.

2. Ammons, J. C., Lofgren, C. B., & McGinnis, L. F. (1985). A large scale machine loading problem in flexible assembly. Annals of Operations Research, 3, 319 – 332.

3. Atmaca, E., & Erol, S. (2000). Goal programming model for loading and routing problems in flexible manufacturing systems. ICMIT 2000 (IEEE), 843–847.

4. Basnet, C. (2012). A hybrid genetic algorithm for a loading problem in flexible manufacturing systems. International Journal of Production Research, 50(3), 707–718.

5. Berrada, M., & Stecke, K. E. (1986). A branch and bound approach for machine load balancing in Flexible Manufacturing Systems. Management Science, 32(10), 1316–1335.

6. Bilgin, S., & Azizoglu, M. (2006). Capacity and tool allocation problem in flexible manufacturing systems. Journal of the Operational Research Society, 57(6), 670–681.

7. Biswas, S., & Mahapatra, S. S. (2007). Machine Loading in Flexible Manufacturing System : A Swarm Optimization Approach. In Eighth International Conference on Operations & Quant. Management (October 17-20, 2007) (pp. 621–628).

8.   Biswas, S., & Mahapatra, S. S. (2008). Modified particle swarm optimization for solving machineloading problems in flexible manufacturing systems. International Journal of Advanced Manufacturing Technology, 39, 931–942.

9.   Bretthauer, K. M., & Venkataramanan, M. A. (1990). Machine loading and alternate routing in a flexible manufacturing system. Computers and Industrial Engineering, 18(3), 341–350.

10. Chan, F. T. S., Swamkar, R., & Tiwari, M. K. (2004). A random search approach to the machine loading problem of an FMS. In Proceedings of the 2004 IEEE, Intemational Symposium on Intelligent Control

11. (pp. 96–101). Taipei, Taiwan, September 2-4,2004.

12. Chen, J., & Ho, S. (2005). A novel approach to production planning of flexible manufacturing systems using an efficient multi-objective genetic algorithm. International Journal of Machine Tools & Manufacture, 45, 949–957.

13. Co, H. C., Biermann, J. S., & Chen, S. K. (1990). A methodical approach to the flexible manufacturing system batching, loading and tool configuration problems. International Journal of Production Research, 28(12), 2171–2186.

14. Dobson, G., & Nambimadom, R. S. (2001). The batch loading and scheduling problem. Operations Research, 49(1), 52–65.

15. Farkas, A., Koltai, T., & Stecke, K. E. (1999). Workload Balancing Using the Concept of Operation Types. doi:10.2139/ssrn.160288

16. Gamila, M. A., & Motavalli, S. (2003). A modeling technique for loading and scheduling problems in FMS. Robotics and Computer Integrated Manufacturing, 19, 45–54.

17. Goswami, M., & Tiwari, M. K. (2006). A reallocation-based heuristic to solve a machine loading problem with material handling constraint in a flexible manufacturing system. International Journal of

18. Production Research, 44(3), 569–588.

19. Greene, T. J., & Sadowski, R. P. (1986). A mixed integer programming for loading and scheduling multiple manufacturing cells. European Journal of Operation Research, 24(3), 379–386. = Hsu, V. N., & De-Matta, R. (1997). An efficient heuristic approach to recognize the infeasibility of a loading problem. The International Journal of Flexible Manufacturing Systems, 9, 31–50.

20. Jahromi, M. H. M. A., & Tavakkoli-Moghaddam, R. (2012). A novel 0-1 linear integer programming model for dynamic machine-tool selection

and operation allocation in a flexible manufacturing system. Journal of Manufacturing Systems, 31(2), 224–231. doi:10.1016/j.jmsy.2011.07.008

21. Kato, K., Oba, F., & Hashimot, F. (1993). A GT-based heuristic approach for machine loading and batch iformation in flexible manufacturing systems. Control Engineering Practice, 1(5), 845–850.

22. Kim, H., Yu, J., Kim, J., Doh, H.-H., Lee, D.-H., & Nam, S.-H. (2012). Loading algorithms for flexible manufacturing systems with partially grouped unrelated machines and additional tooling constraints.

23. International Journal of Advanced Manufacturing Technology, 58, 683–691.

24. Kim Yeong-Dae and Yano Candace A. (1987). A branch and bound approach for the loading problem

25. in flexible manufacturing systems: an unbalancing case (No. Technical Report 87-18). Ann Arbor,

26. Michigan.

27. Kosucuoglu, D., & Bilge, U. (2012). Material handling considerations in the FMS loading problem with

28. full routing flexibility. International Journal of Production Research, 50(22), 6530–6552.

29. Kouvelis, P., & Lee, H. L. (1991). Block angular structures and the loading problem in flexible

30. manufacturing systems. Institute for Operations Research and the Management Sciences (INFORMS),

31. 39(4), 666–676.

32. Kumar, A., Prakash, Tiwari, M. K., Shankar, R., & Baveja, A. (2006). Solving machine-loading problem

33. of a flexible manufacturing system with constraint-based genetic algorithm. European Journal of

34. Operational Research, 175, 1043–1069.

35. Kumar, N., & Shanker, K. (2000). A genetic algorithm for FMS part type selection and machine loading. International Journal of Production Research, 38(16), 3861–3887.

36. Kumar, N., & Shanker, K. (2001). Comparing the effectiveness of workload balancing objectives in FMS loading. International Journal of Production Research, 39(5), 843–871.

37. Kumar, P., Singh, N., & Tewari, N. K. (1991). A nonlinear goal programming model for multistage,multiobjective decission problems

with application to grouping and loding problems in a flexible manufacturing system. European Journal of Operational Research, 53, 166–171.

38. Kumar, V. M., Murthy, A. N. N., & Chandrashekara, K. (2012). A hybrid algorithm optimizationapproach for machine loading problem in flexible manufacturing system. Journal of IndustrialEngineering International, 8(3), 1–10.

39. Lee, D.-H., & Kim, Y.-D. (1998). Iterative procedures for multi-period order selection and loading problems in flexible manufacturing systems. International Journal of Production Research, 36(10), 2653–2668.

40. Lee, D.-H., & Kim, Y.-D. (2000). Loading algorithms for flexible manufacturing systems with partially grouped machines. IIE Transactions, 32, 33–47.

41. Lee, D.-H., Lira, S.-K., Lee, G.-C., Jun, H.-B., & Kim, Y.-D. (1997). Multi-period part selection and loading problems in flexible manufacturing systems. Computers and Industrial Engineering, 33(3-4), 541–544.

42. Mahmudy, W. F., Marian, R. M., & Luong, L. H. S. (2012). Solving part type selection and loading problem in flexible manufacturing system using real coded genetic algorithms – Part II : optimization. World Academy of Science, Engineering and Technology, 69, 778–782.

43. Mandal, S. K., Pandey, M. K., & Tiwari, M. K. (2010). Incorporating dynamism in traditional machine

44. loading problem : an AI-based optimisation approach. International Journal of Production Research, 48(12), 3535–3559.

45. Maria, A. (1997). Introduction to modeling and simulation. In S. Andradottir, K. J. Healy, D. H. Withers, & B. L. Nelson (Eds.), Proceedings of the 1997 Winter Simulation Conference (pp. 7–13).

46. Mgwatu, M. I. (2011). Interactive decisions of part selection , machine loading , machining optimisation and part scheduling sub-problems for flexible manufacturing systems. International Transaction Journal of Engineering, Management, & Applied Sciences & Technologies, 2(1), 93–109.

47. Mukhopadhyay, S. K., Midha, S., & Murlikrishna, V. (1992). A heuristic procedure for loading problem in flexible manufacturing systems. International Journal of Production Research, 30(9), 2213–2228.

48. Mukhopadhyay, S. K., Singh, M. K., & Srivastava, R. (1998). FMS machine loading: A simulated annealing approach. International Journal of Production Research, 36(6), 1529–1547.

49. Murat, A., & Erol, S. (2012). A hybrid simulated annealing-tabu search algorithm for the part selection and machine loading problems in flexible manufacturing systems. International Journal of Advanced Manufacturing Technology, 59, 669–679.

50. Nagarjuna, N., Mahesh, O., & Rajagopal, K. (2006). A heuristic based on multi-stage programming approach for machine-loading problem in a flexible manufacturing system. Robotics and ComputerIntegrated Manufacturing, 22, 342–352.

51. Nayak, G. K., & Acharya, D. (1998). Part type selection, machine loading and part type volume determination problems in FMS planning. International Journal of Production Research, 36(7), 1801–1824.

52. Ozdamarl, L., & Barbarosoglu, G. (1999). Hybrid heuristics for the multi-stage capacitated lot sizing and loading problem. Journal of the Operational Research Society, 50(8), 810–825.

53. Ozpeynirci, S., & Azizoglu, M. (2010). A Lagrangean relaxation based approach for the in flexible allocation capacity problem manufacturing systems. Journal of the Operational Research Society (20, 61(5), 872–877.

54. Petrovic, D., & Akoz, O. (2008). A fuzzy goal programming approach to integrated loading and scheduling of a batch processing machine. Journal of the Operational Research Society, 59(9), 1211–1219.

55. Ponnambalam, S. G., & Kiat, L. S. (2008). Solving machine loading problem in flexible manufacturing systems using particle swarm optimization. World Academy of Science, Engineering and Technology, 2, 14–19.

56. Prakash, A., Khilwani, N., Tiwari, M. K., & Cohen, Y. (2008). Modified immune algorithm for job selection and operation allocation problem in flexible manufacturing systems. Advances in Engineering Software, 39, 219–232.

57. Rahimifard, S., & Newman, S. T. (2000). Machine loading algorithms for the elimination of tardy jobs in flexible batch machining applications. Journal of Materials Processing Technology, 107, 450–458.

58. Rai, R., Kameshwaran, S., & Tiwari, M. K. (2002). Machine-tool selection and operation allocation in FMS: Solving a fuzzy goal-programming model using a genetic algorithm. International Journal of Production Research, 40(3), 641–665.

59. Rajamani, D., & Adil, G. K. (1996). Machine loading in flexible manufacturing systems considering routeing flexibility. International Journal of Advanced Manufacturing Technology, 11, 372–380.

60. Roh H.-K. and Kim Yeon-D. (1997). Due-date based loading and scheduling methods for a flexible manufacturing system with an automatic tool transporter. International Journal of Production Research, 35(11), 2989–3004.

61. Sarin, S. C., & Chen, C. S. (1987). The machine loading and tool allocation problem in a flexible manufacturing system. International Journal of Production Research, 25(7), 1981–1094.

62. Shankar, K., & Srinivasulu, A. (1989). Some selection methodologies for loading problems in a flexible manufacturing system. International Journal of Production Research, 27(6), 1019–1034.

63. Shankar, K., & Tzen, Y. J. (1985). A loading and dispatching problem in a random flexible manufacturing system. International Journal of Production Research, 23(3), 579–595.

64. Stecke, K. E. (1981). Linearized nonlinear MIP formulation for loading a Flexible Manufacturing System. The University of Michigan, Working Paper No 278.

65. Stecke, K. E. (1983a). A hierarchical approach to solving machine grouping and loading problems of FMS. The University of Michigan, Working Paper No. 331-C.

66. Stecke, K. E. (1983b). Formulation and solution of nonlinear integer production planning problems for flexible manufacturing systems. Management Science, 29(3), 273–288.

67. Stecke, K. E. (1986). A hierarchical approach to solving machine grouping and loading problems of flexible manufacturing systems. European Journal of Operational Research, 24, 369–378.

68. Stecke, K. E., & Brian, T. F. (1995). Heuristics for loading flexible manufacturing systems. In Raouf A. and Ben-Daya M. (Ed.), Flexible manufacturing systems: recent developments (pp. 171–180).

69. Elsevier Science B.V. Stecke, K. E., & Talbot, F. B. (1983). Heuristic loading algorithms for Flexible Manufacturing System. The University of Michigan, Working Paper No 348.

70. Sujono, S., & Lashkari, R. S. (2007). A multi-objective model of operation allocation and material handling system selection in FMS design. International Journal Production Economics, 105, 116– 133. doi:10.1016/j.ijpe.2005.07.007

71. Swarnkar, R., & Tiwari, M. K. (2004). Modeling machine loading problem of FMSs and its solution methodology using a hybrid tabu search and simulated annealing-based heuristic approach. Robotics and

Computer-Integrated Manufacturing, 20, 199–209.

72. Taboun, S. M., & Ulger, T. (1992). Multi-objective modelling of operation-allocation problem in flexible manufacturing systems. Computers and Industrial Engineering, 23(1-4), 295–299.

73. Tadeusz, S. (2004). Loading and scheduling of a flexible assembly system by mixed integer programming. European Journal of Operational Research, 154, 1–19.

74. Tiwari, M. K., Saha, J., & Mukhopadhyay, S. K. . (2007). Heuristic solution approaches for combinedjob sequencing and machine loading problem in flexible manufacturing systems. International Journal of Advanced Manufacturing Technology, 31, 716–730.

75. Tiwari, M. K., & Vidyarthi, N. K. (2000). Solving machine loading problems in a flexible manufacturing system using a genetic algorithm based heuristic approach. International Journal of Production Research, 38(14), 3357–3384.

76. Turkcan, A., Akturk, M. S., & Storer, R. H. (2007). Due date and cost-based FMS loading , scheduling and tool management. International Journal of Production Research, 45(5), 1183–1213.

77. Ventura, J. A., Frank, C. F., & Leonard, M. S. (1988). Loading tools to machines in flexible manufacturing systems. Computers & Industrial Engineering, 15(1-4), 223–230.

78. Wilson J. M. (1992). Approaches to machine load balancing in flexible manufacturing systems. The Journal of the Operational Research Society, 43(5), 415–423.

79. Yang, H., & Wu, Z. (2002). GA-based integrated approach to FMS part type selection and machineloading problem. International Journal of Production Research, 40(16), 4093–4110.

80. Yaqoub, D. Z. H., & Abdulghafour, D. A. B. (2012). Development of Job Scheduling and Machine Loading System in FMS. Engineering & Technology Journal, 30(7), 1173–1186.

81. Yogeswaran, M., Ponnambalam, S. G., & Tiwari, M. K. (2009). An efficient hybrid evolutionary heuristic using genetic algorithm and simulated annealing algorithm to solve machine loading problem in FMS. International Journal of Production Research, 47(19), 5421–5448.

82. Yusof, U. K., Budiarto, R., & Deris, S. (2011). Harmony search algorithm for flexible anufacturing system (FMS) machine loading problem. In 3rd IEEE Conference on Data Mining and Optimization (DMO), 28-29 June 2011 (pp. 26–31). Selangor, Malaysia.

83. Yusof, U. K., Budiarto, R., & Deris, S. (2012). Constraint-chromosome genetic algorithm for flexible manufacturing system machine-loading problem. International Journal of Innovative Computing, Information and Control, 8(3), 1591–1609.

84. Yusof, U. K., Budiarto, R., & Venkat, I. (2011). Machine loading optimization in flexible manufacturing system using a hybrid of bio-inspired and musical-composition approach. In Sixth International Conference on Bio-Inspired Computing: Theories and Applications (IEEE) (pp. 89–96).

Chapter 6

A PRODUCTION INVENTORY MODEL WITH FLEXIBLE MANUFACTURING, RANDOM MACHINE BREAKDOWN AND STOCHASTIC REPAIR TIME

S.R. Singh[a] and Leena Prasher[b]

[a]Department of Mathematics, D. N. College, Meerut, 250103,India

[b]Centre for Mathematical Sciences, Banasthali University, Banasthali, Rajasthan 304022, India

ABSTRACT

This paper derives a production inventory model over infinite planning horizon with flexible but unreliable manufacturing process and the stochastic repair time. Demand is stock dependent and during the period of sale it depends on reduction on selling price. Production rate is a function of demand and reliability of the production equipment is assumed to be exponentially decreasing function of time. Repair time is estimated using uniform probability density function. The objective of the study is to determine the optimal policy for production system, which maximizes the total profit subject to some constraints under consideration. The results are discussed with a numerical example to illustrate the theory.

INTRODUCTION

Classical economic production inventory model assumes that manufacturing systems are perfectly reliable. This assumption, however, does not hold for many real systems. Even the best and the most modern production systems face the situation of sudden machine breakdown, and the time taken in the repair of machine also sometimes depend on the type of injury occurred. Reliability of the production equipment is a crucial factor for keeping the synchronization in the production system, and may harm the organization if the existing uncertainty of the production equipment is not taken into account and is planned, accordingly. In this study, the production system is taken as flexible to produce as per the demand but is not reliable. The production equipment may breakdown at any random time and the repair time is also assumed to be stochastic in nature. During a production run, it may shift from incontrol state to out-of control state, and the production process may have to be stopped at any random time. The objective of this study is to determine the expected optimal production run time with a view to maximizing the expected profit per unit time.

LITERATURE REVIEW

The classical production inventory model assumes that manufacturing system is perfectly reliable. Such an assumption appears impractical in real system. Researchers, therefore, have been attracted towards machine breakdown effects on production inventory problem. The effects of machine breakdown and corrective maintenance were studied by Groenevelt et al. (1992). They studied two production control policies to deal with stochastic machine breakdowns. The first one assumes that the production of the interrupted lot is not resumed after a breakdown. While the second policy considers that the production of the interrupted lot will be immediately resumed after the breakdown if the current on-hand inventory level is below a certain threshold level. Incorporated preventive maintenance to production inventory model was done by Cheung and Hausman (1997). They developed a mathematical model with random machine breakdowns and considered preventive maintenance and safety stock. Wang (2004) developed an EPQ mathematical model where production shifts from an in-control state to an out-of control state with a general shift distribution.

Giri et al. (2005) developed EMQ model with machine failure and general repair time. They proposed a model to determine the production rate and production lot size to minimize the expected total cost. Giri and Dohi (2005) developed EMQ model with random variables, corrective and preventive repair. They proposed solution procedure and computational algorithms to find the optimal production rate and lot size. Lin and Gong (2006) developed EPQ model deteriorating inventory model with machine breakdown and fix repair time. Chiu et al. (2007) derived an economic production quantity (EPQ) model with scrap, rework, and stochastic machine breakdowns, assuming some portion of the defective items to be scrapped and the other part to be repairable. Leung (2007) derived a generalized geometric programming solution to an EPQ model with flexibility and reliability considerations. Chakraborty et a,. (2008) developed an EPQ model considering production system that may shift from in-control state to out-of control state or may breakdown at any random time during a production period. Ferik (2008) developed an

EPQ model for unreliable manufacturing facility. Similar research for EPQ model with imperfect process has been done by Liao et al. (2009). Singh and Singh (2010) worked on supply chain model with stochastic lead time under imprecise partially backlogging for expiring items. Widyadana and Wee (2011) developed production inventory model with random machine breakdown and stochastic repair time. They proved that stochastic repair model tends to have larger optimal cost than fixed repair time model.

An increase in the shelf space can influence more customers. In this connection, the observations made by Levin et al. (1972) and Silver and Peterson (1985) was worth noting, that the presence of greater quantity of the same item tends to attract more customers. The reason behind this fact is a typical psychology of the customers. They may have the feeling of obtaining a wide range for selection when a large amount is stored/displayed. Gupta and Vrat (1986) developed models for stock dependent consumption rate. Mandal and Phaujdar (1989) developed an inventory model for deteriorating items and stock dependent consumption rate. Schweitzer and Seidmann (1991) established optimizing processing rate for flexible manufacturing systems. Giri and Chaudhuri (1998) developed deterministic model of perishable inventory with stock-dependent demand rate and nonlinear holding cost and proved that the nonlinear holding cost affected the total average cost. Sana et al. (2006) established a production-inventory model for a deteriorating item with trended demand and shortages. Teng and Chang (2005) proposed economic production model for deteriorating item with price and stock dependent demand. Singh and Jain (2009) worked on reserve money for an EOQ model in an inflationary environment under supplier credits. Singh (2010) gave an inventory model for deteriorating items with shortages and stock-dependent demand under inflation for two-shops under one management. Yadav et al. (2012) developed an inventory model of deteriorating items with stock dependent demand using genetic algorithm in fuzzy environment. Dem and Singh (2012) investigated an EPQ model for damageable items with multivariate demand and flexible manufacturing. Dem and Singh (2013) developed a production model for imperfect production process under volume flexibility.

Goyal et al. (2013) explored an inventory system with variable demand as well as production under partially backordered shortages.

ASSUMPTIONS AND NOTATIONS

The following assumptions and notations are used throughout the model.

Assumptions

- The production rate is a function of demand $P = \ell D(q), \ell > 1$.

- The demand rate D q( ) is a function of on hand inventory level in the interval (0,) m and is given by $D(q) = \alpha + \beta q, \ 0 < \beta < 1, \ q \geq 0$ where β denotes the shape parameter and is a measure of responsiveness of the demand to changes in the level of on hand inventory and a denotes the scale parameter.

- After t = m , the sale starts and demands rate $D(q) = (\alpha + \beta q)(ab^t), 0 \leq r \leq 1$ is taken as

- The time horizon of the inventory system is infinite. Only a typical planning schedule of expected length E (T) is considered, all remaining cycles are identical.

- Machine repair time is independent of machine breakdown.

Notations

$q(t)$: On hand inventory level of products

$D(q)$: Demand rate,

P : Production rate $P = \ell\, D(q)$ where ℓ is a scale parameter, $P > D(q), \ell > 1$

K : Set up cost

S : Selling price per item

r : Reduction (in %) of selling price of products

h : Holding cost per unit of item per unit time

T_1 : Time when production stops

T_p : Time when machine breakdown occurs

μ : Time when sale of products starts

T_2 : Time when inventory of products vanishes and shortages start to accumulate which causes lost sales

$E(T)$: Expected Duration of a production cycle

$E(PC)$: Expected production cost

$E(HC)$: Expected holding cost in the production cycle

$E(SR)$: Expected sales revenues from items in the production cycle

$E(LSC)$: Expected lost sales cost

$E(TAP)$: Expected total profit per unit time from the production cycle

Formulation of Model

We consider a system in which the manufacturing process is flexible as long as machine is working efficiently and hence can produce as per the demand rate. Generally, reliability of the machine is assumed to be an exponentially decreasing function of time and therefore the probability density function for machine breakdown is assumed as $f(T_p) = \lambda e^{-\lambda T_p}$. The demand function for the products is assumed as stock dependent, which is $D(q) = \alpha + \beta q$ where β where b denotes the shape parameter and is a measure of responsiveness of the demand to changes in the level of on hand inventory and a is the deterministic factor. Many of the organizations decide to sell their goods at reduced prices after predestined time, which is normally known as sale period. The demand function during the sale period is assumed as function of stock displayed as well as discounted price for items and so the demand function during sale period is assumed as $(\alpha + \beta q)(ab^r)$, $a > 0, b > 1, 0 \le r \le 1$, where r

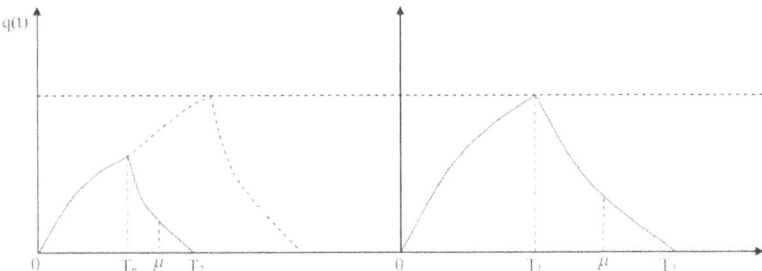

Figure 1: Production system with lost sales.

The production cycle begins with zero inventory and starts at t= 0 . As time advances, if the machine does not breakdown in the production period $[0, T_1]$ inventory level of products pile up even after meeting market demand in the interval $0 \le t \le T_1$, as shown by Eq. (1). Feasibility of this assumption implies $r-1$ must be greater than zero. Production is stopped at time $t = T_1$ and the nventory level of products decreases due to demand in the interval $[T_1, \mu]$ as depicted by Eq (2). The sale period starts at t = m and the inventory of products further decreases due to demand and reaches zero at time t = T_2 as shown by Eq.(3). Since the machine has a possibility of breakdown, the machine may not work the whole T_1 period. When the machine breakdown occurs, the production period stops at p t = T_p and machine requires repair time. As repair time is also stochastic, production may not always be possible and lost sales may occur.

Case I: When m is less than the cycle length

$$\frac{dq}{dt} = P - D(q), \qquad q(0) = 0, \qquad\qquad 0 \le t \le T_1$$

$$\text{(1s)}$$

$$\frac{dq}{dt} = -(\alpha + \beta q), \quad q(T_1^+) = q(T_1^-), \qquad T_1 \le t \le \mu$$

$$\text{(2s)}$$

$$\frac{dq}{dt} = -(\alpha + \beta q)(ab^r), \quad q(T_2) = 0, \qquad \mu \le t \le T_2$$

$$\text{(3s)}$$

Solving Eq. (1), Eq. (2) and Eq. (3) using appropriate boundary conditions, the inventory levels in various intervals are calculated as below

$$q(t) = \frac{\alpha\left(e^{((\ell-1)\beta t} - 1\right)}{\beta}, \qquad 0 \leq t \leq T_1$$

(4)

$$q(t) = \frac{\alpha}{\beta}\left(e^{(\ell T_1 - 1)\beta} - 1\right), \quad T_1 \leq t \leq \mu$$

(5)

To find the relation between the variables using the Taylor series expansion of solutions and using continuity condition, we have $T_2 = \frac{\mu(ab^r - 1) + \ell T_1}{ab^r}$.

For the feasibility of practical situation assumed in case I, T_2 must be greater than m which in turn implies $\ell T_1 - \mu > 0$ Total Inventory in the complete production cycle is calculated as below

$$\int_0^{T_1} \frac{\alpha}{\beta}\left(e^{(\ell-1)\beta t} - 1\right)dt + \frac{\alpha}{\beta}\int_{T_1}^{\mu}\left(e^{(\ell T_1 - t)\beta} - 1\right)dt + \int_{\mu}^{T_2}\frac{\alpha}{\beta}\left(e^{ab^r(T_2 - t)\beta} - 1\right)dt$$

$$= \frac{\alpha}{\beta}\left[\frac{\ell e^{(\ell-1)T_1\beta} - 1}{(\ell-1)\beta} + \frac{(1 - ab^r)e^{(\ell T_1 - \mu)\beta} - 1}{(ab^r\beta)} - \frac{(\mu(ab^r - 1) + \ell T_1)}{ab^r}\right]$$

If machine breakdown occurs at $t = T_{p2}$, then (4) is formulated as below

$$E(I) = \begin{cases} \frac{\alpha}{\beta}\left[\frac{\ell e^{(\ell-1)T_p\beta} - 1}{(\ell-1)\beta} + \frac{(1 - ab^r)e^{(\ell T_p - \mu)\beta} - 1}{(ab^r)\beta} - \frac{\mu(ab^r - 1) + \ell T_p}{ab^r}\right] \dots If \dots T_p < T_1 \\[2em] \frac{\alpha}{\beta}\left[\frac{\ell e^{(\ell-1)T_1\beta} - 1}{(\ell-1)\beta} + \frac{(1 - ab^r)e^{(\ell T_1 - \mu)\beta} - 1}{(ab^r)\beta} - \frac{\mu(ab^r - 1) + \ell T_1}{ab^r}\right] \dots If \dots T_p > T_1 \end{cases}$$

Using the machine breakdown probability density function of T_p, $f(T_p) = \lambda e^{-\lambda T_p}$, $T_p > 0$, the expected inventory is calculated as:

$$\int_{T_p=0}^{T_p=T_1} E(I)\lambda e^{-\lambda T_p}dT_p + + \int_{T_p=T_1}^{\infty} E(I)\lambda e^{-\lambda T_p}dT_p$$

$$\int_{T_p=0}^{T_p=T_1} \frac{\alpha}{\beta}\left[\frac{\ell e^{(\ell-1)T_p\beta} - 1}{(\ell-1)\beta} + \frac{(1 - ab^r)e^{(\ell T_p - \mu)\beta} - 1}{(ab^r)\beta} - \frac{\mu(ab^r - 1) + \ell T_p}{ab^r}\right]\lambda e^{-\lambda T_p}dT_p + \int_{T_p=T_1}^{\infty} E(I)\lambda e^{-\lambda T_p}dT_p$$

$$= \frac{\lambda\alpha}{\beta}\left[\frac{\ell\left(e^{((\ell-1)\beta-\lambda)T_1} - 1\right)}{(\ell-1)\beta((\ell-1)\beta-\lambda)} + \frac{e^{-\lambda T_1} - 1}{\lambda(\ell-1)\beta} + \frac{e^{-\lambda T_1} - 1}{(ab^r)\beta\lambda} + \frac{(1 - ab^r)\left(e^{((\ell\beta-\lambda)T_1 - \mu\beta)} - e^{-\mu\beta}\right)}{(ab^r)\beta(\ell\beta-\lambda)} + \right.$$

$$\left. \frac{\mu(ab^r - 1)(e^{-\lambda T_1} - 1)}{(ab^r)\lambda} - \frac{\ell}{ab^r}\left(\frac{-T_1 e^{-\lambda T_1}}{\lambda} + \frac{1 - e^{-\lambda T_1}}{\lambda^2}\right)\right]$$

$$+ \frac{\alpha}{\beta}\left[\frac{\ell e^{(\ell-1)T_1\beta} - 1}{(\ell-1)\beta} + \frac{(1 - ab^r)\left(e^{(\ell T_1 - \mu)\beta}\right) - 1}{(ab^r)\beta} - \frac{\mu(ab^r - 1) + \ell T_1}{ab^r}\right]e^{-\lambda T_1}$$

Expected holding cost in complete Production cycle, E (HC)

$$= h \left[\frac{\alpha / e^{((i-1)/\beta - \lambda)T_i}}{\beta((i-1)\beta - \lambda)} - \frac{\lambda\alpha I}{(i-1)\beta^2((i-1)\beta - \lambda)} + \frac{I\alpha(1-ab^i)e^{((i-1)\lambda)T_i - \alpha I}}{(ab^i)\beta(i\beta - \lambda)} - \frac{(1-ab^i)e^{-i\beta^i}\lambda\alpha}{(ab^i)\beta^2(i\beta - \lambda)} \right.$$

$$\left. - \frac{\alpha}{\beta^2}\left(\frac{1}{i-1} + \frac{1}{(ab^i)} \right) - \frac{\mu(ab^i-1)\alpha}{(ab^i)\beta} - \frac{I\alpha}{(ab^i)\beta}\left(\frac{1-e^{-\lambda T_i}}{\lambda} \right) \right]$$

$$\tag{5}$$

Lost sales occur when repair time exceeds T_2. Assuming that machine repair time t is a random variable and is uniformly distributed over the interval [0,c].The probability density function f (t) for the repair time is given by

$$f(t) = \begin{cases} \frac{1}{c} & 0 \le t \le c \\ 0 & otherwise \end{cases}$$

For feasibility of the practical situation considered in the model, T_2 must be less than c, otherwise there will be no lost sales interval as lost sales occur when repair time exceeds T_2. Substituting the uniform probability density function of repair time and machine breakdown probability density function, expected lost sales cost is obtained as below Expected Lost sales cost,

$$E\,(LSC) = \frac{S\alpha}{c}\int_{T_p=0}^{T_2}\int_{t=T_2}^{c}(t-T_2)\lambda e^{-\lambda T_p}dt dT_p = \frac{\lambda S\alpha}{2c}\int_0^{T_2}\left(c - \frac{\mu(ab^i-1)+tT_2}{ab^i} \right)^2 e^{-\lambda T_p}dT_p$$

$$\frac{\lambda S\alpha}{2c}\left[\frac{A^2(1-e^{-\lambda T_i})}{\lambda} + \frac{t^2}{(ab^i)^2}\left(\frac{-T_i^2 e^{-\lambda T_i}}{\lambda} - \frac{2T_i e^{-\lambda T_i}}{\lambda^2} + \frac{2(1-e^{-\lambda T_i})}{\lambda^3} \right) - \frac{2At}{ab^i}\left(\frac{-T_i e^{-\lambda T_i}}{\lambda} + \frac{1-e^{-\lambda T_i}}{\lambda^2} \right) \right]$$

where $A = c - \dfrac{\mu(ab^i-1)}{ab^i}$ $\tag{6}$

Next, we calculate production cost as below

$$C_p\int_0^t Pdt = C_p\int_0^t ID(q)dt = C_p la\left(\frac{e^{(i-1)\beta t_p}-1}{(i-1)\beta} \right) \tag{7}$$

When machine breakdown occurs at p t $= T_p$, then Eq. (7) can be formulated as

$$PC = \begin{cases} C_p la\left(\dfrac{e^{(i-1)\beta t_p}-1}{(i-1)\beta} \right) & If\ldots T_p < T_1 \\[3mm] C_p la\left(\dfrac{e^{(i-1)\beta T_i}-1}{(i-1)\beta} \right) & If\ldots T_p > T_1 \end{cases}$$

Using the above formulation, expected production cost can be obtained as

$$E(PC) = \int_{T_p=0}^{T_p=t} C_p la\left(\frac{e^{(i-1)\beta t_p}-1}{(i-1)\beta} \right)\lambda e^{-\lambda T_p}dT_p + \int_{T_p=t}^{T_p=c} C_p la\left(\frac{e^{(i-1)\beta T_i}-1}{(i-1)\beta} \right)\lambda e^{-\lambda T_p}dT_p$$

$$= C_p la\left(\frac{e^{(i-1)\beta - \lambda)T_i}-1}{(i-1)\beta - \lambda} \right) \tag{8}$$

The expected total inventory cost consists of set up cost, expected holding cost, expected lost sales cost and expected production cost

Expected total cost, $E(TC) = K + E(HC) + E(LSC) + E(PC)$ (9)

Next we calculate sales revenue in the complete production cycle,

$$= S\left(\int_{T_1}^{T_p}(\alpha + \beta q)dt + \int_{t_1}^{t_2}(\alpha + \beta q)dt \right) + S(1-r)\left(\int_{t_1}^{T}(\alpha + \beta q)(ab^t)dt \right) = S\alpha \left(\frac{e^{(\ell-1)\beta T_1}-1}{(\ell-1)\beta} + \frac{r\left(1 - e^{(\ell T_1 - \mu)\beta}\right)}{\beta} \right)$$

If machine break down occurs at $pt = T_p$, then sales revenues are formulated as

$$SR = \begin{bmatrix} S\alpha \left(\dfrac{e^{(\ell-1)\beta T_p}-1}{(\ell-1)\beta} + \dfrac{r\left(1 - e^{(\ell T_p - \mu)\beta}\right)}{\beta} \right) & ....If.....T_p < T_1 \\[4mm] S\alpha \left(\dfrac{e^{(\ell-1)\beta T_1}-1}{(\ell-1)\beta} + \dfrac{r\left(1 - e^{(\ell T_1 - \mu)\beta}\right)}{\beta} \right) & ....If.....T_p > T_1 \end{bmatrix}$$

Using the probability density function of machine breakdown time T_p, the expected sales revenues of complete production cycle are obtained as

$$E(SR) = \int_0^{T_1} S\alpha \left(\frac{e^{(\ell-1)\beta T_p}-1}{(\ell-1)\beta} + \frac{r\left(1 - e^{(\ell T_p - \mu)\beta}\right)}{\beta} \right) \lambda e^{-\lambda T_p}dT_p + \int_{T_1}^{\infty} S\alpha \left(\frac{e^{(\ell-1)\beta T_1}-1}{(\ell-1)\beta} + \frac{r\left(1 - e^{(\ell T_1 - \mu)\beta}\right)}{\beta} \right) \lambda e^{-\lambda T_p}dT_p$$

$$= \frac{S\alpha \ell \left(e^{((\ell-1)\beta - \lambda)T_1} - 1 \right)}{(\ell-1)\beta - \lambda} + \frac{S\alpha r\lambda e^{-\mu\beta}}{(\ell\beta - \lambda)\beta} - \frac{S\ell\alpha re^{(\ell\beta-\lambda)T_1-\mu\beta}}{(\ell\beta-\lambda)} + \frac{S\alpha r}{\beta}$$ (10)

The expected total replenishment time is the sum of expected production up time period, non production period and expected repair time after $t = T_2$ Therefore, Expected total time, $E(T) = E(T_2) +$ Expected Repair time

$$\int_{T_p=0}^{T_1} T_2\lambda e^{-\lambda T_p}dT_p + \int_{T_p=T_1}^{T_1} T_2\lambda e^{-\lambda T_p}dT_p + \int_{T_p=0}^{T_1}\int_{t=1}^{T_1} (t-T_2)f(t)\lambda e^{-\lambda T_p}dT_p$$

$$= \frac{\mu(ab^t-1)}{ab^t} + \frac{\ell}{ab^t}\frac{(1-e^{-\lambda t_1})}{\lambda} + \frac{\lambda}{2c}\left[\frac{A^2(1-e^{-\lambda t_1})}{\lambda} + \frac{\ell^2}{(ab^t)^2}\left(\frac{-1^2e^{-\lambda t_1}}{\lambda} - \frac{2T_1}{\lambda^2}e^{-\lambda t_1} + \frac{2}{\lambda^3}(1-e^{-\lambda t_1}) \right) \right.$$

$$\left. - \frac{2A\ell}{ab^t}\left(\frac{-T_1e^{-\lambda T_1}}{\lambda} + \frac{1-e^{-\lambda T_1}}{\lambda^2} \right) \right]$$ (11)

Substituting all the values from Eqs. (5-11), the expected total profit, $E(TP) = E(SR) - E(TC)$ is calculated as follows,

$$\left\{\frac{Sae\left(e^{((\ell-1)\beta-\lambda)T_1}-1\right)}{(\ell-1)\beta-\lambda}+\frac{Sar\lambda e^{-\mu\beta}}{(\ell\beta-\lambda)\beta}-\frac{S\ell are^{(\ell\beta-\lambda)T_1-\mu\beta}}{(\ell\beta-\lambda)}+\frac{Sar}{\beta}\right\}$$

$$-\left\{K+h\left[\frac{\alpha\ell e^{((\ell-1)\beta-\lambda)T_1}}{\beta((\ell-1)\beta-\lambda)}-\frac{\lambda\alpha\ell}{(\ell-1)\beta^2((\ell-1)\beta-\lambda)}+\frac{\ell\alpha\left(1-ab'\right)e^{(\ell\beta-\lambda)T_1-\mu\beta}}{(ab')\beta(\ell\beta-\lambda)}-\frac{\left(1-ab'\right)e^{-\mu\beta}\lambda\alpha}{(ab')\beta^2(\ell\beta-\lambda)}\right]\right\}$$

$$+\left\{\frac{h\alpha}{\beta^2}\left(\frac{1}{\ell-1}+\frac{1}{(ab')}\right)+\frac{h\mu\left(ab'-1\right)\alpha}{(ab')\beta}+\frac{h\ell\alpha}{(ab')\beta}\left(\frac{1-e^{-\lambda T_1}}{\lambda}\right)-C_p la\left(\frac{e^{((1-1)\beta-\lambda)T_1}-1}{(1-1)\beta-\lambda}\right)\right\}$$

$$-\frac{\lambda S\alpha}{2c}\left\{\frac{A^2\left(1-e^{-\lambda T_1}\right)}{\lambda}+\frac{\ell^2}{(ab')^2}\left(\frac{-T_1^2 e^{-\lambda T_1}}{\lambda}-\frac{2T_1 e^{-\lambda T_1}}{\lambda^2}+\frac{2\left(1-e^{-\lambda T_1}\right)}{\lambda^3}\right)-\frac{2A\ell}{ab'}\left(\frac{-T_1 e^{-\lambda T_1}}{\lambda}+\frac{1-e^{-\lambda T_1}}{\lambda^2}\right)\right\}$$

$$\tag{12}$$

$$\frac{dE(TP)}{dT_1}=Sa\ell e^{-\lambda T_1}\left(e^{(\ell-1)\beta T_1}-re^{(\ell T_1-\mu)\beta}\right)-\left\{h\left(\frac{\alpha\ell e^{((\ell-1)\beta-\lambda)T_1}}{\beta}+\frac{\ell\alpha\left(1-ab'\right)e^{(\ell\beta-\lambda)T_1-\mu\beta}}{(ab')\beta}-\frac{\ell\alpha e^{-\lambda T_1}}{(ab')\beta}\right)\right.$$

$$+C_p\ell\alpha e^{((\ell-1)\beta-\lambda)T_1}+\frac{\lambda S\alpha}{2c}\left(A^2 e^{-\lambda T_1}+\frac{\ell^2 T_1^2 e^{-\lambda T_1}}{(ab')^2}-\frac{2A\ell T_1 e^{-\lambda T_1}}{ab'}\right)\right\}$$

$$\tag{13}$$

$$\frac{d^2E(TP)}{dT_1^2}=\left\{Sa\ell\left((\ell-1)\beta e^{((\ell-1)\beta-\lambda)T_1}-r\ell\beta e^{(\ell\beta-\lambda)T_1-\mu\beta}\right)\right\}-\left\{h\alpha\ell e^{-\lambda T_1}\left((\ell-1)e^{((\ell-1)\beta)T_1}+\frac{\ell\left(1-ab'\right)e^{(\ell\beta-\mu)\beta}}{(ab')}\right)\right.$$

$$C_p l(l-1)\beta\alpha e^{((l-1)\beta-\lambda)T_1}+\frac{\lambda S\alpha}{c}\left(\frac{\ell^2 T_1 e^{-\lambda T_1}}{(ab')^2}-\frac{A\ell e^{-\lambda T_1}}{ab'}\right)\right\}$$

$$\tag{14}$$

Using renewal reward theorem, Expected total average profit, E (TAP)

$$E(TAP)=\frac{E(TP)}{E(T)}=\frac{E(SR)-E(TC)}{E(T)}\tag{15}$$

OPTIMAL SOLUTION PROCEDURE:

Our objective is to determine the expected optimal value of T_1 so that E (TAP) is maximized. The necessary condition for E (TAP) to be maximized is $\frac{d(E(TAP))}{dT_1}=0$ and $\frac{d^2E(TAP)}{dT_1^2}<0$. The expected total profit per unit time is concave where $0\leq T_2\leq c$. (see Appendix A, Appendix B and Appendix C for detailed calculations)

The optimal value of T_1 has been obtained numerically using Newton Raphson Method.

The following solution procedure is used to derive the optimal values of T_1

We find the root of the equation $\frac{d(E(TAP))}{dT_1}=0$ using C^{++}

Step 1:

Define $f(T_1) = \dfrac{d(E(TAP))}{dT_1}$

Define $f'(T_1) = \dfrac{d^2(E(TAP))}{dT_1^2}$

Step 2: Choose initial value of $T_1 = t_0$ and set values for other parameters.

Step 3: Apply do-while loop

Find $p = \dfrac{f(t_0)}{f'(t_0)}$

Store in $t_0 = t_0 - p$

If $|p| < 0.0002$ then declare t_0 to be the required root else repeat Step 3.

Case II : When m is greater than the cycle length

When m is greater than the cycle length then there is no sale period and demand remains stock dependent, the governing equations take the form:

$\dfrac{dq}{dt} = (\ell - 1)(\alpha + \beta q), q(0) = 0, 0 \le t \le T_1$

$\dfrac{dq}{dt} = -(\alpha + \beta q), q(T_2) = 0, T_1 \le t \ge T_2$

As $\mu \to T_1$, a $\to 1$ and $r \to 0$, in case I, we can have the expected total profit in Case II as E (TP) = $E(SR) - E(TC)$

$$\left\{ \left[\frac{S\alpha\ell \left(e^{((\ell-1)/\ell - \lambda)T_1} - 1 \right)}{(\ell-1)\beta - \lambda} \right] - \left\{ K + h \left(\frac{\alpha\ell \left(e^{((\ell-1)\beta - \lambda)T_1} - 1 \right)}{\beta((\ell-1)\beta - \lambda)} - \frac{\ell\alpha}{\beta} \left(\frac{1 - e^{-\lambda T_1}}{\lambda} \right) \right) + C_p \ell\alpha \left(\frac{e^{((\ell-1)\beta - \lambda)T_1} - 1}{(\ell-1)\beta - \lambda} \right) \right\} \right.$$
$$\left. - \frac{s\alpha}{2c} \left\{ c^2 - (c - \ell T_1)^2 e^{-\lambda T_1} + \frac{2\ell}{\lambda} \left((c - \ell T_1) e^{-\lambda T_1} - c \right) - \frac{2\ell^2}{\lambda^2} \left(e^{-\lambda T_1} - 1 \right) \right\} \right\}$$

Expected total time, $E(T) = E(T_2) +$ Expected Repair time

$$= \ell \left(\frac{1 - e^{-\lambda T_1}}{\lambda} \right) + \frac{1}{2c} \left\{ c^2 - (c - \ell T_1)^2 e^{-\lambda T_1} + \frac{2\ell}{\lambda} \left((c - \ell T_1) e^{-\lambda T_1} - c \right) - \frac{2\ell^2}{\lambda^2} \left(e^{-\lambda T_1} - 1 \right) \right\}$$

The expected total profit per unit time is concave where $0 \le T_2 \le c$,

NUMERICAL EXAMPLE

In this part, we have presented computational results obtained by using C++ program of Newton Raphson Method which gives insight about the behavior of expected production cycle time E (T), expected optimal run size E(Q) and the expected total average profit E (TAP). The parametric values in the models are taken as

$\beta = 0.4, h = 2, K = 200, \lambda = 0.2, \alpha = 25, C_p = 10, S = 20, c = 5, \mu = 3, a = 0.5, b = 2$

Table 1 Effect of C_p on optimal values of $E(Q), E(T)$ and $E(TAP)$.

	$t = 1.8$			
C_p	9	10	11	12
T_l	2.652	2.444	2.2103	1.931
$E(Q)$	53.72	50.262	46.64	42.653
$E(T)$	4.679	4.3379	3.9373	3.4305
$E(HC)$	281.131	232.846	185.957	139.670
$E(LSC)$	305.612	305.375	305.267	303.678
$E(PC)$	1706.826	1849.723	1978.426	2087.1456
$E(SR)$	3522.820	3514.2084	3496.94	3467.3443
$E(TAP)$	241.34	236.581	235.514	243.944

- As C_p increases, the production cost also increases. To balance the high production cost, production run time decreases. Consequently, inventory level also decreases and which leads to decrease in holding cost.

- Higher values of C_p leads to lower inventory level. As the demand rate is based on the inventory level and sales revenues are based on the demand, therefore, decrease in inventory level decreases the sales revenues, which also causes fall in the expected total profit.

- As C_p increases, expected cycle length, which includes repair time also decreases. It is reasonable that decrease in repair time decreases the lost sales cost also.

Table 2: Effect of h on optimal values of $E(Q), E(T)$ and $E(TAP)$

h	2.5	2	1.5	1
C_p	1.8142	2.3023	2.5263	2.782
T_l	47.505	57.748	63.161	69.963
$E(T)$	3.6193	4.602	5.023	5.485
$E(HC)$	223.097	305.534	285.737	240.78
$E(LSC)$	277.071	277.992	278.863	282.302
$E(PC)$	1321.98	1457.541	1524.323	1604.305
$E(SR)$	2597.82	2666.54	2684.62	2692.99
$E(TAP)$	214.315	135.83	118.595	103.119

- As h decreases, the productions run time increases. Rise in production run time raises the production cost.

- Lower values of h leads to higher production run time. Consequently, there is a rise in the inventory level. As the demand rate is based on the inventory level and sales revenues are based on the demand, therefore, increase in inventory level increases the sales revenues. Although there is rise in the sales revenues but total average profit falls in view of rise in some other costs.

- As h decreases, expected cycle length, which includes repair time also increases. It is reasonable that increase in repair time increases the lost sales cost also.

Table 3: Effect of l on optimal values of Q T, and E (TAP)

ℓ	1.6	1.8	2
T_l	2.609	2.444	2.3023
E(Q)	43.005	50.262	57.748
E(T)	4.0601	4.3379	4.602
E(TAP)	826.639	236.581	135.8304

Table 3: Effect of l on optimal values of Q T, and E(TP)

ℓ	1.6	1.8	2
T_l	2.609	2.444	2.3023
E(Q)	43.005	50.262	57.748
E(T)	4.0601	4.3379	4.602
E(TAP)	826.639	236.581	135.8304

- As l increases, the production runtime decrease, which is very genuine result to expect. Although in view of the increase in production rate there is increase in the inventory level.
- As production rate increases, there is increase in the cycle length.
- Higher values of l leads to lower expected total average profit.
- Higher production run time and lower production rate give higher profit.

CONCLUSION

The model developed above addressed some expected realistic features that usually arose while working on the optimal production policy for stochastic models that maximized the expected profit. It was very important to take the production system as unreliable as uncertainty was very expected feature of a real system. If machine broke down, it was always not certain that it could be repaired in a fixed time period. Normally, in the supermarkets the demand was influenced by the stock displayed on their shelves. In view of the highly competitive situation in the real business, the production system could not afford to be inflexible. Keeping in view all the issues raised, we took the production system unreliable and flexible, taking production rate as a function of demand. In addition, we took a machine repair time stochastic and derived an optimal production policy for the stochastic model which could maximize the expected profit. It was observed that increase in the production rate decreased the production run time and decrease in the production rate increased the production run time. Further, it was observed that longer production up time and lower production rates provided higher expected profits as compared to shorter production uptime and higher production rates in stochastic model. It was also observed that the higher production cost per unit decreased the expected profit. Further research on the problem could be extended to consider more realistic assumptions into the proposed model, for example, imperfect quality of products, reverse manufacturing, trade credit, etc.

REFERENCES

1. Chakraborty, T., Giri, B. C., & Chaudhuri, K. S. (2008). Production lot sizing with process deterioration and machine breakdown. European Journal of Operational Research, 185(2), 606-618.

2. Cheung, K. L., & Hausman, W. H. (1997). Joint determination of preventive maintenance and safety stocks in an unreliable production environment. Naval Research Logistics (NRL), 44(3), 257-272.

3. Chiu, S. W., Wang, S. L., & Chiu, Y. S. P. (2007). Determining the optimal run time for EPQ model with scrap, rework, and stochastic breakdowns. European Journal of Operational Research, 180(2), 664-676.

4. Dem, H., & Singh, S. R. (2012, January). Production scheduling for damageable items with demand and cost flexibility using genetic algorithm. In Proceedings of the International Conference on Soft Computing for Problem Solving (SocProS 2011) December 20-22, 2011 (pp. 747-759). Springer India.

5. Dem, H., & Singh, S. R. (2013). A production model for ameliorating items with quality consideration. International Journal of Operational Research, 17(2), 183-198.

6. El-Ferik, S. (2008). Economic production lot-sizing for an unreliable machine under imperfect age-based maintenance policy. European Journal of Operational Research, 186(1), 150-163.

7. Giri, B. C., & Chaudhuri, K. S. (1998). Deterministic models of perishable inventory with stock-dependent demand rate and nonlinear holding cost. European Journal of Operational Research, 105(3), 467-474.

8. Giri, B. C., Yun, W. Y., & Dohi, T. (2005). Optimal design of unreliable production–inventory systems with variable production rate. European Journal of Operational Research, 162(2), 372-386.

9. Goyal, S. K., Singh, S. R., & Dem, H. (2013). Production policy for ameliorating/deteriorating items with ramp type demand. International Journal of Procurement Management, 6(4), 444-465.

10. Groenevelt, H., Pintelon, L., & Seidmann, A. (1992). Production lot sizing with machine breakdowns. Management Science, 38(1), 104-123.

11. Groenevelt, H., Pintelon, L., & Seidmann, A. (1992). Production batching with machine breakdowns and safety stocks. Operations Research, 40(5), 959-971.

12. Gupta, R., & Vrat, P. (1986). Inventory model for stock-dependent consumption rate. Opsearch, 23(1), 19-24.

13. Konstantaras, I., & Skouri, K. (2010). Lot sizing for a single product recovery system with variable setup numbers. European Journal of Operational Research, 203(2), 326-335.

14. Levin, R. I., Mclaughlin, C. P., Lamone, R. P., & Kottas, J. F. (1972). Production, Operations Management: Contemporary Policy for Managing Operation System. NewYork: McGraw-Hill.

15. Leung, K. N. F. (2007). A generalized geometric-programming solution to "An economic production quantity model with flexibility and reliability considerations". European Journal of Operational Research, 176(1), 240-251.

16. Liao, G. L., Chen, Y. H., & Sheu, S. H. (2009). Optimal economic production quantity policy for imperfect process with imperfect repair and maintenance.European Journal of Operational Research, 195(2), 348-357.

17. Lin, G.C., & Gong, D.C. (2006). On a production inventory system of deteriorating items subject to random machine breakdowns with a fixed repair time. Mathematical Computational Modeling, 43(7), 920-932.

18. Mandal, B.N., & Phaujdar, S. (1989). An inventory model for deteriorating items and stock dependent consumption rate. Journal Operational Research Society, 40, 483–488.

19. Peterson, R., & Silver, E. A. (1979). Decision systems for inventory management and production planning (pp. 799-799). New York: Wiley.

20. Sana, S., Goyal, S.K., & Chaudhuri K.S. (2006). An imperfect production process in a volume flexible inventory model. International Journal of Production Economics, 105,548-559.

21. Schweitzer, P. J., Seidmann, A. (1991). Optimizing, processing rate for flexible manufacturing systems. Management Science, 37, 454-466.

22. Singh, S. R., Kumar, N., & Kumari, R. (2010). An inventory model for deteriorating items with shortages and stock-dependent demand under inflation for two-shops under one management. Opsearch, 47(4), 311-329.

23. Singh, S. R., & Saxena, N. (2012). An optimal returned policy for a reverse logistics inventory model with backorders. Advances in Decision Sciences.

24. Singh, S. R., & Jain, R. (2009). On reserve money for an EOQ model in an inflationary environment under supplier credits. Opsearch, 46(3), 303-320.

25. Singh, S. R., & Singh, C. (2010). Supply chain model with stochastic lead time under imprecise partially backlogging and fuzzy ramp-type demand for expiring items. International Journal of Operational Research, 8(4), 511-522.

26. Teng, J. T., & Chang, C. T. (2005). Economic production quantity models for deteriorating items with price-and stock-dependent demand. Computers & Operations Research, 32(2), 297-308.

27. Wang, C. H. (2004). The impact of a free-repair warranty policy on EMQ model for imperfect production systems. Computers & Operations Research, 31(12), 2021-2035.

28. Widyadana, G. A., & Wee, H. M. (2011). Optimal deteriorating items production inventory models with random machine breakdown and stochastic repair time. Applied Mathematical Modelling, 35(7), 3495-3508.

29. Yadav, D., Singh, S. R., & Kumari, R. (2012). Inventory model of deteriorating items with two-warehouse and stock dependent demand using genetic algorithm in fuzzy environment. Yugoslav Journal of Operations Research ISSN: 0354-0243 EISSN: 2334-6043, 22(1).

Chapter 7

A MEASUREMENT METHOD OF ROUTING FLEXIBILITY IN MANUFACTURING SYSTEMS

F. Zammori[a], M. Braglia[a], M. Frosolini[a]

[a]Dipartimento di Ingegneria Meccanica, Nucleare e della Produzione Via Bonanno, 25/A 56126 Pisa, Italia

ABSTRACT

This paper focuses on routing flexibility, which is the ability to manufacture a part type via several routes and/or to perform different operations on more than one machine. Specifically, the paper presents a comprehensive method for the measurement of routing flexibility, in a generic manufacturing system. The problem is approached in a modular way, starting from a basic set of flexibility indexes. These are progressively extended to include more comprehensive and complex routing attributes, such as: the average efficiency, the range and the homogeneous distribution of the alternative routes. Two procedures are finally proposed to compare manufacturing systems in terms of routing flexibility. The first one uses a vectorial representation of the previously defined indexes and the second one is based on data envelopment analysis, a multi-criteria decision making approach. The paper concludes with a numerical example, supported by discrete event simulation, which validates the proposed approach.

INTRODUCTION

Nowadays the industrial field is characterized by a continuous strain to enhance manufacturing flexibility, which has become one of the main levers to succeed in an ever-changing market. Manufacturing flexibility can be generally defined as the ability of a productive system to quickly react to the changes occurring in its internal and/or external environment, and is especially important for flexible manufacturing systems (FMS) and for flexible assembly systems (FAS). Indeed, this significant attribute distinguishes advanced manufacturing systems from the traditional high-volume process-dedicated production systems, like flow shops and/or automated transfer lines (Chan,

2001; Borenstein & Rohde, 2005). In addition, FMS and FAS generally require conspicuous initial investments that are difficult to be justified, unless the tangible and intangible benefits arising from an increase of flexibility can be fully captured and quantified. As observed by Gupta and Goyal (1989), Sethi and Sethi (1990), Sarker et al. (1994) and Zhang et al. (2002), measuring flexibility is a major concern, because the use of financial evaluation methods (i.e. Net present value, internal rate of return and payback period) is seldom appropriate to take decisions on the acquisition of advanced manufacturing technologies.

Unfortunately, there is not a unified framework to quantify and measure flexibility (Slack, 1987; Beskese et al. 2004) and some authors (Gupta & Buzacott 1989) even raise doubts on the possibility of measuring it, by means of quantitative attributes only. This is due to the high number of potential environmental changes (machine breakdowns, volume and/or mix changes, introduction of new products, etc.) that make it hard or even impossible to include all the critical aspects of flexibility into a single metric (Azzone & Bertelè, 1991). Nonetheless, useful managerial insights can be obtained by dividing flexibility into independent elementary concepts and limiting the analysis to few distinctive features of a manufacturing system. As a matter of fact a great effort has been made to build quantitative analytical tools to measure different types of flexibility, such as: sequencing flexibility, machine flexibility, routing flexibility, volume flexibility, product mix flexibility, layout flexibility and labour flexibility (Brill & Mandelbaun, 1989; Hutchinson & Sinha, 1989; Taymaz, 1989; Kochikar & Narendran, 1992; Nagarur, 1992; Roll et al., 1992; Chen & Chung, 1996; Das, 1996; Chang, 2004).

The present work focuses on routing flexibility, which is the ability to process a part type via several routes. This choice is motivated by the fact that routing flexibility has been recognized as a fundamental competitive lever for advanced manufacturing systems (Caprihan & Wadhwa, 1997; Yu & Green, 2000; Chang, 2007), since it simplifies the scheduling and the balancing of the machines and facilitates the fulfilment of the customers' requirements (Sethi & Sethi, 1990). Furthermore, in technical literature there seems to be a lack of multi dimensional approaches capable to fully describe this important parameter (Chang, 2007). Therefore, the objective of this paper consists in the development of a comprehensive methodology to measure and to capture the main aspects of routing flexibility.

ROUTING FLEXIBILITY

In this paper routing flexibility follows the frequently adopted definition given by Browne et al. (1984) and it is considered as "the ability to handle breakdowns and to continue producing the given set of part types. This ability exists if either a part type can be processed via several routes or, equivalently, if each operation can be performed on more than one machine". In other words, the concept of routing flexibility is strictly related to the capacity to handle breakdowns by means of alternative routes, and becomes evident if production has no meaningful and dramatic downtimes.

$$RF = \frac{\sum_l \sum_m X_{lm}}{M(M-1)},$$
(1)

where X_{lm} equals one if the l-th and the m-th machine (with $l \neq m$) are connected and zero otherwise, and $M(M-1)$ is the total number of potential routes in a system consists of M machines. An equivalent approach was proposed by Kochikar and Narendran (1992), who directly addressed the measurement of routing flexibility in FMSs with the introduction of the Producibility index $\rho_p(M)$ of the p-th part type with respect to the M-th machine set. This index represents the ratio of the number of available routing options at each stage, over the total number of options existing at that stage.

All the previously mentioned approaches consider a single dimension of routing flexibility; however, to get a more comprehensive measure, Das (1996) stressed the necessity to create a more complex analytical tool capable to capture the differences of alternative routes in terms of disparity and efficiency. Disparity should account for the level of machinery and equipment communality, while efficiency should address the difference in processing time between an alternative route and the shortest one. The author also observed that each flexibility dimension should be conveniently evaluated by multiple measures, rather than by a single one. Indeed a multi level approach makes it possible to separate and discriminate the different aspects of flexibility in a more precise and reliable way. A first attempt to address multiple dimensions of routing flexibility was made by Chen and Chung (1996), who proposed a set of indexes relative to a generic set of part types. In doing so the authors introduced and distinguished the concepts of potential routing flexibility, actual routing flexibility, and routing flexibility utilisation.

Recently, Chang et al. (2001) argued that a comprehensive model for the measurement of routing flexibility should consider, at least, two important attributes of a manufacturing system: the efficiency E and the versatility V (i.e. the numerousness) of its routes. Starting form this work, Chang (2007) proposed a three dimensional framework, which also considers routing variety D, an attribute that quantifies the differences of the alternative routes available for a part type p. Specifically, the variety D is evaluated as follows,

$$D_p = \frac{1}{R_p(R_p - 1)} \sum_{r=1}^{R_p} \sum_{\substack{s=1 \\ s \neq r}}^{R_p} d_{(rs),p,}$$

(2)

where R_p is the number of routes (of the p-th part type) and $d_{(rs),p}$ is the ratio of the number of different machines to the total number of machines of the r-th and s-th route.

To compute the efficiency E_p, several operating parameter are combined by means of data envelopment analysis (DEA), a multi-criteria decision making technique based on linear programming (Cooper et al., 2004). In doing so, an output oriented CCR model (Charnes et al., 1978) is used, taking manufacturing costs, set up and processing times as the input decision variables, and the output quantity and quality (of the routes) as the output decision variables. Solving a CCR model for each route, one obtains R_p basic efficiencies e_{rp}, whose average value gives the overall efficiency E_p:

$$E_p = \frac{1}{R_p} \sum_{r=1}^{R_p} e_{rp}.$$

(3)

To evaluate the versatility Vp, the author proposed a method based on the entropy approach, which was firstly introduced by Shannon (1948) in information theory and then adapted to the measurement of flexibility by Yao (1985) and Kumar (1987). As demonstrated by Eq. (4), this choice is motivated by the observation that the entropy approach satisfies both the versatility and the uniformity requirement of a flexible system, as its value increases with the number of alternative routes and with the even distribution of efficient routes among different part types.

$$V_p = -\sum_{r=1}^{R_p} v_{rp} \log(v_{rp}),$$

(4)

where v_{rp} represents the normalized efficiency of the r-th alternative route.

Finally, if P part types are manufactured, the routing flexibility of the whole system is measured as:

$$ROFLX = \frac{1}{P} \sum_{p=1}^{P} E_p \cdot V_p \cdot D_p.$$

(5)

In this paper, an alternative framework for the assessment of routing flexibility is presented, with the objective to incorporate several crucial performance factors into a consistent set of routing flexibility indexes. To this aim, the problem is approached in a modular way, starting from basic flexibility indexes, which are progressively extended in order to include more detailed and complex routing aspects in a set of metrics, suitable at a different level of complexity and characterized by an easy analytical shape. The main aspects that have been considered are: (i) the efficiency (evaluated in term of cost and time), (ii) the number and (iii) the homogeneous distribution of the alternative routes. Additionally, the basic set of indexes has been expanded to capture other meaningful parameters that can affect the routing flexibility of a manufacturing system. These are: the covering degree, the production quality, the backtracking probability and the availability of the equipment installed in the plant.

To complete the work and to improve its practical utility, two approaches, that make it possible to compare manufacturing systems in terms of routing flexibility, are also presented. The first one uses a vectorial representation to visualize the global routing flexibility in a three dimensional space, and to compare it with that of an ideal manufacturing system. The second one is based on a multi-output single-input DEA model, and can be used to compare two or more manufacturing systems in relative terms. Both approaches are clearly explained and validated by means of a numerical example supported by a discrete event simulation model.

OPERATING DATA

In the following part of the paper we will consider a production system characterized by P part types (i.e. jobs) and M machines. Each product must be associated to a binary Part-Machine matrix $\tilde{M} = |x_{m,r}|$ that encodes all its feasible routes. In the matrix, routes are listed in columns, machines are listed in rows and a generic element x_{mr} is one if the r-th processing route requires the m-th machine, and zero otherwise. The convention is used to assign the first column of $\tilde{M}$ to the standard route and to use the subscript $r = 0$ to denote the standard route.

For the sake of simplicity, all the Part-Machine matrixes can be assembled in a three dimensional structure $M = |x_{m,r,p}|$, whose third dimension refers to the part type. In this way, a generic entry x_{mrp} is one if the r-th processing route of the p-th part type requires the m-th machine. An example (Matrix 1) of an M matrix is built considering three part types (p_1) manufactured using five different machines (m_1).

$$M = \begin{array}{c} \\ m_1 \\ m_2 \\ m_3 \\ m_4 \\ m_5 \end{array} \begin{array}{|ccccc|} p_1 & & p_2 & & p_3 \\ 1 & 1 & 1 & 1 & 0 \\ 0 & 1 & 0 & 1 & 1 \\ 0 & 1 & 0 & 1 & 1 \\ 1 & 0 & 0 & 1 & 1 \\ 1 & 1 & 1 & 0 & 0 \\ \end{array}$$

$$r_{01} \quad r_{11} \quad r_{02} \quad r_{12} \quad r_{03}$$

Matrix 1: An Example of Three-Dimensional Part-Machine Matrix.

Each column r_{0p} represents the standard route of the p-th part type, while r_{rp} denotes its r-th alternative route. For example, both p_1 and p_2 can be manufactured via an alternative route, r_{11} and r_{12}, respectively.

Replacing all the values x_{mrp} of the M matrix with the corresponding processing times t_{mrp}, a processing time matrix $T = |t_{m,r,p}|$ can be obtained. In this case, the values t_{mrp} should include, at least, the cycle time and the transportation time that are needed to process and move the p-th part along its r-th route. Besides, the time values t_{mrp} could also include the set-up time, especially when the change-over tasks have a meaningful variation in relation with the adopted production sequence.

The operating data that will be used to evaluate routing flexibility include, lastly, the manufacturing cost per time unit. These costs must be computed for each machine and can be arranged within an Mdimensional array $\tilde{C} = [\tilde{c}_1, \tilde{c}_2, ..., \tilde{c}_m, ..., \tilde{c}_M]$, whose generic element $\tilde{c}_m$ represents the cost per unit of time of the m-th machine. Finally, if the production times are multiplied by the corresponding costs, the production cost matrix $C = |c_{m,r,p}| = |t_{m,r,p} \cdot \tilde{c}_m|$ is obtained. From the values of C, the route production cost RC_{rp} can be easily computed as follows:

$$RC_{rp} = \sum_{m=1}^{M} c_{mrp}.$$

$$(6)$$

BASIC ROUTING FLEXIBILITY INDEXES

In this section the basic metrics of alternative processing routes will be defined. Subsequently, these metrics will be aggregated to evaluate the routing flexibility of a productive system, taking into account the number, the efficiency and the homogeneous distribution of all the feasible alternative routes.

Job Routing Average Efficiency

The efficiency of the r-th alternative route of the p-th part type can be expressed in terms of the route production cost RC_{rp}, taking the cost of the standard route RC_{0p} as the reference parameter. In this way, the alternative route efficiency ARE_{rp} is formally defined as:

$$ARE_{rp} = \frac{RC_{0p}}{RC_{rp}} \qquad \forall\, r \neq 0 \tag{7}$$

If a part type has several alternative routes $\tilde{R}_p$, a second index, called job routing average efficiency $JRAE_p$, can be used to evaluate the average efficiency of its alternative routes:

$$JRAE_p = \frac{\sum_{r=1}^{\tilde{R}_p} ARE_{rp}}{\tilde{R}_p}. \tag{8}$$

Job Routing Range

The index $JRAE_p$ has the advantage to be very easy, but unfortunately it does not adequately consider the number of alternative routes deployed for the p-th part type and, for this reason, it can lead to misleading results.

For instance, consider the situation given in Matrix 2, which displays three part types characterized by one, two and three alternative routes, respectively. In this case, a straight comparison of the job routing average efficiencies ($JRAE_1 = 0.9$, $JRAE_2 = 0.8$, $JRAE_3 = 0.63$) would indicate p_1 as the part type with the higher routing flexibility. This result could legitimate some doubts: as a matter of fact the routing flexibility of p_3 should be greater than that of p_1 because p_3 has three alternative routes and one of them has the same efficiency as r_{11}, which is the only alternative route of p_1.

$$C =
\begin{array}{c|c|c|c|c|c|c|c|c}
2 & 2 & & 1 & & 1 & & 2 & \\
 & 1 & & 2 & 1 & 1 & 1 & & 4 \\
3 & & 2 & & 2 & & 2 & & 3 \\
 & & 3 & 4 & 2 & & & & \\
2 & & & & & 2 & 1.25 & 2 & 4 \\
2 & 3 & & & 1 & 2 & & 1 & \\
 & 3 & 2 & & & 3 & & 3 & \\
1 & 1 & 2 & & & 2 & 1.4 & 3 & 4 \\
\hline
r_{01} & r_{11} & r_{02} & r_{12} & r_{22} & r_{03} & r_{13} & r_{23} & r_{33}
\end{array}$$

Matrix 2: An example of Cost matrix.

Such problem can be overcome through the introduction of an additional index called job routing range JRR, which is zero for an item without alternative routes. To compute JRR we start by observing that, the greater the number of viable routes, the greater the flexibility. However, the marginal increase of flexibility (due to the addition of an alternative route) also depends on the number of the alternative routes $\tilde{R}_p$ available for the p-th part type. If $\tilde{R}_p$ is low, the capacity to counteract failures is noticeably augmented as the number of alternative routes increases from $\tilde{R}_p$ to $(\tilde{R}_p +1)$. Vice versa if $\tilde{R}_p$ is high, the effect of an additional route has a negligible effect. To fulfil these requirements, JRR_p can be represented as an increasing function of $\tilde{R}_p$:

$$JRR_p(\tilde{R}_p) = \begin{cases} 1 - \left(1 - \dfrac{\tilde{R}_p}{(TR - 1)}\right)^{\alpha_p} & \text{if } TR \geq 2 \\ 0 & \text{otherwise} \end{cases} \tag{9}$$

where TR is the total number of routes of the system and α_p is a shape parameter greater than one.

As clearly shown in Eq. (9), JRR_p is a nonzero positive number, provided that there is, at least, one alternative route. Specifically, JRR_p equals zero if the p-th product can be processed only via its standard route (i.e. $\tilde{R}_p = 0$), whereas JRR_p equals one if the part type shares all the alternative routes deployed in the system (i.e. $\tilde{R}_p = [TR-1]$).

Conversely, when there is a unique processing route (i.e. TR = 1 and $\tilde{R}_p = 0$ ∀p) JRR$_p$ equals zero, because the manufacturing system does not have routing flexibility. This particular condition does not have any practical interest and have been included in Eq. (9) only for the sake of clarity. Indeed, a system with a single route corresponds to a rigid production/assembly line, which cannot be adequately described with the set of indexes proposed in this paper.

To better explain the effect of the shape parameter on the job routing range, Fig. 1 shows the shape of JRR$_p$ when TR equals ten and α_p increases from one to nine.

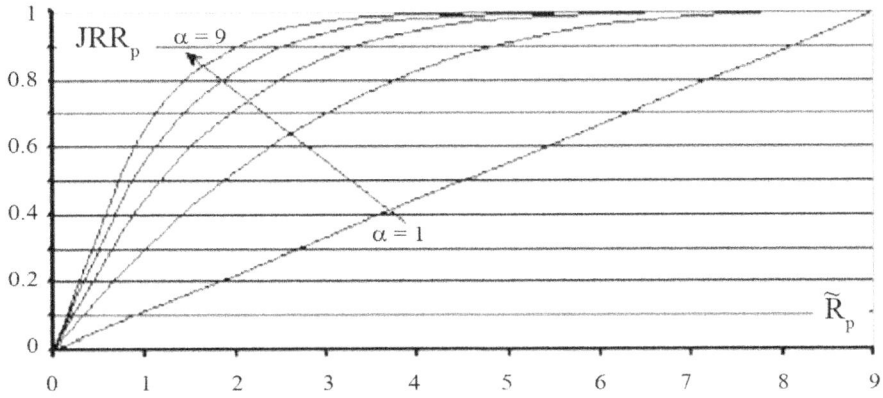

Figure 1: Job routing range evaluation function.

To evaluate JRR$_p$ one needs to define a sensible value for the shape parameter α_p. To this aim, an approach based on the complexity and on the availability of the machines of the standard route can be used. The complexity of the standard production route is an important parameter for the determination of α_p because it influences the easiness to generate new alternative routes. This peculiar parameter can be captured by the standard route complexity index SRC$_p$, defined as the number of machines of the standard route, over the number of machines installed in the plant:

$$SRC_p = \frac{\sum_{m=1}^{M} x_{m0p}}{M},$$

(10)

where M is the total number of machine and x_{m0p} is a generic value of the standard route columns of the M matrix.

In the computation of α_p also machine availability should be taken into account. Indeed, to avoid dramatic production downtimes, it is a good managerial practice to deploy alternative routes for jobs that are processed by machines subject to frequent breakdowns and/or to long lasting periods ofmaintenance. Conversely, if the availability of the machines of the standard route is high, there is no need to introduce many alternative routes.

Under the hypothesis that the likelihood of two simultaneous failures (for the standard and for the alternative route) is negligible, the standard route availability SRA_p can be expressed as:

$$SRA_p = \prod_{m=1}^{M} x_{m0p} A_m$$

(11)

$$A_m = \frac{MTBF_m}{MTBF_m + MTTR_m}$$

(12)

where A is the steady state availability, MTBF and MTTR stand for mean time between failure and mean time to repair, respectively (O'Connor, 2002).

SRC_p and SRA_p can be finally combined to obtain a meaningful value for α_p:

$$\alpha_p = \left(\frac{1}{1 - SRA_p}\right)^{SRC_p}$$

(13)

As shown by Eq. (13) α_p decreases when the route availability SRA_p approaches one. This is because $(1 - SRA_p)$ represents an estimation for the frequency of the p-th job to be processed with an alternative route, and can be seen as an indicator of the necessity to deploy additional routes. Conversely α_p increases as the complexity SRC_p tends to one. Indeed, the capability to arrange alternative routes greatly diminishes with the increase in the complexity of the standard route.

To better explain the use of JRR_p, we re-consider the previous example under the hypothesis that the standard routes for p_1, p_2 and p_3 are characterized by a 90% availability. In this case, remembering that p1, p_2 and p_3 do not share any route (i.e. TR = 9), that M = 8 and that $\tilde{R}_{p=1}$, $\tilde{R}_{p=2}$, $\tilde{R}_{p=3}$, one obtains the following estimates:

$$SRC_1 = SRC_2 = SRC_3 = \frac{\sum_{m=1}^{8} x_{m0p}}{8} = 0.5$$

$$\alpha_1 = \alpha_2 = \alpha_3 = \left(\frac{1}{1 - SRA_p}\right)^{SRC_p} = \left(\frac{1}{0.1}\right)^{0.5} \cong 3.16$$

$$JRR_1 \cong 34.4\%, \qquad JRR_2 \cong 59.7\%, \qquad JRR_3 \cong 77.3\%$$

In accordance with these values, the ranking that was previously obtained by making a straight comparison of the JRAE indexes (i.e. p_1, p_2, p_3) has been completely altered. Indeed p_3 is now ranked as the part type with the highest routing flexibility, followed by p_2 and by p_1, respectively. This seems to be correct because, as previously noted, p_3 has three alternative routes and one of them has the same efficiency as r_{11}, which is the only alternative route of p_1.

Job Routing Versatility

In analogy with the approach proposed by Chang (2001, 2007), the information concerning the efficiency and the range of the routes can be aggregated into a single metric called job routing versatility JRV_p. To this aim, for each part type we introduce two vectors, namely the efficiency vector E_p and the range vector R_p, both of dimension $(TR - 1)$. As shown in Eq. (14), the first $\tilde{R}_p$ elements of E_p are the alternative routing efficiencies of the p-th part type, indexed in a decreasing way. All the remaining $(TR - \tilde{R}_p - 1)$ values are zero.

$$E_p = \left[(ARE_{rp})_1, (ARE_{rp})_2, ..., (ARE_{rp})_i, ..., (ARE_{rp})_{\tilde{R}_p}, 0_{\tilde{R}_p}, ..., 0_{(TR-1)}\right]$$

with $i = 1, ..., (TR - 1)$, $\quad E_p[i] \neq 0 \ \forall i \in [1, \tilde{R}_p]$ and

$$(ARE_{rp})_1 \geq (ARE_{rp})_2 \geq \cdots \geq (ARE_{rp})_i \geq \cdots \geq (ARE_{rp})_{\tilde{R}_p} \qquad (14)$$

Similarly, as shown in Fig. 2, R_p contains the marginal increments of flexibility $\Delta JRR_p(i)$, that can be obtained with the introduction of an additional route for the p-th part type.

$$R_p = \left[\Delta JRR_p(1), \Delta JRR_p(2), ..., \Delta JRR_p(i), ..., \Delta JRR_p(TR - 1)\right]$$

with $i = 1, ..., (TR - 1)$ and $\Delta JRR_p(i) = \left(JRR_p(i) - JRR_p(i - 1)\right)$ $\qquad (15)$

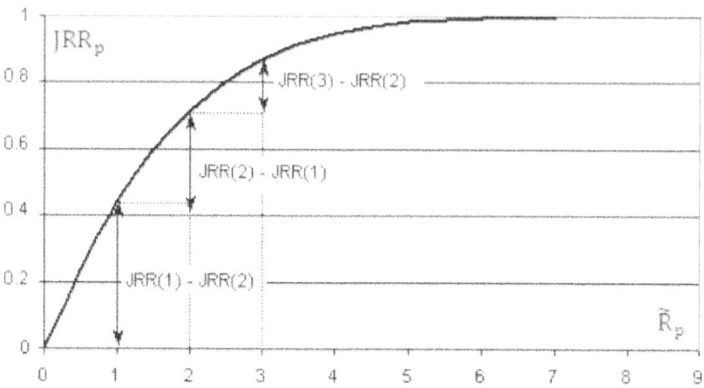

Figure 2: Job Routing range marginal increments.

Since the flexibility rises as the number of alternative routes increases and/ or if the efficient routes are evenly distributed among different part types, the versatility can be obtained as the scalar product of vector E_p and R_p. In this way, the routing flexibility is evaluated as a weighted average of the efficiency of each alternative route, using the marginal increment of flexibility as weighting coefficients.

$$JRV_p = \langle E_p, R_p \rangle .$$

(16)

Since the sum of the elements of R_p equals one (i.e. $\sum_{i=1}^{TR-1} R_p[i] = 1 \ \forall p$), it is easy to see that JRV_p changes between zero and one. Specifically JRV_p can be one only if the p-th part type can be manufactured via (TR-1) alternatives routes with a 100% efficiency. To better explain the concept of JRV_p, let us consider again the numerical example introduced in section 4.2. In this case it results:

$JRV_1 = [0.9,0,0,0,0,0,0,0,0] \times [0.344,0.253,0.176,0.115,0.067,0.033,0.011,0.001] \cong 31\%$

$JRV_2 = [0.8,0.8,0,0,0,0,0,0,0] \times [0.344,0.253,0.176,0.115,0.067,0.033,0.011,0.001] \cong 48\%$

$JRV_3 = [0.9,0.6,0.4,0,0,0,0,0,0] \times [0.344,0.253,0.176,0.115,0.067,0.033,0.011,0.001] \cong 53\%$

This result improves the ranking obtained using the JRR indexes because, as observed at the beginning of section 4.2, the routing flexibility of p_2 and p_3 must be similar and bigger than that of p_1.

GLOBAL ROUTING FLEXIBILITY INDEXES

JRAEp and JRRp can be used, at an aggregate level, to characterize the whole manufacturing system in terms of global routing efficiency (GRE) and global routing range (GRR). These two globalindices are defined in Eqs. (17-18) and give the average efficiency and the average range of the routes of a manufacturing system.

$$GRE = \frac{\sum_{p=1}^{P} JRAE_p}{P},$$

(17)

$$GRR = \frac{\sum_{p=1}^{P} JRR_p}{P}.$$

(18)

Unfortunately, GRE and GRR tend to neglect how evenly routes are distributed among the P part types. As a matter of fact, a system characterised by many alternative routes, but relevant to a few products only, will result less flexible than a system with an identical number of routes, but uniformly distributed among several products. Similarly, a manufacturing system where only a few products can exploit the most efficient processing routes and the majority of products rely on the least efficient ones is, by evidence, less flexible than a system which utilizes both efficient and inefficient routes with higher product uniformity.

To find a remedy to this circumstance we can use JRVp to define an additional global index called global routing versatility (GRV), which should be zero for a system characterized by the worst possible distribution (of the alternative routes) and one in the best case. The first circumstance corresponds to the condition of (P - 1) part types being processed via the standard route and one item alone being the outcome of all the alternative routes listed in the M matrix (i.e. $JRV_1 = JRV_2 = JRV_3 = \ldots = JRV_{(P-1)} = 0$; $JRV_P > 0$). Conversely, the latter circumstance corresponds to a homogeneous distribution of routes among jobs (i.e. $JRV_1 = JRV_2 = \ldots = JRV_P > 0$).

Considering these requirements, to evaluate GRV we can use the Gini concentration index (GCI), which is a common measure of statistical dispersions (Gini, 1921).

$$GCI = \sum_{j=1}^{P} \left(\frac{j}{P} - \frac{\sum_{i=1}^{j} V[i]}{JRV_{TOT}} \right),$$

(19)

where $JRV_{TOT=} \sum_{i=1}^{P} V[i]$, and V[i] is the i-th element of the routing versatility vector V, which contains the values of JRV_p for each part type, indexed in an increasing order:

$$V = \left[\left(JRV_p\right)_1, \left(JRV_p\right)_2, \dots, \left(JRV_p\right)_i, \dots, \left(JRV_p\right)_P\right]$$

with $i = 1, \dots, P$ and $\left(JRV_p\right)_1 \le \left(JRV_p\right)_2 \le \cdots \le \left(JRV_p\right)_i \le \cdots \le \left(JRV_p\right)_P$.   (20)

When a single part type is the outcome of all the alternative routes listed in the M matrix, GCI equals $(P - 1)/2$, which is the maximum value that this index can take. Indeed, in this particular circumstance, since a single product (i.e. the one placed in the P-th position of vector V) benefits of a non null routing flexibility, we have $V[i] = 0 \ \forall i \ne P$ and $JRV_{TOT} = \sum_{i=1}^{P} V[i] = V[P]$. Therefore Eq. (19) can be simplified as follows,

$$GCI = \sum_{j=1}^{P}\left(\frac{j}{P} - \frac{\sum_{i=1}^{j}V[i]}{JRV_{TOT}}\right) == \left(\frac{1}{P} - \frac{0_1}{V[P]}\right) + \left(\frac{2}{P} - \frac{0_1 + 0_2}{V[P]}\right) + \cdots + \left(\frac{P}{P} - \frac{0_1 + \cdots + 0_P + V[P]}{V[P]}\right)$$

$$= \left(\sum_{j=1}^{P-1}\left(\frac{j}{P} + 0\right)\right) + \left(\frac{P}{P} - 1\right) = \frac{1}{P}\sum_{j=1}^{P-1}j = \frac{1}{P}\frac{P(P-1)}{2} = \frac{P-1}{2}.$$

Conversely, GCI is zero if the alternative routes are evenly distributed among part types, so that $V[1] = V[2] = \dots = V[i] = \dots = V[P]$ and $JRF_{TOT} = P \cdot V[i] \ \forall i$. In this case it results:

$$GCI = \sum_{j=1}^{P}\left(\frac{j}{P} - \frac{\sum_{i=1}^{j}V[i]}{JRV_{TOT}}\right) = \sum_{j=1}^{P}\left(\frac{j}{P} - \frac{j \cdot V[i]}{P \cdot V[i]}\right) = 0.$$

Owing to these considerations, to obtain an index that ranges in the interval between zero and one (in correspondence to the worst and to the optimum condition), we can formally define GRV in the following way,

$$GRV = 1 - \frac{2GCI}{(P - 1)}.$$ (21)

ROUTING FLEXIBILITY VECTOR

In order to come up with a single measure of routing flexibility, a three-dimensional picture may be helpful to simultaneously illustrate all the different themes previously discussed. In doing so a global routing flexibility index (GRF) can be conceived as the modulus of a three-dimensional vector $\bar{R}$ whose coordinates are GRE, GRR and GRV, respectively.

$$\bar{R} = GRE\hat{\imath} + GRR\hat{\jmath} + GRV\hat{k} ,$$ (22)

$$GRF = \sqrt{GRE^2 + GRR^2 + GRV^2} ,$$ (23)

with values included between 0 and $\sqrt{3}$.

Although GRF is more synthetic, it is also a bit less selective, since the threefold information concerning number, efficiency and distribution of alternative processing routes gets lost. In other words, GRF is not much relevant by itself. Clearly, what discriminates between the flexibility of two or more alternative systems is the degree of balance among the underlying contributions given by GRE, GRR and GRV.

To maintain the selectiveness of the information, an alternative approach is to consider the cosine similarity ψ between $\bar{R}$ and the best conceivable flexibility vector $\bar{R}_0 = 1\hat{\imath} + 1\hat{\jmath} + 1\hat{k}$. This is a frequently adopted way to compare two vectors, by measuring the cosine of the angle θ between them (Manning et al. 2008). In other words, the cosine similarity determines whether two vectors are pointing in roughly the same direction or not. Therefore, the bigger the angle θ between $\bar{R}$ and $\bar{R}_0$ (i.e. the lower the cosine similarity ψ) the worse the situation. The cosine similarity can be easily computed as shown by Eq. (24):

$$\psi = cos(\theta) = \left(\frac{\langle \bar{R}, \bar{R}_0 \rangle}{|\bar{R}||\bar{R}_0|}\right) = \left(\frac{GRE + GRR + GRV}{GRF\sqrt{3}}\right). \tag{24}$$

Finally GRF and ψ can be conveniently combined in a single metric, with values in the range between zero and one as follows,

$$\vartheta = \frac{GRR}{\sqrt{3}} \cdot \psi \tag{25}$$

Such a framework, based on the above-mentioned indices, is an immediate visual tool to evaluate the flexibility of a manufacturing system and to provide useful indications for corrective actions. Actually, the target should be that to get a balance among the values of GRE, GRR and GRV, which should be as close as possible to one. Suppose for example that an apparently satisfying GRF value comes from the combination of a low GRV and a valuable GRR and GRE. This means that the manufacturing system features a potential machine routing flexibility, which is poorly exploited, since several highly efficient alternative routes are available for a few products only. In this case, priority should be given to redesigning products, rather than keeping on investing in additional flexible machines.

A DEA BASED FLEXIBILITY ASSESSMENT METHOD

In this section, an alternative DEA based approach is presented to combine GRE, GRR and GRV into a global routing flexibility score ϑ. As well known, assessing and comparing the efficiency of N decision making units (DMU) (i.e. operating systems) can be vague and subjective in that it depends on the factors and on the basis for the selected comparison. If a DMU$_j$ is characterized by I inputs x$_{ij}$ and K outputs y$_{kj}$, one could measure its efficiency ϑ as follows,

$$\vartheta_j = \frac{\sum_{k=1}^{K} u_k y_{kj}}{\sum_{i=1}^{I} v_i x_{ij}},$$

(26)

where u$_1$,..., u$_K$ and v$_1$,..., v$_I$ are weighting factors associated with the outputs and the inputs, respectively. Unfortunately, also Eq. (26) generates some problems, since ϑjdepends strongly on the adopted set of weights. At different weights, the efficiency value may undergo relevant variations and it becomes difficult to fix a single structure of weights that might be accepted by all the DMUs.

A DEA approach can solve this problem by evaluating the efficiency of each DMU, through the weights system that is the best for the DMU itself. This implies the solution of N linear programming models to find the system of weights that allows the efficiency of each DMU to be maximized. To this aim models (26-27) are frequently adopted, which are known as the standard output oriented CCR models, expressed in primal and dual form, respectively. In the CCR models the subscript 0 represents the DMU which is being evaluated and s_i^-, s_k^+ are slack variables. DMUs for which the optimal value q* = φ * ≠ 1 are inefficient, whereas, provided that q* = φ * = 1, a DMU is said technically efficient if all the slack variables equal zero and is said weakly efficient if some slack variables are greater than zero. For a further discussion concerning these and similar models, the reader is referred to the clear work by Cooper et al. (2004).

$$\min q = \sum_{i=1}^{I} v_i x_{i0}$$

subject to

$$\sum_{i=1}^{I} v_i x_{ij} - \sum_{k=1}^{K} u_k y_{kj} \geq 0 \quad \forall j \in [1, N]$$

$$\sum_{k=1}^{K} u_k y_{k0} = 1$$

$$u_k, v_i \geq 0 \quad \forall k \in [1, K], \forall i \in [1, I]$$

(27)

$$\max \phi$$
subject to
$$\sum_{j=1}^{N} x_{ij}\lambda_j + s_i^- = x_{i0} \quad \forall i \in [1, I]$$
$$\sum_{j=1}^{N} y_{kj}\lambda_j - s_k^+ = \phi y_{k0} \quad \forall k \in [1, K]$$
$$\lambda_j \geq 0 \quad \forall j \in [1, N]$$

(28)

Unfortunately it is sometimes difficult to recover the explicit input–output relationship among the data, as required by the standard DEA models. This typically occurs when, as in the present case, one wants to combine a set of performance indicator into a global score. In these circumstances data sets are given without inputs (i.e. performance indicators are estimates of the goodness of the outputs), or the original input–output data cannot be easily recovered. As demonstrated by (Lovell & Pastor, 1999), a standard CCR model without explicit inputs and/or outputs cannot be used, since it would rate all DMUs as inefficient. However, a possible solution can be found in the landmark work by Thompson et al. (1986), who suggest using a single constant input CCR model. In the present case, this specific model assumes the dual form given in Eq. (29). Note that in this case, an optimal value $\phi^* = 1$ suggests maximal flexibility, conversely an optimal value $\phi^* > 1$ suggests insufficient flexibility, since it is possible to expand all performance indicators by $100^{(\phi^* - 1)}\%$, without exceeding the best performance observed among the N DMUs. In other words, the larger the value of ϕ^*, the weaker the routing flexibility.

$$\max \phi$$
subject to
$$\sum_{j=1}^{N} GRE_j\lambda_j \geq \phi GRE_0$$
$$\sum_{j=1}^{N} GRR_j\lambda_j \geq \phi GRR_0$$
$$\sum_{j=1}^{N} GRV_j\lambda_j \geq \phi GRV_0$$
$$\sum_{j=1}^{N} \lambda_j \leq 1$$
$$\lambda_j \geq 0 \quad \forall j \in [1, N]$$

(29)

To conclude this section, it is useful to note that, as shown in Liu et al. (2010), after some manipulations the dual form of (29) turns into model (30).

$$\max \vartheta_0 = u_{GRE} GRE_0 + u_{GRR} GRR_0 + u_{GRV} GRV_0$$

subject to

$$u_{GRE} GRE_j + u_{GRR} GRR_j + u_{GRV} GRV_j \leq 1 \ \forall j \in [1, N]$$

$$u_{GRE}, u_{GRR}, u_{GRV} \geq 0 \tag{30}$$

Since the strong duality theorem assures that the optimal value ϕ^* coincides with the optimal value ϑ^*, model (30) shows that, in the proposed approach, the global routing flexibility index ϑ is obtained as a weighed sum of the index GRE, GRR and GRV; where the weightings coefficients are the optimal values obtained by solving a CCR model.

EFFICIENCY ANALYSIS OF ALTERNATIVE ROUTES

In the following part of the paper, a more detailed evaluation of the alternative routing efficiency index ARE_{rp} is presented. To this aim, some critical aspects that may reduce the actual efficiency of alternative routes are analyzed in a more sophisticated way.

The Covering Degree Concept

ARE_{rp} does not take into account some problems that can be better explained through the introduction of the covering degree concept. Covering degree relates to the number of machines of the standard route that can breakdown, without compromising the possibility to access an alternative one. To get a better understanding, consider the situation detailed in Matrix 3.

$$M = \begin{array}{c} \\ m_1 \\ m_2 \\ m_3 \\ m_4 \\ m_5 \end{array} \begin{array}{c} \overbrace{\quad}^{p_1} \\ \left| \begin{array}{cc|cc|cc} 1 & 0 & 0 & 0 & 1 & 1 \\ 0 & 1 & 1 & 0 & 1 & 1 \\ 0 & 1 & 0 & 1 & 0 & 1 \\ 0 & 1 & 0 & 1 & 0 & 0 \\ 1 & 0 & 1 & 1 & 0 & 0 \end{array} \right| \\ \begin{array}{cc} r_{01} \ r_{11} & r_{02} \ r_{12} & r_{03} \ r_{13} \end{array} \end{array}$$

Matrix 3: An example of Part-Machine Matrix.

In this case three items are processed by five machines and each item has two processing routes. Still, in terms of covering degree, r_{11}, r_{12}, r_{13}, behave in a different way. While it is possible to activate the alternative route r11 regardless of machines failure along the standard route r_{01} (i.e. m_1 or m_5), this is not true either for r_{12} or for r_{13}. Actually, if m_5 fails, we cannot use either the standard route or the alternative one for p_2, but also its alternative one cannot be used. At the opposite extreme, all the viable routes of p_3 are conditioned to the availability of both m_1 and m_2.

It is evident that the concept of covering degree largely influences the possibility of accessing alternative routes, so it should be included in the flexibility analysis. This can be done by means of the covering efficiency index CE_{rp}:

$$CE_{rp} = 1 - \frac{B_{rp}}{M_{op}},$$

(31)

where M_{op} is the number of machines used in the standard route and B_{rp} is the number of machines used both in the standard r_{op} and in the alternative route r_{rp} of the p-th part type. The need to consider such an index is urged by the better suitability of a relative rather than an absolute measure. In fact, covering a route that requires several operating machines is much more difficult than covering a route made by a single machine. In the extreme case, if a product requires all the machines to be operating, a real alternative route cannot be established and the problem can only be solved through the adoption of machines, redundancy.

CE_{rp} can be incorporated in the alternative route efficiency, in the following way:

$$ARE_{rp}^C = CE_{rp} \cdot ARE_{rp} = \left(1 - \frac{B_{rp}}{M_{op}}\right) \cdot \frac{RC_{op}}{RC_{rp}}$$

(32)

Note how the optimal solution $ARE_{rp}^C = 1$ is obtained when no machines pertaining to the standard route are shared with that of the r-th alternative route (i.e. maximum degree of covering) and the processing costs of the two routes are identical.

Evidently, starting from ARE_{rp}^C, all the other flexibility indexes can be adjusted to take into account the covering degree concept.

Quality of Routing

Product quality is another critical factor to discriminate between alternative routes. In relative terms, the quality of an alternative route QR_{rp} can be evaluated as:

$$QR_{rp} = \frac{D_{rp}}{D_{0p}},$$

$$(33)$$

where D_{0p} and D_{rp} represent the percentage of defects of the standard and of the alternative route, respectively.

Also QR_{rp} can be easily included as an additional weight in the computation of the routing efficiency index:

$$ARE_{rp}^{QC} = QR_{rp} \cdot CE_{rp} \cdot ARE_{rp} = \frac{D_{rp}}{D_{0p}} \cdot \left(1 - \frac{B_{rp}}{M_{0p}}\right) \cdot \frac{RC_{0p}}{RC_{rp}}$$

$$(34)$$

The implications of Layout Efficiency on Material Handling Management

Several techniques for layout optimization rely on a flow matrix $F = |f_{ij}|$, where f_{ij} is the frequency of items travelling between the i-th and the j-th machine. Typically, the flow matrix F is based on the standard routes and so, if an alternative route is subsequently deployed, the values of F could change making the layout less efficient. To deal with this possibility, it is convenient to measure the layout efficiency in terms of the efficiency of alternative routes, with respect to the existing layout. In many automated manufacturing environment, and especially in the case of FMSs, machines are arranged along a straight track with a material handling device moving jobs from one station to another. In this condition, one can assume, without loss of generality, that materials flow along the line from left to right. If the operations sequence of a job differs from the serial sequence of the machines, the job has to travel to the left (i.e. backward) to be processed and this reverse travel is referred to as backtracking. As noted by Byrne and Chutima (1997), the performance of the system can be exploited if the workloads between machines are balanced and the distances travelled by parts are kept to a minimum. Furthermore, Hassan (1994) and Kouvelis et al. (1995) stated that backtracking has a greater impact on modern manufacturing systems and so, layout should be studied to minimize the total backtracking distance of the material-handling device. An appropriate index to express the worsening performances of the material handling device can be obtained by measuring the increment of backtracking, as compared to the standard route. In this way the layout efficiency index LErp can be computed as:

$$LE_{rp} = \frac{BT_{0p}}{BT_{rp}},$$

$$(35)$$

where BT_{0p} and BT_{rp} represent the backtracking entity of the standard and of the r-th alternative route, respectively.

Again, also LE_{rp} can be used as an additional weight in the computation of the routing efficiency index:

$$ARE_{rp}^{LQC} = LE_{rp} \cdot QR_{rp} \cdot CE_{rp} \cdot ARE_{rp} = \frac{BT_{0p}}{BT_{rp}} \cdot \frac{D_{rp}}{D_{0p}} \cdot \left(1 - \frac{B_{rp}}{M_{0p}}\right) \cdot \frac{RC_{0p}}{RC_{rp}}. \tag{36}$$

NUMERICAL EXAMPLE

This section presents a numerical application concerning two alternative productive systems. It is assumed that ten machines are used to manufacture three products via the standard routes shown in Matrix 4.

		A	p_1	p_2	p_3
	m_1	0.98	1	0	0
	m_2	0.98	1	1	0
	m_3	0.95	0	1	0
	m_4	0.99	1	0	1
	m_5	0.88	0	0	1
$S =$	m_6	0.98	1	1	1
	m_7	0.99	1	1	0
	m_8	0.98	1	0	1
	m_9	0.94	0	1	1
	m_{10}	0.98	0	1	0

Matrix 4: An example of standard route matrix.

The availability of each machine has also been included in the standard route matrix S, as an additional column A. To enhance routing flexibility two alternative configurations have been deployed by the process planner, as shown in the cost matrixes 5 and 6.

Routing Flexibility: A Vectorial Approach

To evaluate the goodness of both solutions, starting from the alternative route efficiency, the whole set of previously introduced metrics has been evaluated, without considering (for the sake of simplicity) either backtracking or the quality of the alternative routes.

Configuration #1

$$C_1 = \begin{array}{cccccccc} p_1 & & p_2 & & & & p_4 & \\ 7 & 7 & & 3 & & & & 1 \\ 3 & 3 & 2 & 2 & 2 & & 3 & \\ & 4 & 4 & & 5 & & 3 & 3 \\ 4 & & & 4 & & 2 & 2 & 3 \\ & 2 & & 5 & 3 & & & \\ 5 & & 5 & & 2 & & & 2 \\ 6 & 3 & & 3 & & 3 & & \\ 5 & 5 & 1 & 5 & 6 & 1 & & \\ & 3 & & & & 5 & 5 & 5 \\ & 7 & 4 & & & 4 & 2 & 2 \end{array}$$

$r_{01} \quad r_{11} \quad r_{02} \quad r_{12} \quad r_{22} \quad r_{03} \quad r_{13} \quad r_{23} \quad r_{13}$

Matrix 5: Cost matrix relative to the first configuration.

Configuration #2

$$C_2 = \begin{array}{cccccccccc} & p_1 & & & p_2 & & & & p_3 & \\ 7 & & 7 & & & & & & & \\ 3 & 4 & 3 & 2 & 4 & & 3 & & 3 & \\ & 6 & & 4 & 5 & 6 & 5 & & & \\ 4 & 5 & & & 6 & 5 & & 2 & 2 & \\ & 3 & 4 & & & & 5 & 3 & & \\ 5 & 6 & 5 & 5 & & 6 & & 2 & & \\ 6 & & 3 & & 5 & 3 & 4 & & 3 & \\ 5 & & 5 & & & & & 1 & 2 & \\ & 7 & & 1 & 2 & 3 & 2 & 5 & & \\ & & 7 & 4 & & & 3 & & 6 & \end{array}$$

$r_{01} \quad r_{11} \quad r_{21} \quad r_{02} \quad r_{12} \quad r_{22} \quad r_{32} \quad r_{03} \quad r_{13}$

Matrix 6: Cost matrix relative to the second configuration.

The results, obtained by means of Eq. (6-25) are listed in Table 1 and Table 2, respectively.

Table 1: Routing flexibility metrics

	Configuration #1						Configuration #2					
	r_{11}	r_{12}	r_{22}	r_{13}	r_{23}	r_{33}	r_{11}	r_{21}	r_{12}	r_{22}	r_{32}	r_{13}
$ARE_{rp} = \dfrac{RC_{op}}{RC_{rp}}$	0.97	0.95	0.905	0.93	0.87	0.81	0.97	0.97	0.86	0.83	0.86	0.81
$CE_{rp} = 1 - \dfrac{B_{rp}}{M_{op}}$	0.50	0.67	0.50	0.60	0.60	0.40	0.50	0.50	0.33	0.33	0.17	0.60
$ARE_{rp}^C = CE_{rp} \cdot ARE_{rp}$	0.48	0.63	0.45	0.56	0.52	0.325	0.48	0.48	0.29	0.275	0.14	0.49
$SRA_{rp} = \prod_{m=1}^{M} A_m x_{mop}$	0.90		0.83		0.79		0.90		0.83			0.79
$SRC_p = \dfrac{\sum_{m=1}^{M} x_{mop}}{M}$	0.60		0.60		0.50		0.60		0.60			0.50
$\alpha_p = \left(\dfrac{1}{1 - SRA_p}\right)^{SRC_p}$	4.1		2.92		2.16		4.1		2.92			2.16
$JRR_p = 1 - \left(1 - \dfrac{\tilde{R}_p}{TR - 1}\right)^{\alpha_p}$	0.42		0.57		0.64		0.69		0.75			0.25
$JRV_p = \langle E_p, R_p \rangle$	0.20		0.315		0.31		0.33		0.19			0.12

Table 2: Aggregate routing flexibility metrics

	Configuration #1	Configuration #2
$GRE = \frac{\sum_{p=1}^{P} JRAE_p}{P}$	0.498	0.402
$GRR = \frac{\sum_{p=1}^{P} JRR_p}{P}$	0.542	0.563
$GRV = 1 - \frac{2GCI}{P-1}$	0.864	0.67
$GRF = \sqrt{GRE^2 + GRR^2 + GRV^2}$	1.135	0.963
$\psi = cos\left(\frac{\langle \overline{R}, \overline{R}_0 \rangle}{|\overline{R}||\overline{R}_0|}\right)$	0.968	0.98
$\vartheta = \frac{GRF}{\sqrt{3}} \cdot \psi$	0.634	0.545

Although both solutions have an equal number of alternative routes, the first one is significantly better than the second one because $\vartheta_1 \approx 0.634$ and $\vartheta_2 \approx 0.545$. The superiority of the first solution is evident also from the three dimensional representation of the routing flexibility vectors $\overline{R}_0, \overline{R}_1, \overline{R}_2$ shown in Fig. 3.

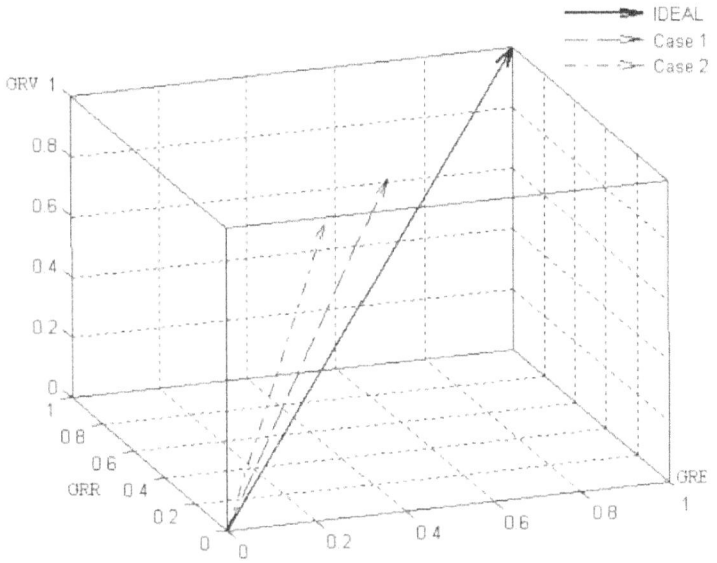

Figure 3: Visual assessment of routing flexibility.

Differences are mainly due to a reduction in the global routing efficiency ($GRE_1 = 0.498$, $GRE_2 = 0.4$) and to an uneven distribution of the alternative routes ($GRV_1 = 0.86$, $GRV_2 = 0.67$). The reasons of these results can be better explained considering that in the second configuration:

- p_2 has three alternative routes, but, due to a poor covering degree (CE_{32} = 0.17), its routing versatility is low (JRV_2 = 0.19);

- there is just an alternative (inefficient) route for p3, and this has a negative impact on its routing versatility (JRV_3 = 0.12). Specifically the standard route r_{03} is characterized by the lowest availability and by the lowest complexity and so, to assure production continuity, more than a single alternative route should be deployed.

Consequently, although the alternative solutions are comparable in terms of the number of alternative routes (i.e. GRR dimension), the first one is preferable, for it gets a significant advantage considering how evenly efficient routes are distributed among the manufactured items (i.e. GRE and GRV dimensions).

This example clearly underlines the fact that the number of available routes is not sufficient to discriminate between alternative solutions and it must be integrated with additional performance factors. It also underlines the utility to keep on separated levels the threefold information concerning the number, the efficiency and the distribution of alternative routes. As a matter of fact, what discriminates between the flexibility of two or more alternative systems is the degree of balance among the underlying contributions GRE, GRR and GRV.

Routing Flexibility: a DEA based approach

A similar result can be obtained using the output oriented CCR model (29) proposed in section 7. In this case, since there are two DMUs, using data of Table 2, the single constant input CCR model (29) assumes the following primal forms, where P_1 and P_2 refer to configuration #1 and #2, respectively.

$(P_1)\max\phi_1$

subject to

$0.498\lambda_1 + 0.402\lambda_2 \geq 0.498\phi_1$

$0.542\lambda_1 + 0.563\lambda_2 \geq 0.542\phi_1$

$0.864\lambda_1 + 0.670\lambda_2 \geq 0.864\phi_1$

$\lambda_1 + \lambda_2 \leq 1, \ \lambda_1,\lambda_2 \geq 0.$

$(P_2)\max\phi_2$

subject to

$0.498\lambda_1 + 0.402\lambda_2 \geq 0.402\phi_2$

$0.542\lambda_1 + 0.563\lambda_2 \geq 0.563\phi_2$

$0.864\lambda_1 + 0.670\lambda_2 \geq 0.670\phi_2$

$\lambda_1 + \lambda_2 \leq 1, \ \lambda_1,\lambda_2 \geq 0.$

Solving models P_1 and P_2 with the dedicated linear programming solver Lindo 6.1 (2002) yields $\phi_1 = \phi_2 = 1$. Therefore, both DMUs appear as efficient, and the DEA model is not capable to discriminate between them. As one usually does in this case, to augment the discrimination power, the cross efficiency method can be used (Cooper et al., 2004). Briefly, this analysis is based on the definition of a cross efficiency matrix $CE_{N \times N} = |\vartheta_{jk}|$ that is a square matrix (with as many rows and columns as there are units being compared), whose generic element ϑ_{jk} represents the efficiency of the j-th DMU evaluated through the optimal weights structure for the k-th one. If DMU$_j$ is efficient (i.e. $\vartheta_{jj} = 1$), but it exhibits a behavior specialized along a given dimension with respect to the other units, ϑ_{jk} will be less than 1 for some value of k. Therefore, an interesting discriminating index $\tilde{\vartheta}_j$ can be obtained taking the average of the values in the k-th row:

$$\tilde{\vartheta}_j = \frac{1}{N} \sum_{k=1}^{N} \vartheta_{jk}$$

(37)

To compute this value, it is convenient to solve the dual problem (30), because this model makes explicit use of the coefficients u$_k$ used to weight the performance indicators of each DMU.

$(D_1) \max v_0 = 0.498u_{GRE} + 0.542u_{GRR} + 0.864u_{GRV}$

subject to

$0.498u_{GRE} + 0.542u_{GRR} + 0.864u_{GRV} \leq 1$

$0.402u_{GRE} + 0.563u_{GRR} + 0.670u_{GRV} \leq 1$

$u_{GRE} \geq 0, \ u_{GRR} \geq 0, \ u_{GRV} \geq 0.$

$(D_2) \max v_0 = 0.402u_{GRE} + 0.563u_{GRR} + 0.67u_{GRV}$

subject to

$0.498u_{GRE} + 0.542u_{GRR} + 0.864u_{GRV} \leq 1$

$0.402u_{GRE} + 0.563u_{GRR} + 0.670u_{GRV} \leq 1$

$u_{GRE} \geq 0, \ u_{GRR} \geq 0, \ u_{GRV} \geq 0.$

Solving these models yields $\vartheta_{12} \approx 1$ and $\vartheta_{21} \approx 0.78$. Therefore $\tilde{\vartheta}_1 \approx$ (1+1)/2 = 1 and $\tilde{\vartheta}_1 \approx (1+1)/2 = 1$ and $\tilde{\vartheta}_2 \approx (1 + 0.78)/2 = 0.89$ and configuration #1 results as the best option. A synthesis of all the results obtained with the DEA based approach is given in Table 3.

Table 3: Results of the output oriented DEA model

	Configuration #1	Configuration #2
ϕ^*	1	1
λ_1	1	0
λ_2	0	1
ϑ^*	1	1
u_{GRE}	0	0
u_{GRR}	0	1.573
u_{GRV}	1.16	0.17
ϑ_{1j}	1	0.775
$\tilde{\vartheta}$	1	0.887

Note that the values obtained with the vectorial and the DEA approaches are similar. Indeed, although their solutions differ (i.e. $\tilde{\vartheta}_1 \approx 1$ and $\vartheta_1 \approx 0.64$), this difference is due to the fact that $\tilde{\vartheta}_1$ expresses the flexibility of the first configuration in relative terms with the second one, whereas ϑ_1 expresses the flexibility of the first configuration with respect to the ideal one. As a matter of fact, computing the percentage difference between $\tilde{\vartheta}_1$ and $\tilde{\vartheta}_2$ and between

ϑ_1 and ϑ_2 one obtains a value of 14% and of 12%, respectively, which are sufficiently aligned values.

VALIDATION THROUGH SIMULATION ANALYSIS

To validate the proposed approach, the operating performance of both configuration #1 and #2 has been assessed via simulation, modelling the two manufacturing systems by means of the dedicated software Simul8 Professional (2006). Making reference to the definition given by Browne et al. (1984), which correlates routing flexibility with the ability to react to manufacturing problems and to assure production continuity, the systems have been compared in terms of their average throughput (i.e. the higher the throughput the more the system is considered to be flexible).

Simulation Models

For the sake of simplicity we considered the same costs (per unit of time) on each machine, that is $\tilde{c}_m = \tilde{C} = 1$ with m = 1,..,10. In this way the processing time matrix T coincides with the cost matrix C and data of Matrix 4 can be directly taken as the average processing time of each machine. Furthermore, processing times are assumed to be normally distributed, with a standard deviation equal to 0.25 of the mean. In a similar way, machines availability (see the first column of Matrix 4) is reproduced using exponential and Erlang distributions to model the time between failures and the repair time, respectively. An example of the modelled systems is given in Fig. 4, which is relative to configuration #1 (a similar model is used for Configuration #2). Note that, to improve the readability of Fig. 4, the main route r_{01} of product p_1 has been highlighted with respect to the other routes deployed in the system.

Most of the elements of Fig. 4 are self explaining, but some ones (i.e. the dummy machines) deserve some further explanations. In particular, to facilitate the attainment of a steady state, and to limit the variability of the throughput, a CONWIP system has been adopted (Hoop & Spearman, 2000). CONWIP stands for constant work-in-process and designates a strategy that uses cards (or other visual control methods) to limit the work in process (WIP) that can accumulate in a manufacturing system. Each job is associated with a card for the whole duration of its manufacturing cycle. As soon as the job is processed by the last machine, its cards are released and can be associated with a new job. Since no job is allowed in the system without a card, the overall amount of WIP equals the number of available cards (Braglia et al., 2010).

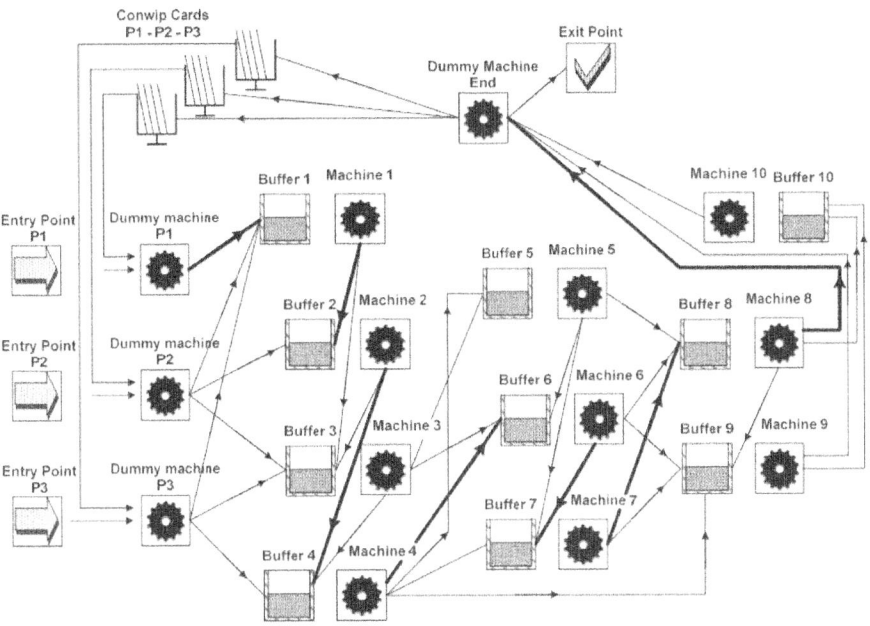

Figure 4: Simulation model.

In the simulation model, this strategy is reproduced by means of four dummy (i.e. zero processing time) machines. Specifically the End machine releases CONWIP cards as soon as a product ends its manufacturing cycle, while machines P_1, P_2 and P_3 check the availability of CONWIP cards before pushing a new job into the system. These dummy machines perform an additional task, since they are used to specify the route (i.e. principal or alternative) assigned to each job. To make this choice, three operating rules have been conceived and implemented. Specifically, after a preliminary check, that identifies the machines (if any) stopped due to failures, jobs are assigned to the remaining available routes based on the following operating rules:

- rule #1: assign the job to the route with the minimum WIP level;
- rule #2: assign the job to the route with the minimum cumulated processing time;
- rule #3: assign the job to the route whose bottleneck is characterized by the lower saturation.

Three systems were compared via simulation, namely base case (i.e. the rigid system with standard routes only), configuration #1 and configuration #2. Operating performance of each one was assessed in five different settings: (i-iii) a single product is manufactured, (iv) all products are manufactured and a homogeneous product mix (i.e. 1:1:1) must be respected, (v) all products are manufactured and any mix can be used, provided that each product accounts, at least, for the 15% of the overallproduction. In each instance, the WIP (i.e. the number of CONWIP cards for each products) that maximises the throughput was determined using the optimization tool (OptQuest) embedded in the simulation environment. Finally, to compare the results, the following scheme was used:

- Output parameter: Daily Throughput [jobs/day];
- Total Simulation time: 48000 [min], that is 100 working days;
- Warm Up period:7200 [min], that is 15 working days;
- Routing Selection: use of rule #3, since a preliminary test demonstrated the superiority of this rule over the other ones;
- Execution scheme: use 30 simulation runs for each analyzed configuration;
- Result analysis: evaluation of the confidence interval for the average Daily Throughput (DT) at a 95% confidence level.

Note that, due to the use of the CONWIP strategy, a warm up period of fifteen working days was considered sufficient (to reach the steady state), as visually demonstrated by Fig. 5, which shows the evolution of the average DT during a simulation run.

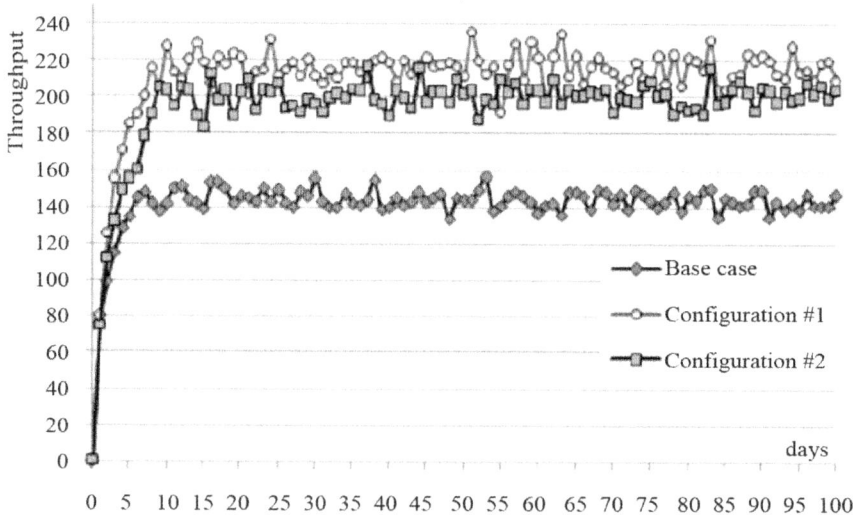

Figure 5: Evolution of the Daily Throughput during a simulation run.

The obtained results, listed in Table 4, demonstrate that, as one could have reasonably guessed, both systems are definitely more flexible that the basic (rigid) configuration. From the data of Table 4 one can also note that Configuration #1 is the best alternative with respect to products p_2 and p_3, but not with respect to p3. Therefore neither solution dominates the other one; this is the reason why, under the constraint of a homogeneous product mix, both configurations perform in a similar way (i.e. there is no statistical evidence that one system is better than the other one).

Table 4: Result of the simulation model

	Base Case				Configuration #1				Configuration #2		
		DT [jobs/day]				DT [jobs/day]				DT [jobs/day]	
	WIP	Low 95%	Avg	High 95%	WIP	Low 95%	Avg	High 95%	WIP	Low 95%	Avg	High 95%
Single product p_1	12	66.8	68.2	69.5	12	67.5	68.7	69.9	12	85.8	88.2	90.6
Single product p_2	12	91.4	93.4	95.4	18	133	135.1	137.2	12	100	102.6	105.2
Single product p_3	6	88.5	90.4	92.3	20	175.6	181.5	187.3	18	165.2	168.4	171.6
All products mix (1 1 1)	47	116	118	120	61	179.8	184.8	189.8	61	182.4	186.1	189.8
All products Max DT	80	138.4	142	145.6	74	209.2	217.2	225.2	80	195.5	200.4	205.3

However, following an approach similar to the DEA one, if we relax the constraints on the homogeneous product mix, and let each system adopt the product mix that maximises its global performance, we reach a clearer situation. Indeed, in this unconstrained setting, Configuration #1 and #2, obtain an average DT of 217±8 and 200±5 [jobs/day], respectively, with a percentage difference of 8.5%. Furthermore, since the DT confidence intervals do not intersect, the superiority of Configuration #1 over Configuration #2 is statistically significant. These results evidently agree with the conclusions obtained in the previous sections of the paper and empirically confirmation of the validity of both approaches presented in the paper.

CONCLUSIONS AND REMARKS

In this paper, a theoretical framework and a set of indexes to evaluate several aspects of routing flexibility have been illustrated. The proposed indexes are intended to measure routing flexibility, by taking into account the number of viable routes and other issues that may compromise the efficiency of the system under analysis. A major peculiarity of the proposed measurement method consists in its flexibility, since a variety of aspects, which could potentially affect the overall effectiveness of alternative routes, are taken into consideration. This allows one to freely decide the extent and the level of details of the indexes included in the flexibility analysis.

Finally, to help practitioners in the planning of the right corrective actions, a graphical and a DEA based approach have also been presented and explained through a meaningful numerical example supported by a simulation analysis.

ACKNOWLEDGMENTS

The authors are grateful to the anonymous referee(s) and to the editor of the journal for their valuable comments. Thanks to their precise and detailed suggestions, both the structure and the technical value of the paper have increased.

REFERENCES

1. Azzone, G., & Bertelè, U. (1991). Techniques for measuring the effectiveness of automation and manufacturing systems. Control and Dynamic Systems, 48, 1-45.

2. Beskese, A., Kahraman, C., & Irani, Z. (2004). Quantification of flexibility in advanced manufacturing systems using fuzzy concept. International Journal of Production Economics, 89, 45-56.

3. Borenstein, D., & Rohde, L.R. (2005). State space representation of routing flexibility. Computer & Industrial Engineering, 49, 537-55.

4. Braglia, M., Frosolini R., Gabbrielli, R., & Zammori, F. (2010). CONWIP card setting in a flow-shop system with a batch processing machine. International Journal of Industrial Engineering Computations, Article in Press.

5. Brill, P.H., & Mandelbaum, M. (1989). On measures of flexibility in manufacturing systems. International Journal of Production Research, 27 (5), 747-756.

6. Browne, J., Dubois, D., Rathmill, K., Sethi, S.P., & Stecke, K.E. (1984). Classification of flexible manufacturing systems. The FMS Magazine, 2 (2), 114-117.

7. Buzacott, J.A. (1982). The fundamental principles of flexibility in manufacturing systems. Proceedings of the 1st International Congress on Flexible Manufacturing Systems, Brighton, England.

8. Byrne, M.D., & Chutima, P. (1997). Real-time operational control of an FMS with full routing flexibility. International Journal of Production Economics, 51 (12), 109-113.

9. Caprihan, R., & Wadhwa, S. (1997). Impact of routing flexibility on the performance of an FMS, a simulation study. International Journal of Flexible Manufacturing Systems, 9 (3), 273-298.

10. Chatterjee, A., Cohen, M.A., Maxwell, W.C., & Miller, L.W. (1987). Manufacturing flexibility: models and measurement, Proceedings of the 1st ORSA/TIMS Special Interest Conference on FMS, Ann Arbor, MI.

11. Chan, F.T.S. (2001). The effect of routing flexibility on a flexible manufacturing system. International Journal of Integrated Computer Manufacturing, 14 (5), 431-445.

12. Chang, A.Y., Whitehouse D.J., Chang S.L., & Hsieh, Y.C. (2001). An approach to the measurement of single-machine flexibility. International Journal of Production Research, 39 (8), 1589-1601.

13. Chang, A.Y. (2004). On the Measurement of Labour Flexibility. Proceeding of the International Engineering Management Conference, IEEE.

14. Chang, A.Y. (2007). On the Measurement of Routing Flexibility: a Multiple Attribute Approach. International Journal of Production Economics, 109 (1), 122-136.

15. Chen, I.J., Calantone, R.J., & Chung, C.H. (1992). The marketing manufacturing interface and manufacturing flexibility. Omega, 20 (4), 431-443.

16. Chen, I.J. & Chung, C.H. (1996). An examination of flexibility measurements and performance of flexible manufacturing systems. International Journal of Production Research, 34 (2), 379-394.

17. Cooper W.W., Seiford, L.M. & Zhu, J. (2004). Handbook on Data Envelopment Analysis. Boston: Kluwer academic publishers.

18. Charnes, A., Cooper W.W., Rhodes, E., (1987) Measuring the efficiency of decision making units. European Journals of Operational Research, 2, 429-444.

19. Das, S.K. (1996). The measurement of flexibility in manufacturing systems. International Journal of Flexible Manufacturing Systems, 8, 67-93.

20. Gerwin, D. (1993). Manufacturing flexibility: a strategic perspective. Management Science, 39 (4), 395-410.

21. Gini, C. (1921). Measurement of Inequality and Incomes. The Economic Journal, 31, 124-126.

22. Gupta, D., & Buzacott, J.A. (1989). A Framework for understanding flexibility of manufacturing systems. Journal of Manufacturing Systems, 8 (2), 89-97.

23. Gupta, Y.P., & Goyal, S. (1989). Flexibility of manufacturing systems: concepts and measurements. European Journal of Operational Research,

43, 119-135.

24. Hassan, M.M.D. (1994). Machine layout problem in modern manufacturing facilities. International Journal of Production Research, 32, 2559-2584.

25. Hopp, J.H., & Spearman, M.L. (2000). Factory Physics. New York: McGraw Hill.

26. Hutchinson, G.K. & Sinha, D. (1989). A quantification of the value of flexibility. Journal of Manufacturing Systems, 8, 47-57.

27. Kochikar, V.P., & Narendran, T.T. (1992). A framework for assessing the flexibility of manufacturing system. International Journal of Production Research, 30 (12), 2873-2895.

28. Kouvelis, P., Chiang, W. C., & Yu, G. (1995). Optimal algorithms for row layout problems in automated manufacturing systems. IIE Transactions, 27 (1), 99-104.

29. Kumar, V. (1987). Entropic measurement of manufacturing flexibility, International Journal of Production Research, 25 (7), 957–966.

30. Liu, W.B., Zhang, D.Q., Meng, W., Li, X.X. & Xu, F. (2011). A study of DEA models without inputs, Omega, Article in Press.

31. Lovell C.A.K., Pastor, J.T. (1999). Radial DEA models without inputs or without outputs. European Journal of Operational Research, 118, 46-51.

32. Manning C.D., Raghavan, P. & Schütze, H. (2008). Introduction to information retrieval, Cambridge: Cambridge University Press.

33. Nagarur, N. (1992). Some performance measures of flexible manufacturing systems. International Journal of Production Research, 30 (4), 799-809.

34. O'Connor. P.D.T. (2004). Practical Reliability Engineering, Chichester: Wiley, 4th edition. Roll, Y., Karni, R., & Arzi, Y. (1992). Measurement of processing flexibility in flexible manufacturing cells. Journal of Manufacturing Systems, 11, 258-268.

35. Sarker, B., Krishnamurthy S., & Kuthethur, G. (1994). A survey and critical review of flexibility measures in manufacturing systems. Production Planning and Control, 5, 512-523.

36. Schrage, L. (2002). LINDO Release 6.1, LINDO System, Inc.

37. Sethi, A.K., & Sethi, S.P. (1990). Flexibility in manufacturing: a survey, International Journal of Flexible Manufacturing Systems, 2 (4), 289-328.

38. Shannon, C.E. (1948). A mathematical theory of communication, Bell System Technical Journal, 27, 379-423.

39. Slack, N. (1987). The flexibility of manufacturing systems. International

Journal of Operations and Production Management, 7 (3), 35-45.

40. SIMUL8, (2000). Version 6.3, Manual and Simulation Guide, SIMUL8 Corporation, Herndon, Virginia.

41. Taymaz, E. (1989). Types of flexibility in a single-machine production system. International Journal of Production Research, 27 (8), 1891-1899.

42. Thompson, R.G., Singleton, Jr., F.D., Thrall, R.M. & Smith, B.A. (1986). Comparative site evaluation for locating a high energy physics lab in Texas. Interfaces, 16, 35-49.

43. Yao, D.D. (1985). Material and information flows in flexible manufacturing systems. Material Flows, Special Issue on Flexible Manufacturing System, 143–149

44. Yu, M.C, & Greene, T.J. (2000). The effects of routing flexibility in a multi-stage pull-type system. International Journal of Production Research, 38 (16), 3725-3746.

45. Zhang, Q., Vonderembse, M.A., & Lim, J.S. (2003). Manufacturing flexibility: defining and analysing relationships among competence, capability, and customer satisfaction. Journal of Operations Management, 21, 173-191.

Chapter 8

INTEGRATION OF PART SELECTION, MACHINE LOADING AND MACHINING OPTIMISATION DECISIONS FOR BALANCED WORKLOAD IN FLEXIBLE MANUFACTURING SYSTEM

Mussa I. Mgwatu

Department of Mechanical and Industrial Engineering, University of Dar es Salaam, P.O. Box 35131, Dares Salaam, Tanzania

ABSTRACT

This paper demonstrates the importance of incorporating and solving the machining optimisation problem jointly with part selection and machine loading problems in order to avoid unbalanced workload in the FMS. Unbalanced workload renders to ineffective FMS such that some machines on the manufacturing shop floor become more occupied than others. Since CNC machine tools employed in the FMS are rather expensive, it is mostly important to balance the workload so that all machines can be effectively utilised. Therefore, in this study, two mathematical models are presented and solved in efforts to balance the workload and improve the performance of the FMS. A two-stage sequential approach is adopted whereby the first stage deals with the maximum throughput objective while the second stage deals with the minimum production cost objective. The results show that when part selection, machine loading and machining optimisation problems are jointly solved, more practical decisions can be made and a wide range of balanced workload in the FMS can be realised with minimum production cost objective. The results also show that the available machine time and tooling budget have enormous effects on throughput and production cost.

INTRODUCTION

Intensive market competition has forced manufacturing industries to focus on flexible manufacturing systems (FMSs). This is because FMSs offer a rapid and timely response to market demands thus helping the manufacturing industries to win competitive advantage. FMSs are highly automated manufacturing systems that consist of a group of processing workstations, usually computernumerical control (CNC) machine tools, interconnected by automated material handling and storage systems altogether interfaced via a central computer. They are designed to combine the flexibility of a low-production-volume job shop and the efficiency of a high-production-volume flow shop to best suit the batch production of mid-volume and mid-variety of products. Job shop and flow shop are two conventional manufacturing systems that are associated with a traditional arrangement of machine tools on a manufacturing shop floor. In a job-shop system, machines are grouped together to perform similar operations for different parts. In a flow-shop system, machines are arranged together to process the parts as they flow from one machine to the next through the sequence of operations.

Although the installation of FMSs requires a greater capital investment, their expected benefits are substantial. The benefits include increased machine utilisation, less machines, reduced floor space, greater responsiveness to changes, reduced inventories, lower manufacturing lead times, and higher labour productivity. In order to justify these benefits, the FMS resources should be well planned and then effectively utilised to meet the production requirements, achieve the operational objectives and yet provide the payback more quickly. Various FMS planning problems have been addressed at different stages of the FMS's life cycle. Stecke (1985) divided the FMS planning problems into five subproblems of part type selection, machine grouping, machine loading, production ratio, and resource allocation. Hwang (1986) found that among the five planning subproblems, part selection and machine loading are the most important in the FMS. However, the decisions of part selection and machine loading problems in the FMSs may not be effective because they often lead to unbalanced workload. As reported by Liang (1994), some planning policies such as maximising throughput may cause extremely unbalance workloads thereby causing bottleneck in some machines involved in the manufacturing process. The source for unbalanced workload decisions in part selection and machine loading problems may be the fact that the values of machining parameters are always predetermined and not optimised during FMS planning process as exhibited in Liang (1994), Yang and Wu (2002), and Choudhary, et al (2006).

The optimisation of machining parameters such as cutting speed, feed rate, and depth of cut is an essential step in achieving high machining efficiency and effective FMS. Ermer and Kromodihardjo (1981) and Manna and Salodkar (2008) optimised machining parameters for single machine problems while Agapiou (1992) optimised machining parameters for a conventional multi-stage machining system for effective machine utilisation. It is however learnt that previous studies on the optimisation of machining parameters mainly relied on either standalone machines or traditional multi-stage manufacturing systems rather than on FMSs. In order to utilise the machines in the FMS more effectively, it is necessary to determine the optimum machining parameters during the production planning stage. It is the purpose of this study to integrate the decisions of part selection, machine loading and machining optimisation problems in an attempt to balance the workload while improving the performance of the FMS.

THEORETICAL FRAMEWORK

Mathematical Notations

The following mathematical notations are used for providing the theoretical concepts and in formulating the mathematical models for the FMS:

Indices

j	index for part type, $j=1,...,J$
k	index for machines, $k=1,...,K$
o	index for operations of part types, $o,...,O_j$
t	index for tool type, $t=1,...,T$

Decisions variables

x_j	$= 1$, if part type j is selected, 0 otherwise
x_{jotk}	$= 1$, if operation o of part j is processed using tool t on machine k, 0 otherwise
y_{tk}	$= 1$, if tool type t is assigned to machine k, 0 otherwise
v_{jotk}	cutting speed for combination j, o, t, k (m/min)
f_{jotk}	feed rate for combination j, o, t, k (mm/rev or mm/tooth)
h	machine processing time (min)

Parameters

α_{jt} tool life constant of the cutting speed for tool t on part j

γ_{jt} tool life constant of the depth of cut for tool t on part j

β_{jt} tool life constant of the feed rate for tool t on part j

λ_{jt} tool life constant of the number of tool teeth in milling operations for tool t on part j

ω_{jt} tool life constant of the tool diameter in milling operations for tool t on part j

δ_{jt} tool life constant of the width of cut in milling operations for tool t on part j

a_{jot} depth of cut for operation o on part j using tool t (mm)

A_k available time at machine k (min)

B available tooling budget ($)

C_e tool cost of machining a single part ($)

C_m machining cost ($)

C_r tool replacement cost ($)

C_t cost per edge of tool t ($)

D_{jot} tool diameter for operation o on part j (mm)

E_{jt} tool life constant for tool t on part j

F^L_{jotk} lower feed rate limit for combination j, o, t, k (mm/rev or mm/tooth)

F^U_{jotk} upper feed rate limit for combination j, o, t, k (mm/rev or mm/tooth)

G_k number of slots on tool magazine of machine k

L_{jo} length of cut for operation o on part j (mm)

M_{jot}, N_{jot} machining constants for operation o on part j using tool t

Q_j production quantity of part type j

q_r number of parts per tool replacement

R_t replacement time for tool t (min)

S_t number of slots required by tool type t

T_{jt} tool life for part j and tool t combination

t_m machining time (min)

t_p processing time of a single part (min)

t_r tool replacement time distributed to each part (min)

u_j value coefficient of part j

V^d_{jotk} lower cutting speed limit for combination j,o,t,k (m/min)

$V^{d/}_{jotk}$ upper cutting speed limit for combination j,o,t,k (m/min)

W_j width of cut on part j (mm)

Z_t number of teeth of tool t

Theoretical Concepts and Main Assumptions

The concept of solving part selection, machine loading and machining optimisation problems jointly comes with the need to define the relationship among their entities. In reality, processing time and tooling cost form a major linkage among part selection, machine loading and machining optimisation problems. On one hand, the selected parts for processing in the FMS depend on the available machine time. Similarly, due to higher tooling cost, the number of tools that can be loaded on the tool magazine depends on the available tooling budget. On the other hand, machining parameters, processing time and tooling cost are strongly coupled. For instance, at higher cutting speeds, machining time and the related cost are less but more cutting tools are consumed thus increasing tooling cost. Major components of processing time are machining time and tool replacement time. Machining time is the actual time required to process a part on a machine described as (Narang and Fischer, 1993):

$$t_m = N_{jot} v^{-1}_{jotk} f^{-1}_{jotk} \tag{1}$$

where

$$N_{jot} = \frac{\pi D_{jot} L_{jo}}{1000}, \quad \text{for drilling and tapping/reaming operations, and} \tag{2a}$$

$$N_{jot} = \frac{\pi D_{jot} L_{jo}}{1000 Z_t} \quad \text{for milling operations.} \tag{2b}$$

Machining cost is the product of machining time and operating cost rate. Denoting U_o as the operating cost rate, the machining cost can be written as:

$$C_m = N_{jot} v^{-1}_{jotk} f^{-1}_{jotk} U_o \tag{3}$$

Tool replacement time is the time needed to remove a worn tool from the machine and replace with a new tool. Tooling cost is the cost of purchasing new cutting tools. Both the tool replacement time and tooling cost are functions of the Taylor's tool life equation. The extended tool life equation reported by Lambert and Walvekar (1978) can be written in the following form:

$$T_{jt} = \frac{E_{jt}}{v_{jotk}^{\alpha_{jt}} f_{jotk}^{\beta_{jt}} a_{jot}^{\gamma_{jt}}}$$

(4)

The tool life is reached and the tool can be replaced when several parts have been processed on a machine. It follows that the number of parts at each tool replacement is the ratio of tool life in Eq. (4) to unit machining time in Eq. (1) expressed as:

$$q_r = \frac{E_{jt}}{N_{jot} v_{jotk}^{\alpha_{jt}} f_{jotk}^{\beta_{jt}} a_{jot}^{\gamma_{jt}}}$$

(5)

Substituting Eqs. (2a) and (2b) into Eq. (5) and rearranging the terms, the tool replacement time and tooling cost distributed to each part can be respectively presented as:

$$t_r = M_{jot} v_{jotk}^{\alpha_{jt}-1} f_{jotk}^{\beta_{jt}-1} R_t$$

(6)

and,

$$C_e = M_{jot} v_{jotk}^{\alpha_{jt}-1} f_{jotk}^{\beta_{jt}-1} C_t$$

(7)

M_{jot} is a machining constant which is defined by Wang and Liang (2005) as:

$$M_{jot} = \frac{\pi D_{jot}^{1-\omega_{jt}} L_{jo}}{1000 E_{jt}} \quad \text{for drilling operations}$$

(8a)

and,

$$M_{jot} = \frac{\pi D_{jot}^{1-\omega_{jt}} L_{jo} a_{jot}^{\gamma_{jt}}}{1000 E_{jt}} \quad \text{for reaming/tapping operations}$$

(8b)

The machining constant for milling operations defined by Shnumugam, et al (2002) and Wang and Liang (2005) is:

$$M_{jot} = \frac{\pi D_{jot}^{1-\omega_{jt}} L_{jo} a_{jot}^{\gamma_{jt}} W_j^{\delta_{jt}} Z_t^{\lambda_{jt}-1}}{1000 E_{jt}}$$

(8c)

Correspondingly, the tool replacement cost is the product of tool replacement time and operating cost rate and is defined as:

$$C_r = M_{jot} v_{jotk}^{\alpha_{ji}-1} f_{jotk}^{\beta_{ji}-1} R_t U_o$$

$$(9)$$

Taking the sum of machining time in Eq. (1) and tool replacement time in Eq. (6), the processing time can be represented as:

$$t_p = N_{jot} v_{jotk}^{-1} f_{jotk}^{-1} + M_{jot} v_{jotk}^{\alpha_{ji}-1} f_{jotk}^{\beta_{ji}-1} R_t$$

$$(10)$$

In similar manner, adding up the machining cost in Eq. (3), tool cost in Eq. (7), and tool replacement cost in Eq. (9), the total production cost can be obtained as:

$$C_p = N_{jot} v_{jotk}^{-1} f_{jotk}^{-1} U_o + M_{jot} v_{jotk}^{\alpha_{ji}-1} f_{jotk}^{\beta_{ji}-1} R_t U_o + M_{jot} v_{jotk}^{\alpha_{ji}-1} f_{jotk}^{\beta_{ji}-1} C_t$$

$$(11)$$

As shown earlier, both processing time and tooling cost make a significant link among part selection, machine loading and machining optimisation problems. Therefore, Eq. (7), Eq. (10) and Eq. (11) are accordingly useful for the formulation of mathematical models in the subsequent sections. Before formulating the models, the main assumptions are defined as follows:

- The manufacturing parameters are deterministic and static. Breakdowns on machines are not
- considered and parameters are considered as constant and hence do not change over time.
- The depths of cut of part-tool-machine combinations are known and given.
- The compatibilities of part-machine and tool-part combinations in the FMS are given.
- A part type is selected or rejected for processing before the beginning of the production period and remains unchanged during this period.
- The CNC machining centres are identical and can perform different operations on any part type. One machining centre can process one operation at a time.
- Machining, loading and unloading workstations have sufficient input and output buffer spaces.
- Machining centres, raw materials, parts and tools are simultaneously available in the beginning of the production period.
- The material handling system, pallets and fixtures are fully available.
- Setting-up of parts is performed offline and the times required for transferring parts between machining centres are negligible.

METHODOLOGY

A theoretical framework in Section 2 provided a relational base of the FMS variables aimed at formulating throughput and production cost models in attempts to integrate the decisions of part selection, machine loading and machining optimisation problems and to achieve balanced workload in the FMS. The model formulation process adopts a two-stage sequential approach as follows. The first stage maximised the throughput of the FMS and the decisions of part selection, machine loading and machining optimisation are simultaneously made. In the second stage, the production cost is minimised while making the decisions of machine loading and machining optimisation but retaining the maximum throughput obtained in the first stage. The two-stage sequential approach was chosen primarily to avoid the complexity of meeting two different objectives concurrently in one aggregate planning problem.

Computational experiments are designed for sensitivity analysis, which involves varying the values of some control factors and examining how the observed results change with variations in control factors. This would help to test out the range of validity of models and data, and also to get more insights of the results and consequently make easier to evaluate and interpret the results. A two-factor full factorial design was adopted with the available machine time and maximum tooling budget as control factors each assigned to 5 levels. Therefore, a total of $5^2 = 25$ computational experiments were conducted using Extended LINGO 11 software. The two FMS models are formulated in the following subsections:

Maximisation of Throughput

The primary objective in the first stage was to maximise the FMS throughput within the boundaries of the operational requirements. The decisions of the integrated part selection, machine loading (operation allocation, part routing and tool assignment) and machining optimisation (cutting speed and feed rate) problem can be obtained when the model is solved. The model for maximising the FMS throughput was formulated by Mgwatu, et al (2009) as follows:

$$\max \sum_{j=1}^{J} Q_j u_j x_j \tag{12}$$

Subject to

$$\sum_{i=1}^{I} \sum_{k=1}^{K} x_{jotk} = x_j , \ \forall (j, o) \tag{13}$$

$$\sum_{j=1}^{J}\sum_{o=1}^{O_j}\sum_{t=1}^{T} x_{jotk} \geq 1, \ \forall k \tag{14}$$

$$\sum_{j=1}^{J}\sum_{o=1}^{O_j} x_{jotk} \leq y_{tk}, \ \forall (t, k) \tag{15}$$

$$\sum_{j=1}^{J}\sum_{o=1}^{O_j} x_{jotk} \leq y_{tk}, \ \forall (t, k) \tag{16}$$

$$\sum_{t=1}^{T} S_t y_{tk} \leq G_k, \ \forall k \tag{17}$$

$$\sum_{j=1}^{J}\sum_{o=1}^{O_j}\sum_{t=1}^{T}\left(N_{jot} v_{jotk}^{-1} f_{jotk}^{-1} + M_{jot} v_{jotk}^{\alpha_n-1} f_{jotk}^{\beta_n-1} R_t\right) Q_j x_{jotk} \leq A_k, \forall k \tag{18}$$

$$\sum_{j=1}^{J}\sum_{o=1}^{O_j}\sum_{t=1}^{T}\sum_{k=1}^{K} M_{jot} v_{jotk}^{\alpha_n-1} f_{jotk}^{\beta_n-1} C_t Q_j S_t y_{tk} \leq B \tag{19}$$

$$V_{jotk}^{L} \leq v_{jotk} \leq V_{jotk}^{U}, \ \forall (j, o, t, k) \tag{20}$$

$$F_{jotk}^{L} \leq f_{jotk} \leq F_{jotk}^{U}, \ \forall (j, o, t, k) \tag{21}$$

$$x_{jotk} = 0 \ \text{or} \ 1, \ \forall (j, o, t, k) \tag{22}$$

$$x_j = 0 \ \text{or} \ 1, \ \forall j \tag{24}$$

$$y_{tk} = 0 \ \text{or} \ 1, \ \forall (t, k) \tag{25}$$

The objective function (12) maximises the FMS throughput where the level of importance of part types at production quantity Q_j can be in terms of dollar or due-date value coefficients u_j. If part type j is selected, $x_j = 1$, and 0 if otherwise. Constraint (13) states that the total proportion of a part processed at all alternative machines using all feasible tools should be the same for all operations, either $x_{jotk} = 0$ or 1. The variable $x_{jotk} = 1$, if operation o of part j is processed using tool t on machine k, and 0 if otherwise. To ensure that all machines are utilised in the shop floor, constraint (14) binds every machine to perform at least one operation on a part. Constraint (15) prevents recirculation of parts on machines and maintains the flexibility of the system. Constraint (16) ensures that if a part is allocated to a machine, the required tool y_{tk} should be assigned to that machine. If tool type t is assigned to machine k, $y_{tk} = 1$, and

0 if otherwise. Constraint (17) restricts the total number tool types needed on tool slots St not to exceed tool magazine capacity G_k. Constraint (18) forces the total processing time at each machine not to exceed the available machine time A_k on the shop floor. Constraint (19) assures that the total tooling cost is not beyond the available tooling budget B. Constraints (20) and (21) give the lower and upper bounds for cutting speed $(1^{L}{}_{jotk}, 1^{U}{}_{jotk})$ and feed rate $(F^{L}{}_{jotk}, F^{U}{}_{jotk})$ respectively. Constraints (22) through (24) represent binary restrictions on the decision variables.

Minimisation of Production Cost

The model in the second stage was formulated to minimise the production cost. The selected parts already obtained in the first phase were maintained and used among other input data in the second stage model. There was an attempt to re-assign the selected parts and reallocate the tools to machines, and also to re-optimise the cutting speed and feed rate for the purpose of exploring the most possible minimum production cost. Once the second stage model is solved, the decisions of machine loading (operation allocation, part routing, tool assignment), machining optimisation (cutting speed and feed rate), and processing times of parts on machines can be made. The formulation of the production cost minimisation model is presented as:

$$\min \sum_{j \in J_s} \sum_{o=1}^{O_j} \sum_{t=1}^{T} \sum_{k=1}^{K} \left(N_{jot} v_{jotk}^{-1} f_{jotk}^{-1} + M_{jot} v_{jotk}^{\alpha_{jt}-1} f_{jotk}^{\beta_{jt}-1} R_t \right) U_o Q_j x_{jotk}$$

(25)

$$+ \sum_{j \in J_s} \sum_{o=1}^{O_j} \sum_{t=1}^{T} \sum_{k=1}^{K} M_{jot} v_{jotk}^{\alpha_{jt}-1} f_{jotk}^{\beta_{jt}-1} C_t Q_j S_t y_{tk} \quad , \quad J_s = \{ j \mid x_j = 1 \}$$

subject to

$$\sum_{t=1}^{T} \sum_{k=1}^{K} x_{jotk} = 1, j \in J_s, \forall o$$

(26)

$$\sum_{j \in J_s} \sum_{o=1}^{O_j} \sum_{t=1}^{T} x_{jotk} \geq 1, \forall k$$

(27)

$$\sum_{o=1}^{O_j} \sum_{t=1}^{T} x_{jotk} \leq 1, j \in J_s, k$$

(28)

$$\sum_{j \in J_s} \sum_{o=1}^{O_j} x_{jotk} \leq y_{tk}, \forall (t, k)$$

(29)

$$\sum_{j \in J_s} \sum_{o=1}^{O_j} \sum_{t=1}^{T} \left(N_{jot} v_{jotk}^{-1} f_{jotk}^{-1} + M_{jot} v_{jotk}^{\alpha_{jt}-1} f_{jotk}^{\beta_{jt}-1} R_t \right) Q_j x_{jotk} = h, \forall k$$

(30)

$$h \leq A_k, \ \forall k \tag{31}$$

$$\sum_{o=1}^{O_j} \sum_{t=1}^{T} \left(N_{jot} v_{jotk}^{-1} f_{jotk}^{-1} + M_{jot} v_{jotk}^{\alpha_{jt}-1} f_{jotk}^{\beta_{jt}-1} R_t \right) Q_j x_{jotk} = p_{jk}, \ j \in J_s, \ \forall k \tag{32}$$

$$\sum_{j \in J, o=1}^{O_j} \sum_{t=1}^{T} \sum_{k=1}^{K} M_{jot} v_{jotk}^{\alpha_{jt}-1} f_{jotk}^{\beta_{jt}-1} C_t Q_j S_t y_{tk} \leq B \tag{33}$$

$$V_{jotk}^{L} \leq v_{jotk} \leq V_{jotk}^{U}, \ j \in J_s, \ \forall (o, t, k) \tag{34}$$

$$F_{jotk}^{L} \leq f_{jotk} \leq F_{jotk}^{U}, \ j \in J_s, \ \forall (o, t, k) \tag{35}$$

$$x_{jotk} = 0 \text{ or } 1, \ j \in J_s, \ \forall (o, t, k) \tag{36}$$

Constraint(17) $\sum_{t=1}^{T} S_t y_{tk} \leq G_k, \ \forall k,$ and Constraint(24) $y_{tk} = 0$ or $1, \ \forall (t, k)$

The objective function (25) minimises the production cost of the selected parts. Constraints (26)-(29) and Constraints (33)-(36) are equivalent to Constraints (13)-(16) and Constraints (19)-(22), respectively. Constraint (30) assigns equal processing time h at all machines to guarantee balanced workload in the FMS. In Constraint (31), the processing times at machines are bound within the available machine time. Constraint (32) specifies the processing time pjk of each part at different machines.

RESULTS AND DISCUSSIONS

Both the formulated throughput and production cost models are integer nonlinear programming (INLP) problems and were computed using Extended LINGO 11 software. LINGO is a nonlinear programming software package which has the capability to solve nonlinear programming problems with unlimited number of linear and nonlinear constraints as well as unlimited number of integer, nonlinear and global variables (LINDO Systems Inc., 2008). The computations were conducted using the numerical data summarised in Table 1-Table 5. In the data, the tool-operation and tool-machine compatibilities are pre-specified. Tool life constants were taken from Shnumugam, et al (2002) and Wang and Liang (2005) while the tool costs per edge were obtained from McMaster-Carr Supply Company (2008). Limits of cutting speeds and feed rates were found in Chapman (2002).

Table 1: General industrial data

S/N	Description	Data
1	Number of identical CNC machines	4
2	Tool magazine capacity in number of slots	10
3	Number of slots per each tool type	1-4
4	Operating cost per hour	$30
5	Tool replacement time	1 min
6	Available machine time	5 days, 24 hours, 3 shifts
7	Estimated maximum tooling budget for a weekly production plan	$25,000
8	Number of tool types	20
9	Number of part types	10
10	Production quantities for part types 1 through 10 respectively	450, 900, 480, 1300, 2000, 700, 1500, 2500, 1000, 850
11	Value coefficients of part types	1
12	Tool materials	HSS, Carbide
13	Tool availability	100%
14	Part materials	Grey cast iron, carbon steel
15	Raw material availability	100%
16	Operation types	Milling, drilling, reaming, tapping
17	Number of operations on parts	2-3
18	Machining width, length and depth (mm)	Varied
19	Tool sizes (mm) and number of tool teeth	Varied
20	The status of tool life	Computer monitored

Table 2: Tool and empirical data

Part Type	Operation No.	Tool Type	C_t ($)	S_t	α_{jt}	β_{jt}	D_{jot} (mm)	Z_t	γ_{jt}	δ_{jt}	ω_{jt}	λ_{jt}	E_{jt}
	1	1	315	3	3.12	1.09	75	5	0.47	0.62	0.62	0	1.62E+08
		2	325	3	3.12	1.09	90	5	0.47	0.62	0.62	0	1.62E+08
1	2	3	210	1	3.12	1.09	30	4	0.47	0.62	0.62	0	1.62E+08
		4	220	1	3.12	1.09	40	4	0.47	0.62	0.62	0	1.62E+08
	3	5	60	2	2.50	1.25	26	2			1.25		11640
	1	6	40	2	3.03	1.51	25	4	1.51	0.3	1.36	0.3	148880
		7	80	2	3.03	1.51	30	6	1.51	0.3	1.36	0.3	148880
2	2	8	25	4	3.03	1.51	16	4	1.51	0.3	1.36	0.3	148880
		9	35	4	3.03	1.51	20	4	1.51	0.3	1.36	0.3	148880
	3	10	20	1	3.03	1.51	12	2	1.51	0.3	1.36	0.3	148880
	1	1	315	3	3.12	1.09	75	5	0.47	0.62	0.62	0	1.62E+08
		2	325	3	3.12	1.09	90	5	0.47	0.62	0.62	0	1.62E+08
3	2	11	25	1	2.50	1.25	10.2	2			1.25		11640
	3	12	30	1	3.33	1.67	12		0.33		0.67		7774
4	1	6	40	2	3.03	1.51	25	4	1.51	0.3	1.36	0.3	148880
		7	80	2	3.03	1.51	30	6	1.51	0.3	1.36	0.3	148880
	2	13	15	2	3.03	1.51	10	2	1.51	0.3	1.36	0.3	148880
	1	14	90	1	2.50	1.25	38	2			1.25		11640
5	2	15	140	1	3.33	1.67	39		0.33		0.67		7774
6	1	1	315	3	3.12	1.09	75	5	0.47	0.62	0.62	0	1.62E+08
		2	325	3	3.12	1.09	90	5	0.47	0.62	0.62	0	1.62E+08
	2	5	60	2	2.50	1.25	26	2			1.25		11640
	3	16	75	1	3.33	1.67	27		0.33		0.67		7774
	1	8	25	4	3.03	1.51	16	4	1.51	0.3	1.36	0.3	148880
7		9	35	4	3.03	1.51	20	4	1.51	0.3	1.36	0.3	148880
	2	13	15	2	3.03	1.51	10	2	1.51	0.3	1.36	0.3	148880
8	1	17	85	1	2.50	1.25	36	2			1.25		11640
	2	18	160	1	3.33	1.67	39		0.33		0.67		7774
	1	6	40	2	3.03	1.51	25	4	1.51	0.3	1.36	0.3	148880
9		7	80	2	3.03	1.51	30	6	1.51	0.3	1.36	0.3	148880
	2	8	25	4	3.03	1.51	16	4	1.51	0.3	1.36	0.3	148880
		9	35	4	3.03	1.51	20	4	1.51	0.3	1.36	0.3	148880
10	1	19	235	1	3.12	1.09	50	4	0.47	0.62	0.62	0	1.62E+08
		20	245	1	3.12	1.09	60	4	0.47	0.62	0.62	0	1.62E+08
	2	5	60	2	2.50	1.25	26	2			1.25		11640

A set of results for the throughput model including the selected parts, maximum throughput, average workload, number of selected tools for different operations, and the status of the FMS are summarised in Table 6. The results indicate that not all parts and tools were selected for immediate and simultaneous processing on the machines symbolizing that the parts and tools were competing for production resources. The workloads on machines in the FMS were either balanced or unbalanced. For unbalanced workload, some bottleneck machines were identified while others had slack times. The reason for unbalanced workload could be that the parts had different processing requirements including the differences in their machining operations, cutting tools, and machining parameters.

Table 3: Part and machining data

Part	Operation	Tool	W_j (mm)	L_t (mm)	a_{tot}(mm)	N_{tot}	M_{tot}
	1	1	68	540	5	25.45	3.15E-07
		2	68	540	5	30.54	3.37E-07
1	2	3	26	180	8	4.24	6.36E-08
		4	26	180	8	5.65	7.10E-08
	3	5		32		2.61	3.82E-06
	1	6	20	600	10	11.78	1.20E-04
		7	20	600	10	9.42	8.44E-05
2	2	8	12	105	10	1.32	2.11E-05
		9	12	105	10	1.65	1.95E-05
	3	10	12	80	5	1.51	1.02E-05
	1	1	68	630	4	29.69	3.31E-07
		2	68	630	4	35.63	3.54E-07
3	2	11		45		1.44	6.80E-06
	3	12		45	0.9	1.70	3.99E-05
	1	6	20	500	8	9.82	7.12E-05
		7	20	500	8	7.85	5.02E-05
4	2	13	10	210	6	3.30	3.55E-05
	1	14		50		5.97	5.44E-06
5	2	15		50	0.5	6.13	6.77E-05
	1	1	68	280	2.5	13.19	1.18E-07
		2	68	280	2.5	15.83	1.26E-07
6	2	5		80		6.53	9.56E-06
	3	16		80	0.5	6.79	6.94E-04
	1	8	12	400	3	5.03	1.31E-05
7		9	12	400	3	6.28	1.20E-05
	2	13	10	160	5	2.51	2.06E-05
8	1	17		40		4.52	4.41E-06
	2	18		40	0.75	4.90	4.92E-05
	1	6	20	420	4	8.25	2.10E-05
		7	20	420	4	6.60	1.48E-05
9	2	8	12	360	4	4.52	1.81E-05
		9	12	360	4	5.65	1.67E-05
	1	19	42	340	3	13.35	1.24E-07
10		20	42	340	3	16.02	1.33E-07
	2	5		50		4.08	5.98E-06

However, this study has revealed the fact that, when machining parameters are optimised with part selection and machine loading problems for maximum throughput objective, they are likely to adjust themselves within their allowable limits and in some cases they can provide balanced workloads in the FMS. Furthermore, the magazine tool slots in the machines were not fully utilised even when the workloads in the FMS were balanced. Although machine tool manufacturers normally try to equip machining centres with larger tool magazines in order to reduce the impact of the magazine capacity constraint, larger tool magazines result in higher spindle idle times during tool change-over between different operations. However, lower tool magazine usage results in using fewer cutting tools and thus spending lower total tooling cost. In addition, better tool magazine utilisation requires the tool slots to be loaded as densely as possible. Therefore, the tooling cost and the penalty cost due to unused tool slots can be compromised basing on the economical benefits that suit the needs of a particular FMS.

Figure 1: and Fig. 2 give more insights of the results summarised in Table 6 and explore the relationships between tooling budget and throughput, and available machine time and throughput. The observations in these figures show that increasing the tooling budget or available machine time results in increased throughput.

Table 4: Upper and lower limits of cutting speeds

Part	Operation	Tool	I^U_{pot1} (m/min)	I^U_{pot2} (m/min)	I^U_{pot3} (m/min)	I^U_{pot4} (m/min)	I^L_{pot1} (m/min)	I^L_{pot2} (m/min)	I^L_{pot3} (m/min)	I^L_{pot4} (m/min)
1	1	1	152			152	91			91
		2	152			152	91			91
	2	3		152	152			91	91	
		4		152	152			91	91	
	3	5		45	45			12	12	
2	1	6	30			30	9			9
		7	30			30	9			9
	2	8		30	30			9	9	
		9		30	30			9	9	
	3	10		30	30			9	9	
3	1	1	152			152	91			91
		2	152			152	91			91
	2	11		45	45			12	12	
	3	12		19	19			6	6	
4	1	6	30			30	9			9
		7	30			30	9			9
	2	13		30	30			9	9	
5	1	14	45			45	12			12
	2	15	15			15	8			8
	1	1	152			152	91			91
		2	152			152	91			91
6	2	5		45	45			12	12	
	3	16	15			15	8			8
7	1	8		30	30			9	9	
		9		30	30			9	9	
	2	13		30	30			9	9	
8	1	17		45	45			12	12	
	2	18	19			19	6			6
	1	6	30			30	9			9
		7	30			30	9			9
9	2	8		30	30			9	9	
		9		30	30			9	9	
10	1	19	152			152	91			91
		20	152			152	91			91
	2	5		45	45			12	12	

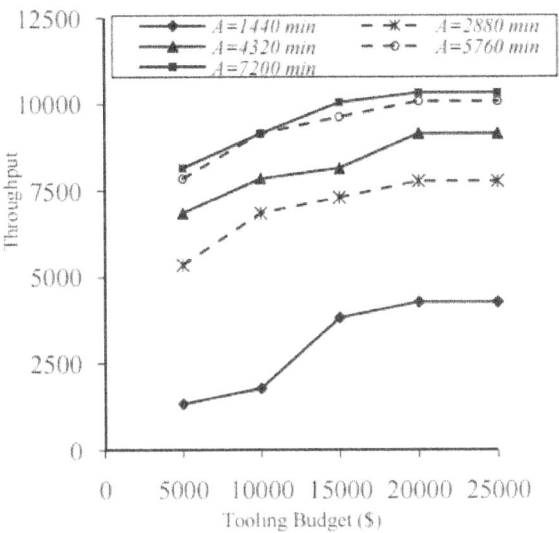

Figure 1: Relationship between tooling budget and throughput.

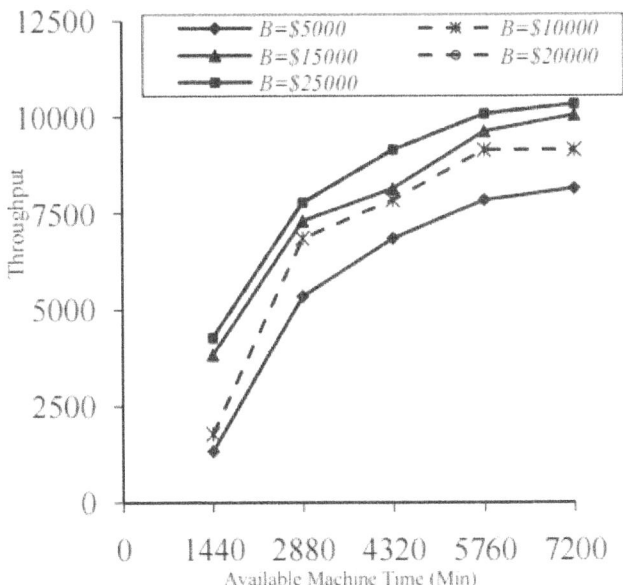

Figure 2: Relationship between available machine time and throughput.

Table 5: Upper and lower limits of feed rates

Part	Operation	Tool	F^U_{iot1}	F^U_{iot2}	F^U_{iot3}	F^U_{iot4}	F^L_{iot1}	F^L_{iot2}	F^L_{iot3}	F^L_{iot4}
	1	1	0.3			0.3	0.075			0.075
		2	0.3			0.3	0.075			0.075
1	2	3		0.3	0.3			0.075	0.075	
		4		0.3	0.3			0.075	0.075	
	3	5		0.5	0.5			0.23	0.23	
	1	6	0.152			0.152	0.102			0.102
		7	0.152			0.152	0.102			0.102
2	2	8		0.127	0.127			0.063	0.063	
		9		0.127	0.127			0.063	0.063	
	3	10		0.089	0.089			0.038	0.038	
	1	1	0.3			0.3	0.075			0.075
		2	0.3			0.3	0.075			0.075
3	2	11		0.3	0.3			0.13	0.13	
	3	12		0.5	0.5			0.15	0.15	
	1	6	0.152			0.152	0.102			0.102
4		7	0.152			0.152	0.102			0.102
	2	13		0.089	0.089			0.038	0.038	
	1	14	0.5			0.5	0.23			0.23
5	2	15	0.5			0.5	0.25			0.25
6	1	1	0.3			0.3	0.075			0.075
		2	0.3			0.3	0.075			0.075
	2	5		0.5	0.5			0.23	0.23	
	3	16	0.5			0.5	0.25			0.25
	1	8		0.127	0.127			0.063	0.063	
7		9		0.127	0.127			0.063	0.063	
	2	13		0.089	0.089			0.038	0.038	
	1	17		0.5	0.5			0.23	0.23	
8	2	18	0.5			0.5	0.15			0.15
	1	6	0.152			0.152	0.102			0.102
9		7	0.152			0.152	0.102			0.102
	2	8		0.127	0.127			0.063	0.063	
		9		0.127	0.127			0.063	0.063	
10	1	19	0.3			0.3	0.075			0.075
		20	0.3			0.3	0.075			0.075
	2	5		0.5	0.5			0.23	0.23	

Feed rates for milling operations in mm/tooth and other operations in mm/rev.

More results of three representative sets of available machine time and tooling budget for the throughput model are presented in Table 7. As can be seen from the table, the decisions of part selection, machine loading (tool assignment, operation allocation and part routes) and machining optimisation (cutting speeds and feed rates) were simultaneously made while the throughput was maximised. The optimum cutting speeds and feed rates were achieved within their recommended limits.

Table 6: General results for maximum throughput model

Available Time (min)	Tooling Budget ($)	Parts Selected	Maximum Throughput	Average Workload (Min)	Total Tools	FMS Status
1440	5000	3,10	1330	1306.75	8	unbalanced
1440	10000	1,3,10	1780	1413.25	14	unbalanced
1440	15000	3,8,10	3830	1440	10	balanced
1440	20000	1,3,8,10	4280	1440	16	balanced
1440	25000	1,3,8,10	4280	1411.8	16	unbalanced
2880	5000	5,7,9,10	5350	2880	17	balanced
2880	10000	5,7,8,10	6850	2878.15	13	unbalanced
2880	15000	1,5,7,8,10	7300	2880	19	balanced
2880	20000	1,3,5,7,8,10	7780	2880	24	balanced
2880	25000	1,3,5,7,8,10	7780	2877.85	24	unbalanced
4320	5000	5,7,8,10	6850	4320	13	balanced
4320	10000	5,7,8,9,10	7850	4317.95	19	unbalanced
4320	15000	4,5,7,8,10	8150	4320	17	balanced
4320	20000	4,5,7,8,9,10	9150	4317.8	23	unbalanced
4320	25000	4,5,7,8,9,10	9150	4320	23	balanced
5760	5000	5,7,8,9,10	7850	5757.675	19	unbalanced
5760	10000	4,5,7,8,9,10	9150	5760	23	balanced
5760	15000	3,4,5,7,8,9,10	9630	5757.675	28	unbalanced
5760	20000	1,3,4,5,7,8,9,10	10080	5758.3	34	unbalanced
5760	25000	1,3,4,5,7,8,9,10	10080	5760	34	balanced
7200	5000	4,5,7,8,10	8150	7200	17	balanced
7200	10000	4,5,7,8,9,10	9150	7200	23	balanced
7200	15000	2,4,5,7,8,9,10	10050	7198.425	30	unbalanced
7200	20000	3,4,5,6,7,8,9,10	10330	7197.675	34	unbalanced
7200	25000	3,4,5,6,7,8,9,10	10330	7200	34	balanced

Table 7: Decisions of part selection, machine loading and machining optimisation for maximum throughput model

Available Time (min)	Tooling Budget ($)	Part	Operation	Tool	Machine	Cutting Speed (m/min)	Feed Rate (mm/tool or mm/rev)	Part routes
		3	1	1	1	96.1	0.103	M1→M3→M2
			2	11	3	28.6	0.131	
1440	5000		3	12	2	6	0.150	
		10	1	19	4	103.7	0.076	M4→M3
			2	5	3	12	0.230	
		4	1	6	1	28.2	0.133	M1→M2
			2	13	2	30	0.089	
		5	1	14	1	41	0.326	M1→M4
			2	15	4	12	0.500	
4320	15000	7	1	8	2	28.6	0.098	M2→M3
			2	13	3	30	0.089	
		8	1	17	3	12.1	0.500	M3→M4
			2	18	4	13.6	0.500	
		10	1	20	4	112.3	0.296	M4→M3
			2	5	3	12.1	0.278	
		3	1	1	1	91	0.300	M1→M2→M3
			2	11	2	41.3	0.300	
			3	12	3	16.9	0.277	
		4	1	7	4	22.5	0.145	M4→M3
			2	13	3	22.7	0.063	
		5	1	14	4	44	0.320	M4→M1
			2	15	1	14.4	0.351	
		6	1	1	4	91	0.192	M4→M3→M1
			2	5	3	21.8	0.365	
7200	25000		3	16	1	8	0.311	
		7	1	8	3	30	0.073	M3→M2
			2	13	2	29	0.069	
		8	1	17	2	31.7	0.366	M2→M4
			2	18	4	9.5	0.486	
		9	1	7	1	29	0.127	M1→M2
			2	8	2	19.7	0.063	
		10	1	19	1	91	0.265	M1→M2
			2	5	2	12.1	0.233	

A set of results for the production cost model are summarised in Table 8 covering the minimum cost of the selected part types, average workload, number of selected tools for different operations, status of the FMS, time savings and tool cost savings. The status of the FMS showed that the workloads were fully balanced for all computational experiments. The values of machining parameters indicate changes where cutting speeds are mostly lower and the feed rates are almost higher for minimum production-cost objective in the second stage than for maximum throughput objective in the first stage. Also, slight tool reassignments and significant part reallocation are observed. It has also been shown that operating machines at higher feed rates within their recommended ranges tend to reduce machining time. It is important to note that the time savings indicated might be caused by reduced machining time, and reassigning of tools and reallocation of parts on machines. These observations designate the fact that by re-allocating parts and re-assigning tools on machines, and re-optimising the machining parameters, there is a great possibility of achieving equal processing times amongst the machines and thus balancing the workload in the FMS. Other observations are that, in some cases, the average workload and tool slot usage decreased with time and tool cost savings as compared to those observed in the first stage.

Table 8: General results for minimum production-cost model

Available Time (min)	Tooling Budget ($)	Minimum Cost ($)	Average Workload (min)	Total Tools	FMS Status	Time Saving (min)	Tool Cost Saving ($)
1440	5000	3329.8	524	8	balanced	916	2717.5
1440	10000	8882.8	830	14	balanced	610	2775.5
1440	15000	15223.1	1440	10	Balanced	–	2656.1
1440	20000	20746.3	1440	16	balanced	–	2132.4
1440	25000	20726.2	1440	16	balanced	–	7152.5
2880	5000	10283.5	476	17	balanced	2404	476.1
2880	10000	12067.8	2880	9	balanced	–	3691.8
2880	15000	15223.3	2880	15	balanced	–	5535.6
2880	20000	22694.2	2880	20	balanced	–	3064.2
2880	25000	22701.8	2880	20	balanced	–	8056.6
4320	5000	10910.8	3493	9	balanced	827	1074.5
4320	10000	13959.9	4320	19	balanced	–	4679.7
4320	15000	19050.0	4320	17	balanced	–	4589.7
4320	20000	24496.1	4320	23	balanced	–	4143.2
4320	25000	24496.1	4320	23	balanced	–	9143.2
5760	5000	14165.2	4664	19	balanced	1096	162.0
5760	10000	19699.8	5760	23	balanced	–	1819.9
5760	15000	22625.9	5760	28	balanced	–	3893.2
5760	20000	28731.0	5760	34	balanced	–	2787.6
5760	25000	28820.1	5760	34	balanced	–	7698.4
7200	5000	17092.8	6047	17	balanced	1153	–
7200	10000	19499.0	6403	23	balanced	797	3306.9
7200	15000	25794.7	7200	30	balanced	–	3604.9
7200	20000	30285.2	7200	34	balanced	–	4113.7
7200	25000	29761.5	7200	34	balanced	–	9637.4

The results for the production cost model also uncover the relationships between tooling budget and production cost, and available machine time and production cost which are well depicted in Figure 3 and Figure 4. These figures illustrate the fact that increasing the tooling budget or available machine time brings about an increased production cost. The results of similar representative sets of available machine time and tooling budget for the production cost model are given in Table 9 and Table 10. As can be noted from the two tables, the decisions of machine loading (including tool assignment, operation allocation and part routes), machining optimisation (cutting speeds and feed rates) and part processing times were concurrently made while minimising the production cost. The optimum cutting speeds and feed rates conformed to their recommended lower and upper limits. It is clearly shown that nearly all cutting speeds and feed rates in the second stage are different from those in the first stage. This is again an indication that the cutting speeds and feed rates were adjusted and significantly contributed to the balanced workload in the FMS. The trend of part routes tends to be different comparing to routes obtained in the first stage because of part reallocations in the second stage.

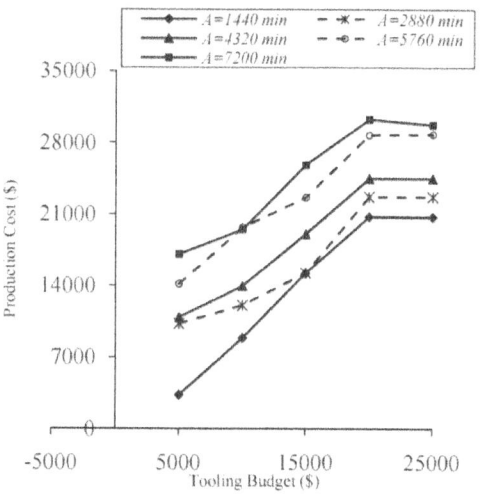

Figure 3: Relationship between Tooling Budget and Production Cost.

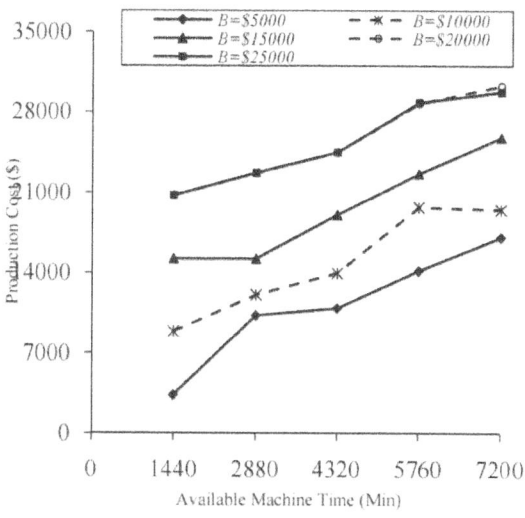

Figure 4: Relationship between available machine time and production cost.

Table 9: Decisions of machine loading and machining optimisation for minimum production-cost model

Available Time (min)	Tooling Budget ($)	Part	Operation	Tool	Machine	Cutting Speed (m/min)	Feed Rate (mm/tool or mm/rev)
		3	1	1	4	91	0.300
			2	11	3	24.4	0.300
1440	5000		3	12	2	6	0.260
		10	1	19	1	91	0.239
			2	5	3	16.2	0.500
		4	1	7	1	20	0.152
			2	13	2	26.5	0.089
		5	1	14	1	45	0.500
			2	15	4	11.5	0.500
		7	1	8	2	23.9	0.127
4320	15000		2	13	3	16.9	0.089
		8	1	17	3	17.6	0.500
			2	18	4	11.4	0.500
		10	1	19	1	91	0.300
			2	5	3	13	0.500
		3	1	1	4	91	0.300
			2	11	2	22.5	0.300
			3	12	3	6	0.500
		4	1	7	1	20	0.152
			2	13	2	17.2	0.089
		5	1	14	4	45	0.500
			2	15	1	8	0.500
		6	1	1	1	91	0.300
			2	5	3	14.2	0.500
			3	16	4	8	0.400
7200	25000	7	1	8	2	15.5	0.127
			2	13	3	18.1	0.089
		8	1	17	3	19.1	0.500
			2	18	4	7.9	0.500
		9	1	7	4	29	0.152
			2	8	3	12.9	0.127
		10	1	19	1	91	0.300
			2	5	2	14.9	0.500

Table 10: Decisions of part processing time and part routes for minimum production-cost model

Available Time (min)	Tooling Budget ($)	Part	Operation	Tool	Machine	Part Processing Time (min)	Part routes
		3	1	1	4	524	M4→M3→M2
			2	11	3	95	
			3	12	2	524	
1440	5000	10	1	19	1	524	M1→M3
			2	5	3	429	
		4	1	7	1	3370	M1→M2
			2	13	2	1829	
		5	1	14	1	533	M1→M4
			2	15	4	2152	
		7	1	8	2	2491	M2→M3
4320	15000		2	13	3	2505	
		8	1	17	3	1283	M3→M4
			2	18	4	2169	
		10	1	19	1	417	M1→M3
			2	5	3	532	
		3	1	1	4	524	M4→M2→M3
			2	11	2	102	
			3	12	3	273	
		4	1	7	1	3368	M1→M2
			2	13	2	2800	
		5	1	14	4	533	M4→M1
			2	15	1	3076	
		6	1	1	1	339	M1→M3→M4
			2	5	3	645	
7200	25000		3	16	4	1520	
		7	1	8	2	3832	M2→M3
			2	13	3	2342	
		8	1	17	3	1182	M3→M4
			2	18	4	3118	
		9	1	7	4	1505	M4→M3
			2	8	3	2758	
		10	1	19	1	417	M1→M2
			2	5	2	466	

CONCLUSIONS

In previous studies, the isolation of machining optimisation from the planning of FMSs resulted in unbalanced workload in the FMSs. This study has made an attempt to integrate the decisions of part selection, machine loading and machining optimisation problems for a more balanced workload and therefore effective FMS. Two integrated models have been formulated and solved based on designed computational experiments. The findings illustrate the following two important implications: (i) the maximum throughput objective in the first stage could not guarantee balanced workload in the FMS for the combined part selection, machine loading and machining optimisation problems. Only some cases would demonstrate balanced workloads in this stage; and (ii) the minimum production-cost objective in the second stage guarantees balanced workload in the FMS without affecting the maximum throughput objective in the first stage. With minimised production cost, the bottleneck machines are eliminated and in some cases, the average workloads are decreased. This is a result of re-adjusting the machining parameters such as cutting speeds and feed rates, and also reallocating parts and reassigning tools on machines. The findings also support the applicability of the adopted approach over a wide range of planning period and tooling budget.

ACKNOWLEDGEMENTS

This research was supported by Sida-SAREC funding at the University of Dar es Salaam in Tanzania and Fulbright Scholarship for research visit at Lehigh University in United States of America.

REFERENCES

1.   Agapiou, J.S. (1992). Optimisation of multistage machining system, part 1: mathematical solution, Journal of Engineering for Industry, 114(4), 524-531.

2.   Chapman, W. (Ed.) (2002). Modern Machine Shop's Handbook for the Metalworking Industries, Hanser Gardner Publications, 1st Edition, Cincinnati, Ohio, USA.

3.   Choudhary, A. K., Tiwari, M.K., & Harding, J.A. (2006). Part selection and operation-machine assignment in a flexible manufacturing system environment: A genetic algorithm with chromosome differentiation-based methodology. Proceedings of the Institution of Mechanical Engineers, Part B: Journal of Engineering Manufacture, 220(5), 677-694.

4.   Ermer, D.S., & Kromodihardjo, S. (1981). Optimisation of multi-pass turning with constraints. Journal of Engineering Industry, 103(3), 462-

468.

5.   Hwang, S. (1986). A constraint-directed method to solve the part selection problem in flexible manufacturing systems planning stage, Proceedings of the Second ORSA/TIMS Conference on Flexible Manufacturing Systems, Ann Arbor, Michigan, USA, 297-309.

6.   Lambert, B.K. & Walvekar, A.G. (1978). Optimisation of Multi-Pass Machining Operations. International Journal of Production Research, 16(4), 259-265.

7.   Liang, M. (1994). Integrating Machining Speed, Part Selection and Machine Loading Decisions in Flexible Manufacturing Systems. Computers in Industrial Engineering, 26(3), 599-608.

8.   LINDO Systems Inc. (2008). LINDO User's Guide, Chicago, Illinois, USA.

9.   Manna, A., & Salodkar, S. (2008). Optimisation of Machining Conditions for Effective Turning of E0300 Alloy Steel. Journal of Materials Processing Technology, 203(1-3), 147-153.

10.   McMaster-Carr Supply Company (2008). Cutting Tool E-Catalog, http://www.mcmaster.com, 1-3675, retrieved on Tuesday, 30th September 2008.

11.   Mgwatu, M.I, Opiyo, E.Z. & Victor, M.A.M. (2009). Integrated Decision Model for Interrelated Sub-Problems of Part Design or Selection, Machine Loading and Machining Optimization, In Proceedings of the American Society of Mechanical Engineers (ASME) International Design Engineering Technical Conferences and Computers and Information in Engineering Conference, San Diego, California, USA, 3-12.

12.   Narang, R.V. & Fischer, G.W. (1993). Development of a Framework to Automate Process Planning Functions and to Determine Machining Parameters. International Journal of Production Research, 31(8), 1921-1942.

13.   Shnumugam, M.S., Reddy, S.V.B., & Narendran, T.T. (2002). Selection of Optimal Conditions in Multi-Pass Face-Milling Using a Genetic Algorithm, International Journal of Machine Tools and Manufacture, 40, 401-414.

14.   Stecke, K.E. (1985). Design, Planning, Scheduling, and Control Problems of Flexible Manufacturing Systems. Annals of Operations research, 3, 3-12.

15.   Wang, P. and Liang, M. (2005). An Integrated Approach to Tolerance Synthesis, Process Selection and Machining Parameter Optimization

Problems. International Journal of Production Research, 43(11), 2237-2262.

16. Yang, H., & Wu, Z. (2002). GA-Based Integrated Approach to FMS Part Type Selection and Machine-Loading Problem. International Journal of Production Research, 40(16), 4093-4110.

Chapter 9

SUFFICIENT CONDITIONS FOR A FLEXIBLE MANUFACTURING SYSTEM TO BE DEADLOCKED

Paul E. Deering

Department Engineering Technology and Management, Ohio University, Athens, OH, USA.

ABSTRACT

In recent years, researchers have been interested in scheduling algorithms to avoid deadlock in Flexible Manufacturing Systems (FMS). FMS are discrete event systems characterized by the availability of resources to produce a set of products. Raw parts, which belong to various product types, enter the system at discrete times and are processed concurrently while sharing a limited number of resources. In such systems, a situation may occur in which parts become permanently block. This is called deadlock. This paper presents the sufficient conditions for deadlock to exist in a FMS; it models a FMS using digraphs to calculate slack, knot, order and space; it identifies three types of circuits that are fundamental in determining if a FMS is in deadlock.

INTRODUCTION

Moving the wrong part in a manufacturing system could place the live (deadlock-free) system into a deadlocked state or dead state. The only recourse would be to manually resolve the deadlock and reset the FMS to a live state. Clearly, avoiding deadlock altogether would lead to increased production and decreased labor costs. To prevent manual deadlock resolution a Deadlock Avoidance Algorithm (DAA) was developed in Deering (2008). The DAA did not allow the system to enter any dead states and proved sufficient conditions for the system to be live. The DAA introduced the idea of space. If space > 0 of all closed paths in the manufacturing then deadlock would be avoided. The only problem was that some live states were detected dead states. See Fig. 1. The DAA in Deering (2008) only proved sufficient conditions for a system to be live.

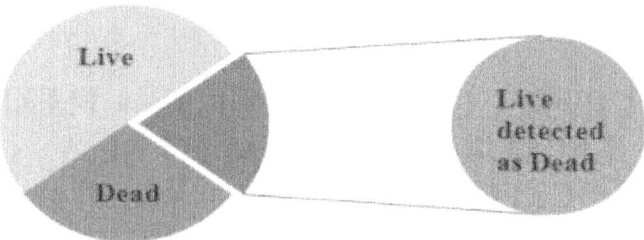

Figure. 1: Live states detected as dead

This paper will prove sufficient condition for the manufacturing system to be dead and is the partial results of Deering (2000). This paper is organized as follows: the first section discusses previous research on deadlock in a FMS; the next section defines a mathematical model of a manufacturing systems; circuit parameter slack, knot order and space is then defined; the next section introduces three types of circuit uses to proves sufficient conditions for a manufacturing systems to be dead.

RELATED RESEARCH

Many researchers use Petri nets Banaszak and Krogh (1990), Barkaoui and Abdallah (1995), Hsieh and Chang (1994), Viswanadham et al. (1990), Zhou and DiCesare (1992), Zhou (1996) and Ezpeleta et al. (1995) as a formalism to describe deadlock in a manufacturing system. Banaszak and Krogh (1990) proposed a deadlock avoidance algorithm (DAA), which developed a restriction policy based on production route information to guarantee that no circular wait situations would occur. Their DAA is sufficient for avoiding deadlocks but is not an optimal solution. Viswanadham et al. (1990) developed a deadlock avoidance algorithm that employed a look-ahead policy. This algorithm did not detect all deadlocked states, and the authors suggested using a recovery mechanism in case of system deadlock. Zhou and DiCesare (1992) and Zhou (1996) generalized the sequential mutual exclusions (SME) and parallel mutual exclusions (PME) concepts and derived the sufficient conditions for a Petri net (PN) containing such structures to be bounded, live, and reversible. In general, PN solutions are suitable for manufacturing systems that contain few resources but become very complicated for larger systems. Another formalism to describe the manufacturing system is to use graphs Cho et al. (1995), Fanti et al. (1996), Judd and Faiz (1995), Judd et al. (1997), Lipset et al. (1997), Zhou (1996), Ezpeleta et al. (1995), Wenle et al. (2003), Deering (2000), Fanti et al. (1995), Wenle et al. (2004), Wenle et al. (2007). In this approach, the vertices represent resources and the edges represent part flows between resources.

Wysk et al. (1991) were the first to develop a specialized directed graphical structure called a wait relation graph (WRG) to model a manufacturing system. They developed a string manipulation procedure that yields a set of control actions to detect and recover from primary deadlock. Cho et al. (1995) used system status graphs to develop the concept of simple and nonsimple bounded circuits with empty and non-empty shared resources to detect part flow deadlock and impending part flow deadlock. This method introduced the concept of a bounded circuit to detect deadlock. The method detected deadlock based on characteristics of this bounded circuit. The methods in references Wysk et al. (1991) and Cho et al. (1995) could only handle single capacity resources. Fanti et al. (1996) used a graph called working procedure digraph and developed a simple graph-theoretic method for deadlock detection and recovery in systems with multiple capacity resources. This algorithm did not prevent deadlock from occurring either, but it suggested a suitable recovery strategy. Judd and Faiz (1995) expanded on the original formulation proposed by reference Wysk et al. (1991) and the first to define slack, order, and space to avoid deadlock. This method provided sufficient conditions for deadlock by satisfying a set of linear inequalities. Lipset et al. (1997) extended Judd et al. (1995) to precisely quantify necessary and sufficient conditions for deadlock to exist. In this research, they redefined the order of a knot, defined a special state called an evaluation state, and defined the concept of order reduction. The approach was to put the system into an evaluation state and then compute the order. Deering (2000) and Deering (2008) improved Lipset et al. (1997) by further refining the order of a knot and evaluation state, as well as eliminating the need for order reduction. Wenle et al. (2003) developed a deadlock avoidance algorithm (DAA) based on Lipset et al. (1997) and Deering (2000), which avoided deadlock and was executed in polynomial time. Wenle et al. (2004) expanded upon Wenle et al. (2003) and Deering (2000) to quantify the sufficient conditions for a system state to be live and derived the liveness necessary and sufficient conditions for an evaluation state. Wenle and Judd (2007) extended Wenle et al. (2004) to allow choice in process flow or flexible part routing.

MODELING A MANUFACTURING SYSTEM

An FMS consists of a set, R, of finite resources, such as robots, buffers, and machines, which produce a finite set, P, of products. Each resource $r \in R$ has a capacity of cap(r) units that can perform the required operations. The capacity function can be extended to a set of resources, that is:

$$\text{cap}(R_1) = \sum_{\forall r \in R_1} \text{cap}(r), \quad \text{for any } R_1 \subseteq R.$$

(1)

For each product $p \in P$, the process plan $\mathrm{plan}(p) = r_1 r_2 \cdots r_m$ defines the sequence of resources that are required to produce p. Resource r_m is the terminal resource for product p. It is assumed that all process plans are fixed, finite, and sequential. A part is an instance of a product that flows through the system. At any given time, a manufacturing system is working on a set Q_o f parts. The function class(q) returns the product p to which part q belongs.

A manufacturing system can be represented by a WRG, $G = (V, A)$. Each vertex represents a resource; that is, V=R. A directed arc is drawn from vertex r_1 to vertex r_2, if r_2 immediately follows r_1 in at least one process plan. Each arc will be labeled with the part(s) that will flow through it. A subgraph $G_1 = (R_1, A_1) \subset G$ of an WRG consists of a subset of the resources and arcs of G, so that all the arcs in A_1 connect resources in R_1. The union (intersection), denoted by $G_1 \cup G_2 (G_1 \cap G_2)$, of two subgraphs is the union (intersection) of the component resource and arc sets. A path $P = (R_p, A_p)$ is a subgraph whose resources and arcs can be ordered in the list $r_1 a_1 r_2 a_2 \cdots a_{n-1} r_n$ where each arc in the list connects the resources on either side. When specifying a path, writing the arcs is redundant. Therefore, only the resources will be enumerated when a path is defined. A simple path is a path with no repeated elements in the ordered list. A closed path is a path with the same first and last element. A simple circuit is a closed path with no repeated elements in the ordered list, except the first and last elements.

The function n(q) returns a positive integer that represents the position in plan[class(q)] of the operation that is currently processing q. When a new part q is added to the system, then $n(q) = 1$. As the part is moved from resource to resource according to its plan, n(q) is incremented until it reaches the end of its plan and exits the system. The state n of a manufacturing system is a vector containing the current $n(q)$ for all $q \in Q$. A state n of a manufacturing system is live if a sequence of part movements exist that will empty the system. A state n of a manufacturing system is dead, or deadlocked, if it is not live. Given a manufacturing system $G = (R, A)$, let $a \in A$ and $r \in R$. Then, the function tail(a) returns the resource at the tail of the given arc; the function head(a) returns the resource at the head of the arc. A unit of the resource $r = \mathrm{tail}(a)$ is said to be committed to arc a if it is processing a part q whose next resource in its process plan is head(a). It is important to note that the number of resource units committed to the outgoing arcs of r can be less than the number of busy units. This happens when some of the busy units are being used for terminal operations. A resource unit is free if it is not committed to an arc; by this definition, a busy unit that is not committed is still termed free. A resource is free if any of its units are free. A resource is empty if it contains no parts. The commitment function com(a, n) returns the number of resource units that are committed to arc a when the system is in state n. The commitment function is extended to a set of arcs as follows:

$$\text{com}(A_1, n) = \sum_{\forall a \in A_1} \text{com}(a, n), \quad \text{for any } A_1 \subseteq A$$

$$(2)$$

A part is enabled if either the next resource in its process plan contains at least one resource unit that is not busy, or the part is in the last step of its process plan. Suppose that the system is in state n_o there exists an arc a such that resource $r_2 = \text{head}(a)$ is free and the part in the resource $r_1 = \text{tail}(a)$ is committed to a. Then, when r_1 finishes its operation, this part can be moved to resource r_2 . This process is called propagation. The symbol n_k is used to denote the state of the system after the kth propagation. A part q in WRG G can be shifted to resource r if it can be propagated to r without propagating any other part in G. A part q in WRG G is said to have a free exit if it can shift its terminal resource r_m in G

SLACK, KNOT, ORDER, AND SPACE

This section will summarize the major concepts and results from Judd and Faiz (1995), Judd et al. (1997), Lipset et al. (1997), and Deering (2000). This section defines the concept of slack, knot, order, and space. The slack is the number of free resource units available for parts to flow on a subgraph.

Definition 1: The slack of any subgraph $G_1 = (R_1, A_1) \subseteq G$ is given by

$$\text{slack}(G_1, n) = \text{cap}(R_1) - \text{com}(A_1, n)$$

$$(3)$$

A closed path c in a WRG G is in primary deadlock in state n if slack(c, n) = 0 .

Definition 2: Let c_1 and c_2 be any two closed paths in a WRG of a manufacturing system. If $c_1 \cap c_2$ consists of exactly one resource with a capacity of one, then this resource is called a knot with respect to $c_1 \cap c_2$

Definition 3: Let c_1 and c_2 be two closed paths in a WRG G. Path c_1 is connected to c_2 if $c_1 \cap c_2 \neq 0$ and a part currently exists in the system that must propagate from c_1 to c_2 without leaving $c_1 \cup c_2$

Definition 4: Given two closed paths c_1 and c_2, then c_1 and c_2 are cross-connected if c_1 is connected to c_2 and c_2 is connected to c_1

Definition 5: Let the closed path c in state n consist of two closed paths, c_1 and c_2 such that $c = c_1 \cup c_2$ and $c_1 \cap c_2 = k$, where k is a knot. The order of knot k with respect to the closed path c in state n is defined as:

$$\text{order}(k,c,n) = \begin{cases} 1, \text{if } c_1 \text{ and } c_2 \text{ are cross connected.} \\ 0, \text{otherwise.} \end{cases}$$

(4)

The order of any simple circuit is zero.

Definition 6: Let c be a closed path in a WRG G in state n that contains m knots. Then, the order of c is given by:

$$\text{order}(c,n) = \sum_{i=1}^{m} \text{order}(k_i,c,n)$$

(5)

Definition 7: Let c be a closed path in a WRG G of a manufacturing system in state n. The free space on a closed path c is the difference between the slack and the order:

$$\text{space}(c,n) = \text{slack}(c,n) - \text{order}(c,n) \quad \forall c \in C_G$$

(6)

where C_G is the set of all closed paths in G. The following theorem proves that if all closed paths of a WRG G have space greater than zero, G is live.

Theorem 1: Let C_G be the set of all closed paths in a non-empty WRG G in state n. If,

$$\text{space}(c,n) > 0 \quad \forall c \in C_G$$

(7)

then G is live. Proof: See Deering (2000).

SUFFICIENT CONDITIONS FOR A SYSTEM TO BE DEAD

The previous section proved sufficient conditions for a manufacturing system to be live; that is, if the space of all closed paths in a manufacturing system is greater than zero, then the system is live. This section will prove sufficient conditions for a manufacturing system to be dead. Unfortunately, this cannot be proven in the general case, since there is insufficient information in the WRG to determine these conditions. However, when the system is in a special system state called an evaluation state, it can be shown that a manufacturing system is dead if one of the closed paths equals zero. The following example will demonstrate this more clearly.

Example 1: Let the WRG G in Figure 2 be in state n. Suppose that the process plans for parts a, b and c appear as presented in Table 1. Assume that the state of the system is $n = [n(a), n(b), n(b)] = [1,1,1]$. Table 2 Table 2 depicts the order and space computations for this system.

Table 1: Process plans for example 1

Part	Process Plan
a	$r_2 r_3 r_4$
b	$r_4 r_3 r_1$
c	$r_1 r_2 r_3 r_1$

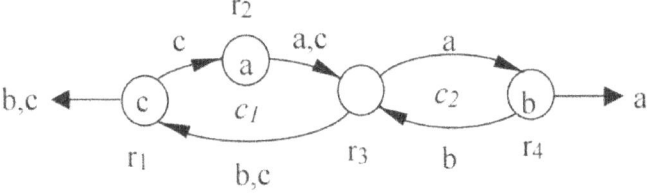

Figure. 2: Manufacturing system for example 1

Table 2: Order and space computations for example 1

Circuit	Order	Space
c_1	0	1
c_2	0	1
$c_1 \cup c_2$	1	0

Since the space of the union between c_1 c and c_2 c is zero, the method previously presented in Deering (2008) cannot conclude whether the system is live or dead. This revised method will show that the order in which parts flow through order-one knots is required to describe sufficient conditions for a dead system. For example, knowing that parts a and b must pass through r_3 before any other parts can leave c_1 and c_2 and that the space of the union between c_1 and c_2 is zero, will allow researchers to know that the system in Fig. 1 to be dead. This section will contain three parts: the first section shows necessary and sufficient conditions which render basic closed path as dead; the next two sections show sufficient conditions for deadlock of chained and complex closed paths.

Basic Closed Paths

Definition 8: A basic closed path c is a closed path in a WRG G in state n such that order(c, n) = 0

Theorem   1:   Given   a   basic   closed   path $c = (R, A)$ in state n. If space$(c, n) = 0$ then c is dead.

Proof: See Deering (2000).

Theorem 1 allows us to conclude that space greater than zero of a basic closed path is necessary and sufficient for the system to be live. The next section addresses a particular closed path that contains order-one knots.

Chained Closed Paths

This section defines a chained closed path and introduces a special state called an evaluation state. A series of definitions, some lemmas and a theorem will prove that if a chained closed path is in an evaluation state and its space is equal to zero, then the chained closed path is dead.

Definition 9: A chained closed path c is a closed path containing one or more order-one knots with respect to c, such that c can be decomposed into a set of basic closed paths which intersect at only the order-one knots.

The following is a simple example of a chained closed path:

Example 2: Consider the manufacturing system in Figure 3. Assume that all a part types flow to the right from c_1 to c_3, and that all b part types flow to the left from c_3 to c_1. In this state, resources r_2 to r_3 are order-one knots. The manufacturing system can be decomposed into three simple circuits, c_1, c_2 and c_3. Let $c = c_1 \cup c_2 \cup c_3$ In this example, c is a chained closed path, since c can be decomposed into basic closed paths so that each circuit intersects each other at only the order-one knots (i.e. $c_1 \cap c_2 = r_2$ and $c_2 \cap c_3 = r_3$).

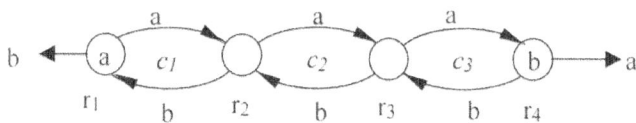

Figure. 3: A chained closed path

The following example will help the reader conceptualize the need for an evaluation state:

Example 3. Suppose that a part exists in all the resources shown in Figure 4 except r_4 Each part is committed to the outgoing arc of its resource. Assume that all part a types must flow to circuit c_2 before completion and parts d_1 and d_2 must flow to circuit c_1 before completion. Call this state n. This state may, or may not, be dead, depending on the ultimate destination of part b in the resource r_7

Case 1. Suppose part b must move to resource r_4 and then to r_5 and exit the system. Clearly, in this case, state n is a live state.

Case 2. Suppose part b must flow to r_4 and commit to circuit c_1 Then state n is a dead state.

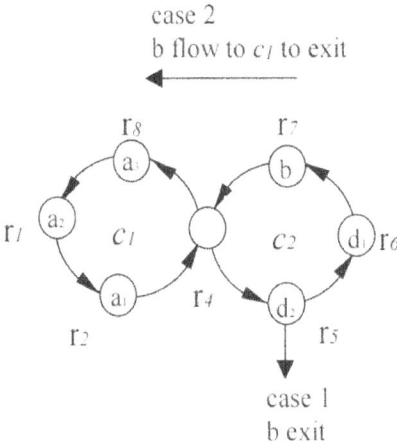

Figure. 4: Manufacturing system for example 3

To distinguish and to evaluate these two cases, the dynamics of the part crossing through the knot should be analyzed more closely. Notice that in both cases, all part a's must cross knot r_4 before any other part on c_1 can leave c_1 But the part crossing dynamics are different on circuit C, in the two cases. Notice that in case 1, part b can leave circuit c_2 before part d_1 must cross knot k. In other words, a resource may become free on c_2 before part d_1 must cross the knot. In this state, we conclude that state n is not in an evaluation state. The method in Deering (2008) cannot determine if deadlock exists by computing the space in state n. Notice that in Case 2, part b must cross knot r_4 before any other part can leave c_2. In this situation, no part can escape c_2 before the crossing must occur. The state of the system in case 2 is considered to be an evaluation state. These ideas motivated the following definitions.

Definition 10: Let C_1 and C_2 be two closed paths in a WRG G such that $c_1 \cap c_2 = k$ where k is an order-one knot. If a part q on c_1 propagates to k and commits to an arc on c_2, then q is said to cross knot k.

Definition 11: A basic closed path in a WRG G is always in an evaluation state.

Definition 12: An empty chained closed path in state n is in an evaluation state.

Definition 13: Let a non-empty chained closed path c be in state n. A chained closed path c can be divided into two closed paths, c_1 and c_2, at any order-one knot k such that $c_1 \cap c_2 = k$. Then chained closed path c is in an evaluation state if

- all order-one knots are empty, and
- for each order-one knot k, two parts, q_1 and q_2 exist, such that

a. part q_1 must cross from c_1 to c_2 before any other part can leave c_1, and c_2 is in an evaluation state after the move; and

b. part q_2 must cross from c_2 to c_1 before any other part can leave c_2, and c_1 is in an evaluation state after the move.

The system in Example 2 is in an evaluation state. Resources r_2 and r_3 are order-one knots. For order-one knot r_2, part a must cross from c_1 to $c_2 \cup c_3$ before any other part can leave C_1, and part b must cross $c_2 \cup c_3$ to c_1 before any other part can leave $c_2 \cup c_3$. For order-one knot r_3, part a must cross from $c_1 \cup c_2$ to c_3 before any other part can leave $c_1 \cup c_2$ and part b must cross c_3 to $c_1 \cup c_2$ before any other part can leave c_3. After moving either part a or part b, both $c_2 \cup c_3$ and $c_1 \cup c_2$ are in evaluation states.

The next lemma will show how the parts are committed when a chained closed path is in an evaluation state.

Lemma 1: Given a chained closed path c = (R, A) in a WRG G that is in an evaluation state n, space(c,n) = 0 if, and only if, all order-one knots are empty in c and all other resources in c are filled and committed to resources on c.

Proof: See Deering (2000).

The next two lemmas are preliminary results that are required to prove the final theorem of this section.

Lemma 2: Given a chained closed path c that is in an evaluation state n, if space(c, n) = 0 then a part q exists such that when it is moved, it will fill an order-one knot and commit an outgoing arc of that knot on c.

Proof: See Deering (2000)

Lemma 3: Given a non-empty chained closed path c that is in an evaluation state n_0, if space$(c, n_0) = 0$ then propagating any part will create a chained closed path c_2 such that $c_2 \subset c$ and space$(c_2, n_1) = 0$.

Proof: See Deering (2000)

Definition 14: If any subgraph in a WRG G is dead, then G is dead.

Theorem 2: Given a non-empty chained closed path c that is in an evaluation state n_o, if space $(c, n_n) = 0$ then c is dead.

Proof: See Deering (2000)

Chained Closed Paths

Closed paths that are not basic closed paths or chained closed paths are classified as complex closed paths. This section will introduce complex closed paths. It will also be shown, if a complex closed is in an evaluation state and it contains a path with space equal to zero, then this is sufficient for determining if the system is dead. We will first define a complex closed path and its various components, then follow these definitions with an example.

Definition 15: A complex closed path is a closed path that contains one or more order one knots that is not a chained closed path.

Definition 16: A complex path can be decomposed into two paths, one being a chained closed path and the other is called the auxiliary closed path. The intersection of the auxiliary closed path intersects and the chained closed path must contain one or more order one knots of the chained path.

Definition 17: A bypass path is the portion of the auxiliary path that does not intersect the chained closed path.

Definition 18: The first arc on the bypass path is a bypass arc.

Consider the following example. Example 4: Suppose that the system in Figure 5 has the following parts and process plans as depicted in Table 3.

Table 3: Process plans for example 4

Part	Process plan
a	$r_1 r_2 r_3 r_4$
b	$r_4 r_2 r_6 r_1$
d	$r_3 r_5 r_6 r_1$

Assume that the system is in state $n = [n(a_1), n(a_2), n(b), n(d_1), n(d_2)] = [3,1,1,3,2]$.

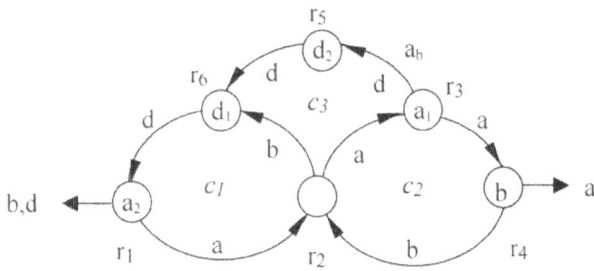

Figure. 5: Complex closed path for example 4

The system consists of three simple closed paths: $c_1 = r_1 r_2 r_6 r_1$, $c_2 = r_2 r_3 r_4 r_2$, and $c_3 = r_2 r_3 r_5 r_6 r_1 r_2$. The order $(r_2, c_1 \cup c_2, n) = 1$ Clearly, the manufacturing system in Figure 5 is not a basic closed path. The system cannot be a chained closed path either since $(c_1 \cup c_2) \cap c_3$ is not a knot. According to Definition 15, the system in Figure 5 is a complex closed path. The complex closed path can be decomposed into a chained closed path, (i.e., $c_1 \cup c_2$, and an auxiliary closed (i.e., c_3). Closed path c_2 is an auxiliary closed path since the intersection of c_3 and the chained closed path $c_1 \cup c_2$ contain the order-one knot r_1. The simple path $r_3 r_5 r_6$ is a bypass path that joins c_1 and c_2 together. Arc a_b on resource r_3 is a bypass arc since it leaves r_3 along the auxiliary closed path c_3. We next define the evaluation state for a complex closed path.

Definition 19: A complex closed path in a WRG G is in an evaluation state if its bypass arcs are not committed.

Definition 20: An empty subgraph that is a complex closed path in a WRG G in state n is in an evaluation state.

Definition 21: A WRG G is in an evaluation state if all closed paths in G are in an evaluation state. The system in Example 4 is in an evaluation state. This is because part a_1 in resource r_3 is not committed to the bypass arc a_b. The chained closed path $c_1 \cup c_2$ is in as evaluation state per Definition 13. The space of the chained closed path $c_1 \cup c_2$ is zero. Clearly, the system is dead. The next theorem proves this concept in general.

Theorem 3: Given a complex closed path c_p that is in an evaluation state n_o, if any chained closed

path $c^* \subset c_p$ has $space(c^*, n_0) = 0$, then c_p is *dead*.

Proof: See Deering (2000).

CONCLUSION

Three types of closed paths were identified to prove sufficient conditions for a manufacturing system to be dead. A special state called an evaluation state was introduced. It was showed that if a basic, chained or complex closed path that is in an evaluation state with space=0 then the system is dead. Unfortunately, determining if a closed path is in an evaluation state is a problem. The problem is there is insufficient information in the WRG to determine if a system is in an evaluation state. This is a topic of future research.

REFERENCES

1. Banaszak, Z., & Krogh, B. (1990). Deadlock avoidance in flexible manufacturing systems with concurrently competing process flows. IEEE Trans. on Robotics and Auto., 6(6), 724–733.

2. Barkaoui, K. and I.B. Abdallah. (1995). A deadlock method for a class of FMS, Proceedings of the 1995 IEEE Int. Conf. On Systems, Man and Cybernetics, 4119–4124.

3. Cho, H., Kumaran, T. K., & Wysk, R. (1995). Graph-theoretic deadlock detection and resolution for flexiblemanufacturing systems. IEEE Trans. on Robotics and Auto., 11(3) 550–527.

4. Deering, E. P. (2000). Necessary and sufficient conditions for deadlock in manufacturing systems. PhD Dissertation, Ohio University.

5. Deering, E. P. (2008). A simple deadlock avoidance algorithm in flexible manufacturing systems. International Journal of Modern Engineering, 9(1) 19-26.

6. Ezpeleta, J., Colom, J., & Martinez, J. (1995). A petri net based deadlock prevention policy for flexible manufacturing systems. IEEE Trans. on Robotics and Automation, 11(2), 173–184.

7. Fanti, M.P., Maione, B., Mascolo, S., &Turchiano, B. (1995). control polices conciliating deadlock voidance and flexibility in FMS resource allocation. IEEE Symposium on Emerging Technologies and Factory Automation, 1, 343–351.

8. Fanti, M., Maione, G., & Turchiano, B. (1996). Deadlock detection and recovery in flexible productionsystems with multiple capacity resources. Industrial Applications in Power Systems Computer Science and Telecommunications Proceedings of the Mediterranean Electrotechnical Conference, 1, 237–241.

9. Hsieh, F., & Chang, S. (1994). Dispatching-driven deadlock avoidance

controller synthesis for flexiblemanufacturing systems. IEEE Transaction on Robotics and Automation, 10(2), 196–209.

10. Judd, R. P., & Faiz, T. (1995). Deadlock detection and avoidance for a class of manufacturing systems. Proceedings of the 1995 American Control Conference, 3637–3641.

11. Judd, R. P., Deering, P., & Lipset, R. (1997). Deadlock detection in simulation of manufacturing systems. Proceedings of the 1997 Summer Computer Simulation Conference, 317–322.

12. Lipset, R., Deering, P., & Judd, R. P. (1997). Necessary and sufficient conditions for deadlock inmanufacturing systems. Proceedings of the 1997 American Control Conference, 2,1022–1026.

13. Lipset, R., Deering, P., & Judd, R. P. (1998). A stack-based algorithm for deadlock avoidance in flexible manufacturing systems. Proceedings of the 1998 American Control Conference.

14. Viswanadham, N., Narahari, Y., & Johnson, T. (1990). Deadlock prevention and deadlock avoidance in flexible manufacturing systems using petri net models. IEEE Transaction on Robotics and Automation,6(6), 713–723.

15. Wysk R., Yang, N., & Joshi, S. (1991). Detection of deadlocks in flexible manufacturing systems. IEEE Transactions Robotics and Automation, 7(6), 853–858.

16. Wenle, Z., Judd, R.P., & Deering, P. (2003). Evaluating order of circuits for deadlock avoidance in a lexible manufacturing system. Proceedings of the 2003 American Control Conference, 3679–3683.

17. Wenle, Z., Judd, R.P., & Deering, P. (2004). Necessary and sufficient conditions for deadlocks in flexible manufacturing systems based on a digraph model. Asian Journal of Controls, 6(2) 217–228.

18. Wenle, Z., & Judd, R. P. (2007). Evaluating order of circuits for deadlock avoidance in a flexible manufacturing system. Asian Journal of Controls, 9(2), 111–120.

19. Zhou, M., & DiCesare, F. (1992). Parallel and sequential mutual exclusion for petri net modeling of manufacturing systems with shared resources, IEEE Transaction on Robotics and Automation, 7(4), 550–527.

20. Zhou, M. (1996). Generalizing parallel and sequential mutual exclusions for petri net synthesis ofmanufacturing systems, IEEE Symposium on Emerging Technologies and Factory Automation, 1 49–55.

Chapter 10

A CENTRALIZED REVERSE CHANNEL STRUCTURE WITH FLEXIBLE MANUFACTURING UNDER THE STOCK OUT SITUATION

S.R. Singh[a] , Leena Prasher[b] and Neha Saxena[a]

[a]Department of Mathematics, D. N. College, Meerut, India
[b]Department of Mathematics, QIFGOI, Mohali, India

ABSTRACT

Keeping in view the concern about environmental protection, the study investigates effects of remanufacturing in an integrated production inventory model consisting of forward and reverse supply chain over infinite planning horizon. This article is developed for the deteriorating products with stock dependent demand under shortages. To make the model more realistic, flexibility of production system has been incorporated during forward manufacturing. We derive total cost function and using the results of calculus, optimum production policy is derived, which minimizes the total cost incurred. The results are discussed with a numerical example to illustrate the theory.

INTRODUCTION

Environmental degradation has emerged as a serious social and economic problem. In fact, several governmental policies also encourage the business organizations to re-use or re-cycle used materials with a view to prevent further environmental degradation. The impact of this consciousness on organizations is forcing them to adopt all such methods and to undertake necessary activities to prevent further degradation of the environment. Reverse manufacturing is one of the popular methods undertaken by the manufacturing organizations to recycle the goods after these have been procured from the customers and their reuse effectively for the same purpose. Re-usable and recycle-able materials/articles are procured from the customers through reverse-distribution channels and reconverted through appropriate processes to appear as new and usable. This paper has been prepared in the backdrop of a very high level of ecological consciousness on the part of the government and society. Our research work also facilitates to include implication of research topics such as flexible manufacturing system. In the present consumerist society and a cut-throat competition in the market, the manufacturers are not only employing newer methods of distribution but also newer formats of distribution. The companies are entering rural markets, semi-urban areas and reaching out to the unexpected segments of potential customers. In addition to generate a spurt in demand, the companies are using innovative marketing strategies and innovative marketing tactics with varying degrees of effectiveness. As far as distribution is concerned new departmental stores, new shopping malls are sprouting up even in the unrepresented geographical areas. Because of all this, the visibility and reach of the brand/product has increased manifolds, which causes sudden fluctuations in demand. There is a strong need for a flexible manufacturing system, which can take care of the above realities and adjusts itself to the realities of the market. For the past few decades, reverse logistics has been receiving much attention. Schrady (1967) first studied the problem on optimal lot sizes for production/procurement and recovery. For issues in the greening process, Nahmias and Rivera (1979) studied an EPQ variant of Schrady's model (1967) with a finite recovery rate. Richter (1996a, 1996b, 1997) and Richter and Dobos (1999) investigated a waste disposal model by considering the returned rate as a decision variable. Dobos and Richter (2003, 2004) investigated a production/remanufacturing system with constant demand that is satisfied by noninstantaneous production and remanufacturing for single and multiple remanufacturing and production cycle. Dobos and Richter (2006) extended their previous model and assumed that the quality of collected returned items is not always suitable for further repairing. Konstantaras and Skouri (2010) presented a model by considering a general cycle pattern in which a variable number of reproduction lots of equal size were followed by a variable number of manufacturing lots of equal size. They also studied a special case where shortages were allowed in each manufacturing and reproduction cycle and similar sufficient conditions, as the non-shortages case, are given. El Saadany and Jaber (2010) extended the models developed by Dobos and Richter (2003, 2004) by assuming that the collection rate of returned items is dependent on the purchasing price and the acceptance quality level of these returns. That is, the flow of used/returned items increases as the purchasing price increases, and decreases as the corresponding acceptance quality level increases. Alamri (2010) developed a general reverse logistics inventory model. Chung and Wee (2011) developed an inventory model on short life-cycle deteriorating product remanufacturing in a green supply chain model.

Singh and Saxena (2012) derived an optimal returned policy for a reverse logistics inventory model with backorders. An increase in the shelf space can influence more customers. In this connection, the observations made by Levin et al. (1972) and Silver and Peterson (1985) should be mentioned. They observed that the presence of greater quantity of the same item tends to attract more customers. The reason behind this fact is a typical psychology of the customers. They may have the feeling of obtaining a wide range for selection when a large amount is stored/displayed.Gupta and Vrat (1986) developed models for stock dependent consumption rate. Mandal and Phaujdar (1989) developed an inventory model for deteriorating items and stock dependent consumption rate. Schweitzer and Seidmann (1991) established optimizing processing rate for flexible manufacturing systems. Giri and Chaudhuri (1998) developed deterministic model of perishable inventory with stock-dependent demand rate and nonlinear holding cost and proved that the non-linear holding cost affects the total average cost. Sana et al. (2004) established a production-inventory model for a deteriorating item with trended demand and shortages. Teng and Chang (2005) proposed economic production model for deteriorating item with price and stock dependent demand. Singh and Jain (2009) worked on reserve money for an EOQ model in an inflationary environment under supplier credits. Singh and Singh (2010) worked on supply chain model with stochastic lead-time under imprecise partially backlogging for expiring items. Singh et al. (2010) contributed on an inventory model for deteriorating items with shortages and stock-dependent demand under inflation for two-shops under one management. Yadav et al. (2012) developed an inventory model of deteriorating items with stock dependent demand using genetic algorithm in fuzzy environment. Singh et al. (2013) developed a supply chain inventory model for shortages with variable demand rate.

This model consists of two systems forward manufacturing and reverse manufacturing. At the beginning of each cycle, the inventory is zero. The production starts at the very beginning of the cycle. As production progresses the inventory of finished goods piles up even after meeting the market demand, deterioration/obsolescence. At the beginning of each cycle, the process of collecting returnable items in a separate store also begins. At a point where the production from the forward manufacturing system stops; the collection process of returnable items also stops at the same point (For simplicity, we assume there is no collection of used items once the remanufacturing of collected items starts). At this very point the remanufacturing of reusable items begin at a constant rate. The accumulated inventory produced from the advanced manufacturing system in the meanwhile starts getting consumed and ultimately becomes nil. The accumulated inventory of remanufacturing products, which are assumed to be as good as the newly produced products is consumed when the shortages from the forward manufacturing system begin to surface. In addition, at this stage, there is no production and inventory of remanufactured items is consumed till it becomes nil. When the inventory of remanufactured items is also nil, inventory shortages begin to accumulate for some time. Thereafter, production starts and shortages are gradually cleared after meeting demand and the cycle ends with zero inventories. Geometrical description is shown in Fig 1.

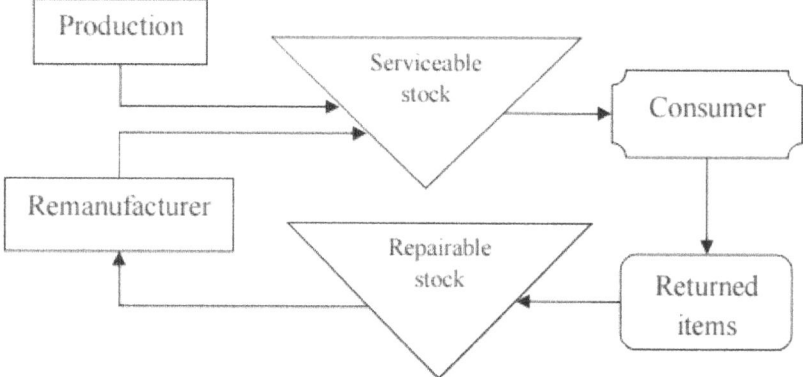

Figure. 1: Flow of inventory in the integrated supply system

ASSUMPTIONS & NOTATIONS

*   Production rate is linear function of demand.
*   The demand rate is deterministic and is a known function of the on hand inventory q. The functional relationship between the demand rate f q( ) and the inventory level q t( ) is given by the following expression:

$f(q) = Dq^{\beta}, D > 0, 0 < \beta < 1, q \geq 0$

Where b denotes the shape parameter and is a measure of responsiveness of the demand to changes in the level of on hand inventory and D denotes the scale parameter.

*   Deterioration rate is constant.
*   Items are returnable and are remanufactured. Remanufactured items are as good as new ones and they are used during the shortage period of forward manufacturing.
*   The time horizon of the inventory system is infinite. Only a typical planning schedule of length T is considered, all remaining cycles are identical.

- Shortages are allowed and are completely backlogged.
- The production time interval for forward production coincides with the collection time interval for reverse manufacturing. (This assumption is not applicable during the period of shortages)

$q(t)$: On hand inventory level at any time t.

$f(q)$: Demand rate, $f(q) = Dq^\beta$, $D > 0$, $0 < \beta < 1$

P : Production rate, $P = If(q)$ where I is a scale parameter, $P > f(q)$, $I > 1$

K : Ordering cost per order

c_h : Holding cost per unit per unit of time during the forward manufacturing.

c_p : Production cost per item.

θ : Deterioration rate.

$c_h^\cdot$: Holding cost per unit per unit of time during the collecting and consuming process for the reverse manufacturing

$c_h^{\cdot\cdot}$: Holding cost per unit per unit of time during the remanufacturing process for the reverse manufacturing.

$q_c(t_c)$: Inventory level during the collecting process for the reverse manufacturing.

$q_1(t)$: Inventory level during the remanufacturing process for the reverse manufacturing

ξ : Fraction of the production lot size $0 < \xi < 1$

Q : Maximum inventory level during forward manufacturing.

R_c : Rate of collection of returnable items.

M : Rate of production of returnable items to be remanufactured.

t_p : Time when production of forward manufacturing stops and also the time when collecting process for reverse manufacturing stops. At this very time remanufacturing of collected items start.

t_s : Time when remanufacturing of returnable items stops and also the time when accumulated inventory of forward manufacturing vanishes.

t_{s1} : Time when accumulated remanufactured inventory vanishes and shortages start.

t_{s2} : Time when production starts again during the period of shortage.

T : Time to complete cycle.

S_1 : Maximum inventory level of remanufactured items.

S : Maximum shortages.

$c_p^{\cdot}$: Cost of purchasing the returnable items per unit.

$c_p^{\cdot\cdot}$: Production cost of remanufactured items per unit.

c_s : Shortage cost per unit per unit of time.

MATHEMATICAL MODELING

There are five stages in the Model (in each cycle as represented in the figure). The governing differential equations are as below: Forward manufacturing process

$$\frac{dq}{dt} + \theta q = (I-1)Dq^\beta \qquad q(0) = 0, \qquad 0 \le t \le t_p \tag{1a}$$

$$\frac{dq}{dt} + \theta q = -Dq^\beta \qquad q(t_p) = Q \qquad t_p \le t \le t_s \tag{1b}$$

Differential equations representing reverse manufacturing collecting time & consuming time.

$$\frac{dq_c(t_c)}{dt_c} = R_c - \theta q_c(t_c) \qquad\qquad q_c(0) = 0 \qquad\qquad 0 \le t_c \le t_p$$

(2a)

$$\frac{dq_c(t_c)}{dt_c} = -M - \theta q_c(t_c) \qquad\qquad q_c(t_p) = B\xi \qquad\qquad t_p \le t_c \le t_s$$

(2b)

where $B = (1-\theta)lD\left(\dfrac{Q}{(l-1)D} + \dfrac{\theta Q^{\alpha+1}}{(l-1)^2 D^2 (\alpha+1)} \right)$ is the production lot size during

the interval $[0, t_p]$ in forward manufacturing system (see Appendix A)

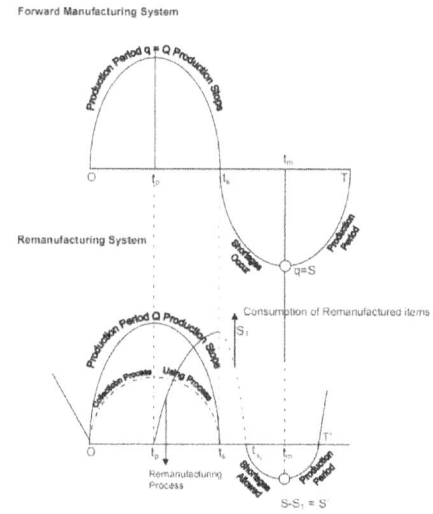

Figure. 1: Flow of inventory

Differential equations representing inventory of remanufactured items.

$$\frac{dq_1}{dt} = M - \theta q_1 \qquad\qquad q_1(t_p) = 0 \qquad\qquad t_p \le t \le t_s$$

(3a)

$$\frac{dq}{dt} + \theta q(t) = -Dq^\beta \qquad\qquad q(t_s) = S_1 \qquad\qquad t_s \le t \le t_{s_1}$$

(3b)

$$\frac{dq}{dt} = -Dq^\beta \qquad\qquad q(t_{s_1}) = 0 \qquad\qquad t_{s_1} \le t \le t_m$$

(3c)

$$\frac{dq}{dt} + \theta q(t) = (l-1)Dq^\beta \qquad\qquad q(t_m) = S = S - S_1 \qquad\qquad t_m \le t \le T$$

(3d)Solving Eq. (1a) and Eq. (1b), we have

$$t = \left(\frac{q^\alpha}{(l-1)\alpha D} + \frac{\theta q^{2\alpha}}{2(l-1)^2 \alpha D^2} + -- \right)$$

$$0 \le t \le t_p$$

(1as)

$$q^\alpha e^{\theta \alpha t} = \frac{-D}{\theta} e^{\theta \alpha t} + \left(\frac{D}{\theta} + Q^\alpha \right) e^{\theta \alpha t_p}$$

$$t_p \le t \le t_s$$

(1bs)

Solving Eq. (2a) and Eq. (2b)

$$q_c = \frac{R_c}{\theta} \left(1 - e^{-\theta t_c} \right)$$

$$0 \le t_c \le t_p$$

(2as)

$$q_c = -\frac{M}{\theta} + \left(\frac{M}{\theta} + B\xi \right) e^{\theta(t_p - t_c)}$$

$$t_p \le t_c \le t_s$$

(2bs)

Now to find holding cost for inventory of collected items during interval [0, t_s], we have

$$= c_h' \int_0^{t_p} q_c (t_c) dt_c + c_h' \int_{t_p}^{t_s} q_c (t_c) dt_c$$

$$= c_h' \left(\frac{B^2 \xi^2}{2R_c} + \frac{B^3 \xi^3 \theta}{3R_c^2} \right) + c_h' \left(\frac{B^2 \xi^2}{2M} - \frac{B^3 \xi^3 \theta}{3M^2} \right) \text{ (See Appendix C)}$$

∴ Total cost of collected items during [0, t_s] (Holding cost + Deterioration cost)

$$= \left(c_h' + \theta c_p' \right) \left(\frac{B^2 \xi^2}{2} \left(\frac{1}{R_c} + \frac{1}{M} \right) + \frac{B^3 \xi^3 \theta}{3} \left(\frac{1}{R_c^2} - \frac{1}{M^2} \right) \right)$$

Inventory of Remanufactured items

Solving (3a),(3b), (3c) & (3d) and using boundary conditions

$$q_1 = \frac{M}{\theta} \left(1 - e^{\theta(t_p - t)} \right)$$

$$t_p \le t \le t_s$$

(3as)

$$q^\alpha = -\frac{D}{\theta} + \left(S_1^\alpha + \frac{D}{\theta} \right) e^{\theta \alpha(t_s - t)}$$

$$t_s \le t \le t_{s_1}$$

(3bs)

$$q^\alpha = D\alpha\left(t_{s_1} - t\right) \qquad\qquad t_{s_1} \le t \le t_m \quad\text{(3cs)}$$

$$q^\alpha = \frac{(l-1)D}{\theta} + \left(S'^\alpha - \frac{(l-1)}{\theta}D\right)e^{\theta\alpha(t_m - t)} \qquad\qquad t_m \le t \le T' \quad\text{(3ds)}$$

The cycle consists of five stages; time for each stage and the cycle time have been calculated as below:

$$t_p = \frac{Q^a}{(l-1)D\alpha}\left(1 + \frac{\theta Q^a}{2(l-1)D}\right) \quad t_{s_1} = t_p + \frac{Q^a}{D\alpha}\left(1 - \frac{\theta Q^a}{2D}\right) \quad t_{s_1} = t_s + \frac{S^a}{D\alpha} - \frac{\theta S_1^{2a}}{2D^2\alpha} \quad t_m = t_{s_1} - \frac{S'^a}{D\alpha}$$

$$T = \frac{Q^a}{(l-1)D\alpha}\left(1 + \frac{\theta Q^a}{2(l-1)D}\right) + \frac{Q^a}{D\alpha}\left(1 - \frac{\theta Q^a}{2D}\right) + \frac{S^a}{D\alpha} - \frac{\theta S_1^{2a}}{2D^2\alpha} + \frac{l}{(l-1)D\alpha}S'^a - \frac{S'^{2a}\theta}{2(l-1)^2 D^2\alpha}$$

The above expression represents time to complete one cycle.

Inventory of remanufactured items during interval $\left(t_p \le t \le t_{s_1}\right)$

$$= \int_{t_p}^{S_1} q_1 dt \qquad = \int_0^{S_1} q_1\left(\frac{1}{M} + \frac{\theta q_1}{M^2}\right)dq_1 \qquad = \frac{S_1^2}{2M} + \frac{\theta S_1^3}{3M^2}$$

Total cost of remanufactured inventory (Holding cost + Deterioration cost)

$$= \left(c_h^{"} + \theta c_p^{"}\right)\left(\frac{\theta S_1^3}{3M^2} + \frac{S_1^2}{2M}\right) \qquad\text{where } t_{s_1} - t_p = \frac{S_1}{M} + \frac{\theta S_1^2}{2M^2}$$

(See Appendix D to find the relation to find S_1 at time t_s). In forward manufacturing system, period of shortage starts at $t = t_s$. It has been assumed that remanufactured items are as good as the new ones and they are used during the shortage period of forward manufacturing.

Holding cost in interval $[t_s, t_{s_1}]$

$$= c_h \int_{t_s}^{t_{s_1}} q dt = c_h \int_{S_1}^0 q\left(-\frac{q^{\alpha-1}}{D} + \frac{\theta q^{2\alpha-1}}{D^2}\right)dq \quad\text{(See Appendix E)}$$

$$hc = c_h\left(\frac{S_1^{\alpha+1}}{D(\alpha+1)} - \frac{\theta S_1^{2\alpha+1}}{D^2(2\alpha+1)}\right)$$

Total cost of remanufactured inventory (i.e. holding cost + deterioration cost) in interval $[t_s, t_{s_1}]$

$$= \left(c_h + \theta c_p \right) \left[\frac{S_1^{\alpha+1}}{D(\alpha+1)} - \frac{\theta S_1^{2\alpha+1}}{D^2(2\alpha+1)} \right)$$

Shortage cost in $\left[t_s, t_m \right]$

$$= -c_s \int_{t_s}^{t_m} q \, dt \qquad = -c_s \int_{0}^{S_1} -\frac{q^\alpha}{D} dq = \frac{c_s S_1^{\alpha+1}}{D(\alpha+1)}$$

Shortage cost in $\left(t_m \leq t \leq T \right)$ (see Appendix F)

$$= -c_s \int q \left(\frac{q^{\alpha-1}}{(I-1)D} + \frac{\theta q^{2\alpha-1}}{(I-1)^2 D^2} \right) dq = c_s \left(\frac{S_1^{\alpha+1}}{(I-1)D(\alpha+1)} + \frac{\theta S_1^{2\alpha+1}}{(I-1)^2 D^2 (2\alpha+1)} \right)$$

This represents shortage cost in $\left[t_m, T \right]$

Total shortage cost

$$= sc \text{ in } \left[t_s, t_m \right] + sc \text{ in } \left[t_m, T \right] = \frac{c_s S_1^{\alpha+1}}{D(\alpha+1)} + \frac{c_s S_1^{\alpha+1}}{(I-1)D(\alpha+1)} + \frac{c_s \theta S_1^{2\alpha+1}}{(I-1)^2 D^2 (2\alpha+1)}$$

Total holding cost and deterioration cost in interval [0,T] (Forward system + Reverse System)

HC + DC = cost in $[0, t_p]$+ cost in $[t_p, t_s]$+ cost in $[t_s, t_m]$

$$= \frac{\left(c_h + \theta c_p \right)}{(I-1)D} \left[\frac{Q^{\alpha+1}}{\alpha+1} + \frac{\theta Q^{2\alpha+1}}{(2\alpha+1)(I-1)D} \right] + \left(c_h + \theta c_p \right) \times \left[\frac{Q^{\alpha+1}}{D(\alpha+1)} - \frac{\theta Q^{2\alpha+1}}{D^2(2\alpha+1)} \right]$$

$$+ \left(c_h + \theta c_p \right) \left[\frac{B^2 \xi^2}{2} \left(\frac{1}{R} + \frac{1}{M} \right) + \frac{B^2 \xi^3 \theta}{3} \left(\frac{1}{R^2} - \frac{1}{M^2} \right) \right]$$

$$+ \left(c_h + \theta c_p \right) \left(\frac{\theta S_1^3}{3M^2} + \frac{S_1^2}{2M} \right) + \left(c_h + \theta c_p \right) \left(\frac{S_1^{\alpha+1}}{D(\alpha+1)} - \frac{\theta S_1^{2\alpha+1}}{D^2(2\alpha+1)} \right)$$

The total inventory cost per unit time is therefore given by

$$TAC \left(Q, S \right) = \frac{K + HC + DC + SC}{T}.$$

Our problem is to find the time to stop the production when q takes optimum value Q and the time to again start the production when maximum shortages accumulate.

$$\frac{\partial}{\partial Q}(TAC) = 0, \frac{\partial}{\partial S}(TAC) = 0, \frac{\partial(TAC)}{\partial S} = 0 \Rightarrow T \left[\frac{\partial SC}{\partial S} + \frac{\partial}{\partial S} DC \right] - (K + HC + DC + SC) \frac{\partial T}{\partial S} = 0$$

Also $T \left[\frac{\partial HC}{\partial Q} + \frac{\partial DC}{\partial Q} \right] - (K + HC + DC + SC) \frac{\partial T}{\partial Q} = 0$

As $\theta \to 0, \alpha \to 1$, we have

$$\frac{\frac{\partial SC}{\partial S}}{\frac{\partial HC}{\partial Q}} = \frac{\frac{\partial T}{\partial S}}{\frac{\partial T}{\partial Q}}$$

As $\theta \to 0, \alpha \to 1$

$$T = \frac{Qt}{(I-1)D} + \frac{I}{(I-1)D} S + \frac{S_1}{D} \qquad \text{(R1)}$$

Using Appendix D, As $\theta \to 0$ $S_1 \to M\left(t_s - t_p\right)$

and using relation $\dfrac{B\xi}{M} = t_s - t_p$ $S_1 \to B\xi$

and also $B \to \dfrac{Ql}{l-1}$ $\therefore S_1 \to \dfrac{Ql\xi}{l-1} = Q\eta \,(say)$

$\therefore T' = \dfrac{Ql}{(l-1)D} + \dfrac{l}{(l-1)D}S' + \dfrac{Ql\xi}{(l-1)D}$

(Also as $\xi \to 0$ i.e. there is no remanufacturing)

[Than $T' \to T$ which is time to complete the cycle in forward manufacturing when reverse manufacturing is not included.]

As $\theta \to 0$, $\alpha \to 1$, we have DC $= 0$ and

$$HC = \frac{c_h l Q^2}{2(l-1)D} + c_h'\frac{B^2\xi^2}{2R_c} + c_h''\frac{S_1^2}{2M} + c_h\frac{S_1^2}{2D} + \frac{B^2\xi^2 c_h'}{2M}$$

and also $S_1 \to B\xi$

$$HC = \frac{c_h l Q^2}{2(l-1)D} + c_h'\frac{B^2\xi^2}{2R_c} + c_h''\frac{B^2\xi^2}{2M} + c_h\frac{B^2\xi^2}{2D} + \frac{B^2\xi^2 c_h'}{2M}$$

As $\theta \to 0$, $\alpha \to 1$

$$SC = \frac{c_s S'^2 l}{2(l-1)D}$$

Substituting all the above values in relation (R1)

$$\frac{\dfrac{c_s S'l}{(l-1)D}}{\dfrac{c_h l Q}{(l-1)D} + \dfrac{c_h' Q\eta^2}{R_c} + \dfrac{c_h'' Q\eta^2}{M} + \dfrac{c_h Q\eta^2}{D} + \dfrac{Q\eta^2 c_h'}{M}} = \frac{l/(l-1)D}{\dfrac{l}{(l-1)D} + \dfrac{\eta}{D}} \qquad \text{where } \eta = \dfrac{l\xi}{l-1}$$

[As $\xi \to 0$, we get $c_s S = -c_h Q$.] If we assume holding cost per unit per unit of time remains same during forward and reverse manufacturing i.e. $c_h = c_h' = c_h''$, We have

$$\frac{c_s S'}{\dfrac{c_h l Q}{l-1}\left(\dfrac{1}{D} + \dfrac{l}{(l-1)}\xi^2\left(\dfrac{1}{R_c} + \dfrac{2}{M} + \dfrac{1}{D}\right)\right)} = \frac{-1}{\dfrac{l}{(l-1)D}(1+\xi)}$$

$\Rightarrow c_s S'(1+\xi) = -c_h Q(1+\eta\xi\zeta D)$ Which gives relation in ' S and Q. Substituting all the values in (R1)

As $\theta \to 0$, $\alpha \to 1$

$$\frac{2k(l-1)D}{l} = Q^2 \left(1+\eta\xi\zeta D\right)c_h\left(1+\frac{c_h\left(1+\eta\xi\zeta D\right)}{\left(1+\xi\right)^2 c_s}\right)$$

As $\xi \to 0$ $Q^* = \sqrt{\dfrac{2k(l-1)Dc_s}{lc_h\left(c_s+c_h\right)}}$

As l increases, production occurs at a more rapid rate. Hence for large l, the model should approach the instantaneous delivery situation of the EOQ model. For large l, $1-\frac{1}{l} \to 1$ Thus as l increases towards infinity, the optimal run size for the model approaches the EOQ when shortages are allowed.

NUMERICAL EXAMPLE

The above theoretical results are illustrated through the numerical verification. Here we are presenting the computational results obtained using Newton Raphson Method which give insight about the behavior of optimal run size Q* , production cycle time ' T and the effects of reverse manufacturing on the total average cost TAC. To illustrate the proposed model, we have considered the following input parameters in appropriate units

$D=2.0, c_h = c_{l_r} = c_{j_r} = 0.5$, $K = 200$, $l = 2$, $c_r = 0.5$ and as we have taken in the last section q ® 0.

Here we derive the optimal solution for the different returned rate and holding cost. Results are presented in Tables 1 as follows,

Table 1: Effects of ξ on (Q^*, T, TAC)

ξ	0.2	0.4	0.6	0.8	1
Q^*	14.825	11.785	9.7823	8.3621	7.303
T	38.54	37.712	37.173	36.792	36.516
Average HC	2.3952	2.302	2.266	2.2239	2.192
Average SC	2.793	2.983	3.115	3.2124	3.288
TAC	10.38	10.606	10.7605	10.8720	10.954

Table 2: Effects of holding cost on (Q* T, TAC)

c_h	0.2	0.4	0.6	0.8	1
Q^*	14.433	8.704	6.29	4.950	4.082
T	46.186	38.298	35.22	33.66	32.64
Average HC	2.706231	2.3738	2.022013	1.74706	1.53
Average SC	1.6238	2.8487	3.6383	4.193	4.59
TAC	8.6603	8.66084	8.6467	8.6593	8.655

Observations

Following observations are made from Table 1:

- As x increases, Q and ' T decreases and holding cost also decreases.
- As x increases, shortage cost increases and total average cost slightly increase.

Observations made from Table2:

- Case 1: When holding cost per unit is lesser than the shortage cost per unit (0.5)
- (a) As holding cost increases, Q & ' T decreases.
- (b) As holding cost increases, there is very slight increase in the total average cost.

· Case 2:When holding cost per unit is greater than the shortage cost per unit (0.5)

- As holding cost increases, Q & ' T decreases but the rate of decreasing is less as compared to Case 1.
- As holding cost increases, total average cost increases & the rate of increasing is more as compared to Case 1.

Comparative observations from Table 1 and Table 2

- When h $c_h = c_s$, total average cost incurred is more as compared to the cost incurred when $c_h \neq c_s$

CONCLUSION

When remanufacturing is undertaken, from the management standpoint there is no perceptible cost difference in terms of total average cost consisting of holding cost, shortage cost, deterioration cost & set-up cost. In view of the governments concern about ecological protection, the management can adopt the system at almost no major incremental costs. As the ratio x increases, there is very slight increase in the total average cost. Further research can be extended to consider the issue of multi objective optimization model, collection of used items during reverse manufacturing period also, inflation and discounting etc.

REFERENCES

1. Alamri, A.A. (2010). Theory and methodology on the global optimal solution to a Reverse logistics inventory model for deteriorating items and time varying rates. Computer & Industrial Engineering, (60), 236-247.

2.   Chung, C. J., & Wee, H. M. (2011). Short life-cycle deteriorating product remanufacturing in a green supply chain inventory control system. International Journal of Production Economics, 129(1), 195-203.

3.   Dobos, I., & Richter, K. (2003). A production/recycling model with stationary demand and return rates. Central European journal of Operations Research, 11(1), 35-46.

4.   Dobos, I., & Richter, K. (2004). An extended production/recycling model with stationary demand and return rates. International Journal of production Economics, 90(3), 311-323.

5.   Dobos, I., & Richter, K. (2006). A production/recycling model with quality considerations. International Journal of Production Economics, 104(2), 571-579.

6.   El Saadany A. M. A., & Jaber M. Y. (2010). A production/remanufacturing inventory model with price and quality dependant return rate. Computers and Industrial Engineering, 58(3), 352–362.

7.   Giri, B. C. & Chaudhuri, K. S. (1998). Deterministic model of perishable inventory with stock-dependent demand rate and nonlinear holding cost. European Journal of Operations Research, 105, 467-474.

8.   Gupta, R., & Vrat, P. (1986). Inventory models for stock dependent consumption rate. Opsearch, 23, 19-24.

9.   Konstantaras, I., & Skouri, K. (2010). Lot sizing for a single product recovery system with variable set upnumbers. European Journal of Operations Research, 203(2), 326-335.

10.   Levin, R. I., Mclaughlin, C. P., Lamone, R. P., & Kottas, J. F. (1972). Production, Operations Management: Contemporary Policy for Managing Operation System. (Newyork: McGraw-hill), 373.

11.   Mandal, B.N., & Phaujdar, S. (1989). An inventory model for deteriorating items and stock dependent consumption rate. Journal Operational Research Society, 40, 483–488.

12.   Nahmias, N., & Rivera, H. (1979). A deterministic model for a repairable item inventory system with a finite repair rate. International Journal of Production Research, 17 (3), 215–221

13.   Richter, K. (1996). The EOQ repair and waste disposal model with variable setup numbers. European Journal of Operational Research, 95(2), 313-324.

14.   Richter, K, (1996b). The extended EOQ repair and waste disposal model. International Journal of Production Economics, 45(1-3), 443-447.

15. Richter, K., & Dobos, I. (1999). Analysis of the EOQ repair and waste disposal model with integer set up numbers. International journal of production economics, 59(1-3), 463-467.

16. Sana, S., Goyal, S.K., Chaudhuri K.S. (2004). A production-inventory model for a deteriorating item with trended demand and shortages. European Journal of Operational Research, 157 (2), 357–371.

17. Schrady, D.A. (1967). A deterministic inventory model for repairable items. Naval Research Logistics Quarterly, 14, 391–398.

18. Schweitzer, P. J., & Seidmann, A. (1991). Optimizing, processing rate for flexible manufacturing systems.Management Science, 37, 454-466.

19. Silver, E. A., & Peterson, R. (1985). Decision systems for inventory management and production lanning,

20. 2nd ed. (Newyork: Wiley).

21. Singh, S.R., Gupta, V., & Gupta, P. (2013).Three stage supply chain model with two warehouse, imperfect production, variable demand rate and inflation. International Journal of Industrial Engineering Computations, 4, 81–92.

22. Singh, S.R., & Jain R. (2009). On reserve money for an EOQ model in an inflationary environment under supplier credits. Opsearch, 46 (2), 352-369.

23. Singh, S. R., Kumar N., & Kumari, R. (2010). An inventory model for deteriorating items with shortagesand stock-dependent demand under inflation for two-shops under one management. Opsearch, 47 (4),311–329.

24. Singh, S. R., & Saxena N. (2012). An Optimal Returned Policy for a Reverse Logistics Inventory Model with Backorders. Advances in Decision Sciences, vol. 2012, Article ID 386598, 21 pages.

25. Singh, S. R. & Singh C. (2010). Supply chain model with stochastic lead time under imprecise partiallybacklogging and fuzzy ramp-type demand for expiring items. International Journal of perational Research, 8(4), 511 - 522.

26. Teng, J. T. & Chang, C. T. (2005). Economic production quantity models for deteriorating items with price and stock-dependent demand. Computational Operations Research, 32 (2), 297-308.

27. Yadav, D., Singh, S. R., & Kumari R. (2012). Inventory model of deteriorating items with two-warehouse and stock dependent demand using genetic algorithm in fuzzy environment. Yugoslav Journal of Operations Research, 22 (1), 51-78.

Chapter 11

A CRITICAL REVIEW OF MACHINE LOADING PROBLEM IN FLEXIBLE MANUFACTURING SYSTEM

Ranbir Singh[1], Rajender Singh[2], B. K. Khan[3]

[1]Research Scholar, Department of Mechanical Engineering, DCRUST Murthal, Sonepat, India

[2]Professor, Department of Mechanical Engineering, DCRUST Murthal, Sonepat, India

[3]Technical Advisor, MSIT, Sonepat, India

ABSTRACT

Production planning is the foremost task for manufacturing firms to deal with, especially adopting Flexible Manufacturing System (FMS) as the manufacturing strategy for production seeking an optimal balance between productivity-flexibility requirements. Production planning in FMS provides a solution to problems regarding part type selection: machine grouping, production ratio, resource allocation and loading problem. These problems need to be solved optimally for maximum utilization of resources. Optimal solution to these problems has been a focus of attention in production and manufacturing, industrial and academic research since a number of decades. Evolution of new optimization techniques, software, technology, machines and computer languages provides the scope of a better optimal solution to the existing problems. Thus there remains a need of research to solve the problem with latest tools and techniques for higher optimal use of available resources. As an objective, the researchers need to reduce the computational time and cost, complexity of the problem, solution approach viz. general or customized, better user friendly communication with machine, higher freedom to select the desired objective(s) type(s) for optimal solution to the problem. As an approach to the solution to the problem, a researcher first needs to go for an exhaustive literature review, where the researcher needs to find the research gaps, compare and analyze the tools and techniques used, number of objectives considered for optimization and need, and scope of research for the research problem. The present study is a review paper analyzing the research gaps, approach and techniques used, scope of new optimization techniques or any other research, objectives considered and validation approaches for loading problems of production planning in FMS.

INTRODUCTION

Manufacturing is the pilot element within the overall enterprise. Possible manufacturing outputs of the firm to meet pre-determined corporate level goals should be known to remain in competition at global market. Manufacturing strategy writes the script to calculate possible manufacturing outputs. Existence of the manufacturing strategy guides daily decisions and activities with clear understanding of decision-goal relationship of the corporation and provides a vision for the firm to remain aligned with the overall business strategy of the firm. The firms having manufacturing strategies for achieving corporate goals survive for long run. A strategy is also a strong communication tool between different levels of management to bring all operations in line with corporate objectives. Custom manufacturing, continuous manufacturing, intermittent manufacturing, flexible manufacturing, just-in-time manufacturing, lean manufacturing and agile manufacturing are major manufacturing strategies revealed in the literature.

FMS is an automated manufacturing system consisting of computer numerical control (CNC) machines with automated material handling, storage and retrieval system. The aim of FMS is to attain the efficiency of mass production while utilizing the flexibility of job shop simultaneously. FMS is adopted for batch production of mid production volume and mid part variety (flexibility) requirements. Since its evolution, researchers are working for optimality of FMS strategy. FMS is a field of great potential hence a numerous complex planning problems need to be solved. Major complex production planning problems are part type selection: machine grouping, production ratio, resource allocation and loading problem (Stecke, 1983). All the production planning problems need to be optimally solved. The present research is the critical literature review for the loading problem of production planning in FMS.

Tooling individual or group of machine(s) to collectively accomplish all manufacturing operations concurrently for all part type in a batch is termed as loading problem. A solution to the problem specifies the machine(s) to which a job has to be routed in sequence for each of its operation(s) with respective tooling under capacity and technological constraint(s) for all jobs in a batch simultaneously to achieve certain objective(s). Loading is a complex combinational planning problem because a batch of jobs is to be machined simultaneously and each job requires unique set of operations effect on manufacturing cost.

To solve the problem, highly experienced and skilled professionals are required. Without the use of some computational or optimization technique, the solution may or may not be optimal. Thus there arises the need of optimal solution with the help of computational methods using optimization techniques. The paper is a critical review paper analyzing the research gaps, approach and techniques used, scope of new optimization techniques or any other research, objectives considered and validation approaches for loading problems of production planning in FMS.

LITERATURE REVIEW OF LOADING PROBLEMS IN FMS

In brief, to solve a problem using optimization techniques and computational analysis, objective(s) are first set, the physical system is modelled using certain technique like mathematical modelling, the solution is then derived under given boundary conditions and constraints to achieve the given objectives, the results are then analysed and the solution approach is then validated. Heuristics has been widely used by the researchers. Table 1 presents the tabulated research review discussing the approach, objectives and results of the loading problems in FMS. Flexible manufacturing is an overall pilot element within an enterprise. Each multinational manufacturing concern has to satisfy business goals to remain in competition with the global market. The manufacturing firm should be aware of the possible manufacturing outputs that will closely match the goals and strategy determined at the corporate level. The existence of a manufacturing strategies guide the daily decisions and activities with clear understanding of how those daily decisions relate to the overall goals of the corporation. The firms having manufacturing strategies for achieving corporate goals survive long. A manufacturing strategy provides a vision to the manufacturing organization for keeping itself aligned with the overall business strategy of the corporation. A strategy is also a strong communication tool between different levels of management to bring all operations in line with corporate objectives. Custom manufacturing, continuous manufacturing, intermittent manufacturing, flexible manufacturing, just-in-time manufacturing, lean manufacturing and agile manufacturing are the major manufacturing strategies which are revealed in literature.

FMS is an automated manufacturing system consisting of numerical control (computer) machines with automated material handling, automated storage and retrieval system. The aim of FMS is to attain the efficiency of mass production while utilizing the flexibility of job shop simultaneously.

Table 1: Review of machine loading problems in FMS based on heuristics approach

			HEURISTIC APPROACH			
Sr.	Year	Researcher Name	Approach	Objectives	Results	Validation approach
1	1983	K. E. Stecke & F. Brian Talbot [2]	Heuristic methods	O Minimizing part movements O Balancing of workload O Unbalancing of workload	Determined how machine tool magazine in a FMS can be loaded to meet simultaneous requirements of a number of different parts	Computational results are presented
2	1985	K. Shankar & Y. J. Tzen [14]	Heuristic methods	O Minimizing system unbalance O Number of late jobs O Balancing of workload	Computational results presented gives improved results	performance is compared with previous results from literature
3	1988	J. A. Ventura, F. F. Chen. & M. S. Leonard [15]	Heuristic algorithms	O Minimizing make span	Improved Performance	The performance of each of the proposed algorithms is evaluated by testing on two hypothetical FMSs.
4	1990	B. Ram. S. Sarin. & C. S. Chen [16]	Fast heuristic algorithms	O Maximizing throughput	FMS loading problem can be solved near optimally in short time	Computational results are produced and compared with previous results
5	1992	S. K Mukhopadhyay, S. Midha, & V. Murlikrishna [17]	Heuristic procedure	O Minimizing system unbalance O Maximizing throughput	Results show that algo developed is very reliable and efficient	tested on ten problems and are compared with existing results
6	1993	K. Kato. F. Oba, & F. Hashimoto [18]	Heuristic approach	O Minimizing total number of cutting tools required O Maximizing utilization rate of each machine	Computational results shows improved effectiveness	Computational results are given to demonstrate the effectiveness of the proposed method
7	1995	E. K. Steeke & F. Brian Talbot [19]	Heuristic Algorithms	O Minimizing part movements O balancing of workload O unbalancing of workload	Results are computationally demonstrated & found improved significantly	Computational results are produced and compared with previous results
8	1997	M. K. Tiwari et al. [20]	Heuristic solution approach	O Maximizing throughput O Minimizing system unbalance	Graphical representation and subsequent model validation	Computational results are produced and compared with previous results
9	1998	G. K. Nayak & D. Acharya [21]	Heuristics and mathematical programming approaches	O Maximizing part types in each batch O Maximizing routing flexibility of batches	Heuristic proposed for part type selection & simple mathematical programs for other two problems	Computational results are compared with existing results
10	2000	D.-H. Lee & Y.-D. Kim [22]	Heuristic algorithms	O Minimizing maximum workload of machines	Results show that suggested algos perform better than existing	Simulation results are compared with existing results
11	2006	N. Nagarjunaa, O. Maheshb, & K. Rajagopal [23]	Heuristic based on multi stage programming approach	O Minimizing system unbalance	Bring together productivity of flow lines and flexibility of job shops	Tested on 10 sample problems available in FMS literature and compared with existing solution methods
12	2006	M. Goswami & M. K. Tiwari [24]	Heuristic-based approach	O Minimizing system unbalance O Maximizing throughput	Loading problem is crucial link between tactical planning and operational decisions	Extensive computational experiments have been carried out to assess the performance of the proposed heuristic and validate its relevance
13	2007	M. K. Tiwari. J. Saha, & S. K. Mukhopadhyay [25]	Heuristic Solution Approaches	O Minimizing system unbalance O Maximizing throughput	GA based heuristic are found more efficient and outperform in terms of solution quality	tested on problems representing three different FMS scenarios from available literature

Most of the researches are focused on increasing the production volume of FMS with increased part varieties. FMS is an interesting field of research to solve the issues and problems encountered by industries. Though FMS has great potential benefits, a numerous control and planning problem need to be taken care of. Kathryn E. Stecke in 1983 described five complex productions planning problems namely part type selection problem, Machine Grouping Problem, production ratio problem, resource allocation problem and loading problem [1] .

Loading means allocation of the operations and required tools to a part types among the set of machine(s), subjected to resource & technological constraints to collectively accomplish all manufacturing operations for each pat type machined concurrently. The allocation of workloads to the existing production facilities for manufacturing products with several constraints in order to perform production activities according to the production plan established, it is essential to adjust the workload for each of the facilities and workers in each time period so they are not assigned work exceeding the given capacity. A solution to this problem specifies the tools which must be loaded in each machine tool magazine and the machine(s) to which a part can be routed for each of its operations before production begins. A variety of products are manufactured simultaneously in FMS, where each part requires potentially unique set of operations, and loading problem is declared as a combinational problem by Kathryn E. Stecke [2] which is highly complex, time-consuming and tedious in nature & requires highly experienced process planners.

Machine loading is one of the most critical production planning problems of FMS. It concerns with the time spend by the job(s) on machine(s) and the manufacturing cost. Manufacturing cost is the sum of fixed and variable costs. Variable cost varies with the level of production output. As output increases, variable cost increases. Once invested, we can't play around the fixed cost; hence to reduce the manufacturing cost, researcher has to minimize the variable cost while maximizing the output. This is done by developing and optimizing a virtual model of manufacturing by some conventional or non-conventional technique for certain number of objectives with their individual weightage accordingly. A researcher has to solve the manufacturing model to minimize the time spent by the job on machine, number of tool used, and movements of tool and job. FMS is a group technology concept hence all the operations on the group jobs are required to be completed at once keeping in view that no machine should be idle or overloaded at any instance of time. Thus the optimized solution of the machine loading problem for certain objectives under technological and capacity constraints is required. The solution to the machine loading problem is to minimize the manufacturing cost as a whole.

Increasing part varieties with raised productivity is necessary to be in competition and to maintain the demand of the product, which is possible by continuous research and optimized solutions to each of the production planning problem. This paper presents a research review of the optimization techniques and the objectives for which the machine loading problem in FMS has been solved, and scope of the research in the field.

Before presenting literature review, an introduction to the optimization techniques and their classification seems necessary to be discussed here for better understanding of the subject. Optimization is the approach for ideal solution. Accuracy of the solution depends on the approach, modeling, computational time and capacity, and nature of the problem. Optimization is classified into six categories: function, and trial and error, single variable and multiple variables, static and dynamics, continuous and discrete, constrained and unconstrained, and random and minimum seeking.

A functional optimization is for theoretical approach where a mathematical formula describes the objective function. Trial-and-error optimization is for experimental optimization with change in the variables which affect output without knowing much about the process. An optimization can be single variable for one dimensional analysis, and multi variables for multi-dimensional analysis. As the number of variables increases, the complexity of the problem also increases. Static optimization is independent of time and dynamic optimization as a function of time. Discrete optimization has a finite number of variables with all possible values, while continuous optimization has infinite number of variables with all possible values. Values are incorporated in equalities and inequalities to an objective of variable function in constrained optimization while the variables can take any value in unconstrained optimization. Random optimization finds sets of variables by probabilistic calculations while minimal seeking is the traditional optimization algorithms which are generally based on calculus methods and minimizes the function by starting from an initial set of variable values.

These optimization approaches can be further sub-categorized as stochastic programming, integer programming, linear programming, nonlinear programming, bound programming, network programming, least squares methods, global optimization, and non-differential optimization.

Most of the researches are focused on solving the machine loading problem by global optimization algorithms. Global optimization algorithms are generally categorized into two approaches: deterministic and probabilistic. Deterministic are sub-categorized into static space search (1992) [3] , branch and bond and algebraic geometry algorithms. Probabilistic is sub-categorized as Monte Carlo algorithms, soft computing and Artificial Intelligence (AI). Monte Carlo algorithms includes two classes, one covers Stochastic (hill climbing) (2002) [4] , Random optimization (1963) [5] , Simulated Annealing (SA) (1953) [6] , Tabu Search (TS) (1989) [7] , Parallel tempering, Stochastic tunneling and Direct Monte Carlo Sampling, and second class includes Evolutionary Computation (EC). EC can be performed by Monte Carlo algorithms or soft computing or AI. EC is further classified as Evolutionary Algorithms (EA), Memetic Algorithms (hybrid Algorithms) (1989) [8] , Harmonic Search (HS), Swarm Intelligence (SI). EA is sub-classified as Genetic Algorithms (GA) (1962) [9] , Learning Classifier System (LCS) (1977) [10] , Evolutionary Programming, Evolution Strategy (ES), Genetic Programming (GP) (1958) [11] . ES includes Differential Evolution (DE), and GP includes Standard GP, Linear GP and Grammar Guided GP. SI includes Ant Colony Optimization (ACO) (1996) [12] and Particle Swarm Optimization (PSO) (1995) [13]. The above discussed classifications scheme will be used for classifying the optimization techniques for solving the machine loading problems of FMS in the paper. Figure 1 shows the evolution of the major optimization techniques along the time axis.

LITERATURE REVIEW OF MACHINE LOADING PROBLEMS IN FMS

An exhaustive research review has been carried out for study of approaches and optimization techniques for machine loading problems in FMS. A. Baveja, A. Jain, A. K. Singh, A. Kumar, A. M. Abazari, A. Murthy, A. Prakash, A. Srinivasulu, A. Turkcan, C. A. Yano, C. Basnet , C.S. Chen, D. Acharya, D. Kosucuoglu, D.H. Lee, F. Brian Talbot, F. F. Chen, F. Guerrero, F. Hashimoto, F. Oba, G. K. Nayak, G.C. Lee, H. C. Co, H. Sattari, H. Yong, H.B. Jun, H.-K. Roh, J. A. Ventura, J. Larranaeta, J. S. Biermann, J. Saha, J. G. Shanthikumar, J. N. D. Gupta, K Chandrashekara, K. E. Stecke, K. Kato, K. M. Bretthauer, K. Rajagopal, K. Shankar, L. H. S. Luong, L. S. Kiat, M. A. Gamila, M. A. Venkataramanan, M. Arıkan, M. Berrada, M. Goswami, M. I. Mgwatua, M. K. Pandey, M. K. Tiwari, Ming Liang, M. M. Aldaihani, M. S. Akturk, M. S. Leonard, M. Savsar, M. Solimanpur, M. Yogeswaran, N. K. Vidyarthi, N. Khilwani, N. Kumar, N. Nagarjunaa, N. K. Vidyarthi, O. Maheshb, Prakash, R. P. Sadowski, R. Budiarto, R. D. Matta, R. H. Storer, R. M. Marian, R. R. Kumar, R. Shankar, R. Swarnkar, S. Biswas, S. Deris, S. Erol, S. G. Ponnambalam, S. K. Mandal, S. K. Mukhopadhyay, S. Kumar, S. Lozano, S. Midha, S. Motavalli, S. P. Dutt, S. Rahimifard, S. S. Mahapatra, S.C. Sarin, S.K. Chen, S.K. Lim, S.T. Newman, T. J. Greene, T. J. Sawik, T. Koltai, T. L. Morin, T. Sawik, U. Bilge, U. K. Yusof, V. H. Nguyen, V. M Kumar, V. Murlikrishna, V. N. Hsu, V. Tyagi, W. F. Mahmudy, Y. Cohen, Y. D. Kim, Y. J. Tzen and Z. Wu are key researchers for solving the loading problem of production planning in FMS.

The tabulated research review discussing the approach, objectives, results and validation approach for machine loading problems in FMS is discussed in Tables 1-3. The literature review is classified into three groups: (1) heuristics; (2) global optimization; and (3) other optimization techniques.

Table 1 presents the review of machine loading problems of FMS based on heuristics approach. The heuristics approach has been significantly used for solving the research problem. Research has gained significant acceleration with the evolution and growth of global optimization techniques.

Table 2 presents the review of machine loading problems in FMS based on global optimization algorithms. Global optimization techniques have been explored rigorously by the researchers. The natural selection techniques have reported good results compared to others. The application of global optimization techniques for solving machine loading problem is increasing with growth of natural optimization techniques. The results reported by natural optimization techniques are more acceptable. Natural optimization techniques, GA and PSO are widely used techniques.

Table 3 presents the review of machine loading problems in FMS based on optimization techniques not falling in the above classification. Since the major focus is on heuristics and global optimization techniques, thus other techniques are grouped in a single table. These techniques have been adopted from time to time for solving the machine loading problem as shown year wise in Table 3.

Optimization techniques and approaches under the classification of global optimization scheme are discussed in Table 2.

Optimization techniques and approaches not falling under the above classifications are discussed in Table 3.

Table 4 has been formulated on regressive analysis of Tables 1-3, for the analysis of the loading objectives to be fulfilled while solving the loading problem. It is a year-wise tabulation and analysis of the loading objectives.

Table 2: Review of machine loading problems in FMS based on global optimization algorithms

			GLOBAL OPTIMIZATION ALGORITHMS		
			a. Deterministic approach		
			1. Branch and bound		
Sr.	Year	Researcher Name	Approach	Objectives	Results	Validation approach
1	1986	M. Berrada & K. E. Stecke [26]	Branch and bound approach	➢ Balancing of workload	Computational results gives fruitful results	Computational results are produced and demonstrated the efficiency of suggested procedures
2	1989	K. Shankar & A. Srinivasulu [27]	Branch & backtrack procedure and Heuristic procedures	➢ Maximizing assigned workload ➢ Maximizing throughput ➢ Minimizing workload unbalance	Each procedure is illustrative by numerical example and results are with improved performance	An illustrative numerical example
3	1994	Y. D. Kim & C. A. Yano [28]	New branch and bond algorithm	➢ Maximizing throughput	Improved efficiency	Computational results are produced and compared with previous results
				2. Algebraic Geometry		
4	1986	T. J. Greene & R. P. Sadowski [29]	Mixed integer programming	➢ Minimizing make span ➢ Minimizing mean flow time ➢ Minimizing mean lateness	Explained simple numeric example	a simple numeric example
5	1987	S.C. Sarin & C.S. Chen [30]	Mathematical model	➢ Minimizing overall machining cost	Computational results are reported	Computational results are compared with literature results
6	1990	K. M. Bretthauer & M. A. Venkataramanan [31]	Linear Integer Programming	➢ Maximizing weighted sum of number of operation to machine assignments	Computational results are satisfactory with improved performance	Computational results are produced
7	1990	H. C. Co, J. S. Biermann, & S K. Chen [32]	Mixed-integer programming (MIP)	➢ Balancing of workloads	Results were found practical	Computational results are produced
8	1990	M. Liang & S. P. Dutt [33]	Mixed-Integer Programming	➢ Minimizing production cost	Demand for change on optimal solution	An example problem is solved
9	1993	Ming Liang [34]	Non-linear programming	➢ Maximizing system output	production cost can be significantly reduced using this approach	Computational results with an illustrative example is demonstrated
10	1994	Ming Liang [35]	Non-linear programming	➢ Maximizing system output ➢ minimizing production cost	Production cost can be significantly reduced using this approach	An illustrative example is solved using the suggested approach
11	1997	V. N. Hsu & R. D. Matta [36]	Lagrangian-based heuristic procedure (MIP problem formulation)	➢ total processing cost	finds a good loading solution	iteratively compared different scenarios
12	1998	T. J. Sawik [37]	Integer programming & approximative lexicographic approach	➢ Balancing workloads ➢ Minimizing total interstation transfer time	Results of computational experiments are reported	illustrative example and some results of computational experiments
13	1999	F. Guerrero, S. Lozano, T. Koltai, & J. Larranaeta [38]	Mixed-integer linear program	➢ Balancing of workload	New approach to loading problem	Computational results are produced
14	2001	N. Kumar & K. Shanker [39]	Mixed integer programming	➢ Balancing of Workload	Results are in agreement with previous findings	Computational results are compared with the previous findings

15	2003	M. A. Gamila & S. Motavalli [40]	mixed integer programming	➢ Minimizing completion time ➢ Minimizing Material handling time ➢ Minimizing total processing time	Results reported increased efficiency and performance of system	Computational results are compared with the previous findings
16	2004	T. Sawik [41]	Mixed integer programming	➢ Minimizing production time	Computational results reported better performance	Numerical examples and some computational results are compared with available literature
17	2011	M. I. Mgwatua [42]	Linear Mathematical Programming	➢ Maximizing throughput ➢ Minimizing make span	More interactive decisions and well-balanced workload of the FMS can be achieved when sub-problems are solved jointly	Compared with results from previous literature
18	2012	A. M. Abazari, M. Solimanpur, & H. Sattari [43]	Linear mathematical programming	➢ Minimizing System unbalance	Genetic algorithm (GA) is proposed and performance of proposed GA is evaluated based on some benchmark problems	Performance is evaluated based on some benchmark problems adopted from the literature

      b.   Probabilistic

   3.   Monte Carlo algorithms

19	1998	S. K. Mukhopadhyay et al. [44]	Simulated annealing (SA) approach	➢ Minimizing system imbalance	Tried to give global optimum solution	Computational results are compared with existing results
20	2004	R. Swarnkar & M. K. Tiwari [45]	Hybrid tabu search and simulated annealing based heuristic approach	➢ Minimizing system unbalance ➢ Maximizing throughput	Results reported better performance	Tested on Standard problems and the results obtained are compared with those from some of the existing heuristics from literature
21	2005	M. M. Aldaihani & M. Savsar [46]	Stochastic model	➢ Minimizing total (FMC) flexible manufacturing cell cost per unit of production	Results reported better performance	Computational results were presented
22	2006	M. K. Tiwari, S. Kumar, S. Kumar, Prakash, & R. Shankar [47]	Constraints-Based Fast Simulated Annealing (SA) Algorithm	➢ Minimizing system unbalance ➢ Maximizing throughput	Proposed algorithm enjoys the merits of simple SA and simple genetic algorithm	The application of the algorithm is tested on standard data sets
23	2012	M. Arikan & S. Erol [48]	Hybrid simulated annealing-tabu search algorithm	➢ Maximizing weighted sum ➢ Minimizing system unbalance ➢ Balancing of workload	Results shows improved system performance compared to earlier results in literature	The results are compared with those developed earlier by the authors

   4.   Evolutionary Computation (EC)

     ✓   Evolutionary algorithms (EA)

24	2000	N. Kumar & K. Shanker [49]	Genetic algorithm (GA)	➢ Maximizing number of part types in a batch ➢ Maximizing number of parts selected a batch ➢ Maximizing mean machine utilization	Results reported reduced computational requirements	comparative study of Computational results
25	2002	H. Yong & Z. Wu [50]	GA-based integrated approach	➢ Balancing of workloads	Results shows that suggested approach perform better than existing	Computational results are compared with the previous findings

26 2006	A. Kumar, Prakash, M. K. Tiwari, R. Shankar, & A. Baveja [51]	Constraint based genetic algorithm (CBGA)	➢ Balancing machine processing time ➢ Minimizing number of movements ➢ Balancing of workload ➢ Unbalancing of workload ➢ Filling the tool magazines as densely as possible ➢ Maximizing sum of operations priorities	The methodology developed here helps avoid getting trapped at local minima	The application of the algorithm is tested on standard data sets from available literature.
27 2007	A. Turkcan, M. S. Akturk, & R. H. Storer [52]	Genetic Algorithm (GA)	➢ Minimizing manufacturing cost ➢ Total weighted tardiness	Approach improves CNC machine efficiency & responsiveness to customer due date requirements	compared with the performance of most commonly used approach in the literature
28 2008	V. Tyagi & A. Jain [53]	Genetic algorithm based methodology	➢ Minimizing system unbalance	For a given number of tool copies of each tool type tool loading is affected by the availability of flexible process plans	An illustrative example
29 2012	U. K. Yusof, R. Budiarto, & S. Deris [54]	Constraint-chromosome genetic algorithm	➢ Minimizing system unbalance ➢ Maximizing throughput	Overall combined objective function increased by 3.60% from previous best result	tested on 10 sample problems available in the FMS literature and compared with existing solution methods
			✓ Memetic (hybrid) Algorithms		
30 2000	M. K. Tiwari & N. K. Vidyarthi [55]	Genetic Algorithm (GA) based (HA) Heuristic Approach	➢ Minimizing system unbalance ➢ Maximizing throughput	Optimal solution to problem	Tested on ten sample problems and the computational results obtained have been compared with those of existing methods
31 2009	M. Yogeswaran, S. G. Ponnambalam, & M. K. Tiwari [56]	Hybrid genetic algorithm simulated annealing algorithm (GASAA)	➢ Minimising system unbalance ➢ Maximising throughput	Results support better performance of GASA over algorithms reported in literature	results compared with reported in the literature
32 2010	S. K. Mandal, M. K. Pandev, & M. K. Tiwari [57]	Genetic algorithm simulated annealing Heuristics approach	➢ Minimizing breakdowns ➢ Minimizing system unbalance ➢ Minimizing make span ➢ Maximizing throughput	Results incurred under breakdowns validate robustness of developed model for dynamic ambient of FMS	Compared with dataset from previous literature
33 2012	V. M Kumar, A. Murthy, & K. Chandrashekara [58]	Meta-hybrid heuristic technique based on genetic algorithm and particle swarm optimization	➢ Minimizing system unbalance ➢ Maximizing throughput	Model efficiency and performance of system is comparable with results compared to literature	Computational results are presented
34 2012	C. Basnet [59]	Hybrid genetic algorithm	➢ Minimizing system unbalance	Better solutions for system unbalance	Computational comparison between the genetic algorithm and previous algorithms is presented
35 2012	D. Kosucuoglu & U. Bilge [60]	Genetic algorithm based mathematical programming (GAMP)	➢ Minimizing total distance travelled by parts during production	GALP integration works successfully for this hard-to-solve problem	tested through extensive numerical experiments
			✓ Swarm Optimization		
36 2007	S. Biswas & S. S. Mahapatra [61]	Swarm Optimization Approach	➢ Minimizing system unbalance	Results reported improved system balance	compared with existing techniques for ten standard problems available in literature representing three different FMS scenarios

37	2008	S. Biswas & S. S. Mahapatra [62]	Modified particle swarm optimization	➤ Minimizing system unbalance	Proposed algorithm produces promising results in comparison to existing methods	comparison to existing methods for ten benchmark instances available in the FMS literature
38	2008	S. G. Ponnambalam & I. S. Kiat [63]	Particle Swarm Optimization (PSO)	➤ Minimizing system unbalance ➤ Maximizing throughput	Performance of PSO is satisfactory compared with heuristics reported in literature	tested by using 10 sample dataset and the results are compared with the heuristics reported in the literature
				✓ Artificial intelligence		
39	2001	N. K. Vidyarthi & M. K. Tiwari [64]	Fuzzy-based Heuristic Approach	➤ Minimizing system unbalance ➤ Maximizing throughput	Substantial improvement in solution quality over some existing heuristic-based approaches	Tested on 10 problems adopted from literatures and computational results are compared with the previous findings
40	2004	R. R. Kumar, A. K. Singh, & M. K. Tiwari [65]	Fuzzy based algorithm	➤ Minimizing system unbalance ➤ Maximizing throughput	Extended neuro fuzzy petri net is constructed	Computational results are compared with standard data set adopted from literature
41	2008	A. Prakash, N. Khilwani, M. K. Tiwari, & Y. Cohen [66]	Modified immune algorithm	➤ Maximizing throughput ➤ Minimizing system unbalance	Good results as compared to best results reported in literature	compared to the best results reported in the literature

The table is showing the list of objectives for which the loading problem is solved. The tick mark ($\sqrt{}$) in the table shows the density for repeatability of the objectives.

Abbreviations used in Table 4:

- Minimizing system unbalance
- Maximizing throughput
- Balancing of workload in the system configured of groups composed of machines of equal size
- Minimizing make span
- Meeting delivery dates
- Minimizing manufacturing cost/Minimizing total processing cost/ Minimizing total flexible manufacturing cell cost per unit of production
- Minimizing tardiness
- Minimizing production cost
- Unbalancing the workload per machine for a system of groups of pooled machines of unequal sizes
- Minimizing part movements
- Maximizing part types in each batch
- Minimizing subcontracting costs
- Maximizing weighted sum of number of operation to machine assignments
- Minimizing flow time
- Minimizing late jobs (number)/ lateness

- Minimizing machine processing time
- Minimizing production time
- Filling the tool magazines as densely as possible
- Maximizing assigned workload
- Maximizing routing flexibility of batches
- Maximizing the sum of operations priorities
- Minimizing material handling time
- Minimizing total distance travelled by parts during production
- Minimizing total number of cutting tools required
- Minimizing workload of machines
- Minimizing breakdowns
- Minimizing earliness

After regressive analysis of the loading objectives of various researchers the optimization approaches and techniques utilized by researchers for problem formulation and its solution are identified and tabulated in Table 3.

Table 3: Review of machine loading problems in FMS based on optimization techniques not falling in the above classification

				OTHER OPTIMIZATION TECHNIQUES		
1	1984	K. E. Stecke & T. L. Morin [67]	Single server closed queueing network model	➤ Balancing of workload	Maximizes expected production of FMS	Results are compared and contrasted with previous models of production systems
2	1986	K. E. Stecke [68]	Hierarchical approach	➤ Maximizing throughput	Nonlinear integer programs models	Ties with some previous results & use of the proposed models to solve realistic loading problems is discussed
3	1986	J. G. Shanthikumar & K. E. Stecke [69]	Dynamic approach	➤ Balancing of workload	Result maximizes expected production	results obtained here complement previous results from literature
4	1993	Y.-D. Kim [70]	Due-Date Based Loading methods	➤ Maximizing throughput	Results reported reduced tardiness and makespan & increased throughput	Computational tests
5	1997	H.-K. Roh & Y-D. Kim [71]	Due-Date Based Loading methods	➤ Minimizing total tardiness	Iterative approach performs better than others	Computational tests on randomly generated problems
6	1997	D. H. Lee, S. K. Lim, G. C. Lee, H. B. Jun, & Y. D. Kim [72]	Iterative algorithms	➤ Minimizing subcontracting costs	Solved part selection and loading problems	computational experiments on randomly generated test problems
7	1997	Y. D. Kim and C. A. Yano [73]	Queueing network model	➤ Maximizing throughput ➤ Maximizing make span ➤ Balancing of workload	Reducing number of machine groups and balancing workloads among machines help to reduce make span	Computational results are produced
8	1998	D.-H. Lee & Y.-D. Kim [74]	Iterative procedures	➤ Minimizing earliness ➤ Minimizing tardiness ➤ Minimizing subcontracting costs	Computational experiments on randomly generated test problems are produced	computational experiments are done on randomly generated test problems and the results are compared with existing results
9	1999	J. N. D. Gupta, J. H. S. Luong, & V. H. Nguyen [75]	Dispatching approach	➤ Minimizing make spans ➤ Minimizing average flow time ➤ Minimizing tardiness	Satisfactory performance of given dispatching algorithm	Simulation results are compared with existing results

10	2000	S. Rahimifard & S.T. Newman [76]	Combined machine loading (CML) algorithms	➤ Meeting delivery dates ➤ Minimising production costs	Adoption of algorithms within an application is dependent on number of manufacturing constraints	Computational results are produced and performance measure is carried out in virtual environment
11	2012	W. F. Mahmudy, R. M. Marian, & L. H. S. Luong [77]	Real coded genetic algorithms (RCGA)	➤ Maximizing throughput ➤ Minimizing system unbalance	RCGA improves FMS performance & minimizes required computational time	Results are compared to the previous literature work

The tick marks (√) shows the density of repetitive occurrence of the optimization techniques and approaches for solving the machine loading problem.

Abbreviations used in Table 5:

- Genetic Algorithm (GA): GA, Hybrid GA, Constraint based GA, Constraint-chromosome GA, Real coded GA, integrated approach based on GA

- Heuristic Algorithm (HA): HA, Fast HA, Fuzzy based HA, GA based HA, Hybrid TS and SA based HA, Lagrangian based HA, GA and PSO based Meta-hybrid HA, multi stage programming approach based HA

- Simulated annealing (SA): SA, Constraints-Based Fast SA, GA based SA, Hybrid GA-SA & SA-TS algorithm

Table 4: Objectives of machine loading in FMS

Sr.	Year	Researcher Name	1	2	3	4	5	6	7	8	9	10	11	12	13	14	15	16	17	18	19	20	21	22	23	24	25	26	27
1	1983	K. E. Stecke & F. B. Talbot [2]		√						√	√																		
2	1984	K. E. Stecke & T. L. Morin [67]		√																									
3	1985	K. Shankar & Y. J. Tzen [14]	√	√													√												
4	1986	M. Berrada & K. E. Stecke [26]		√																									
5	1986	K. E. Stecke [68]			√																								
6	1986	J.G.S. Kumar & K. E. Stecke [69]			√																								
7	1986	T. J. Greene & R. Sadowski [29]					√										√	√											
8	1987	S.C. Sarin & C S. Chen [30]							√																				
9	1988	J. A. Ventura et al. [15]					√																						
10	1989	K. Shankar & A. Srinivasulu [27]	√											√								√							
11	1990	B. Ram et al. [16]					√																						
12	1990	K. M. Bretthauer et al. [31]													√														
13	1990	H. C. Co et al. [32]			√																								
14	1990	M. Liang & S. P. Dutt [33]							√																				
15	1992	Y. D. Kim & C. A. Yano [28]		√																									
16	1992	S. K. Mukhopadhyay et al. [17]	√	√																									
17	1993	K. Kato et al. [18]		√																							√		
18	1993	Ming Liang [34]		√																									
19	1993	Y-D. Kim [70]		√																									
20	1994	Ming Liang [35]		√				√																					
21	1995	E. K. Stecke & F. B. Talbot [19]		√						√	√																		
22	1997	M. K. Tiwari et al. [20]	√	√																									

No	Year	Author											
23	1997	V. N. Hsu & R. D. Matta [36]				√							
24	1997	H.-K. Roh & Y.-D. Kim [71]			√								
25	1997	D. H. Lee et al. [72]						√					
26	1997	Y. D. Kim and C. A. Yano [73]	√	√	√								
27	1998	S. K. Mukhopadhyay et al. [44]	√										
28	1998	D.-H. Lee & Y.-D. Kim [74]			√			√					√
29	1998	G. K. Nayak & D. Acharya [21]					√			√			
30	1998	T. J. Sawik [37]	√			√							
31	1999	F. Guerrero et al. [38]	√										
32	1999	J. N. D. Gupta et al. [75]		√	√			√					
33	2000	N. Kumar & K. Shanker [49]	√			√							
34	2000	D.-H. Lee & Y.-D. Kim [22]									√		
35	2000	S. Rahimifard & S. Newman [76]		√	√								
36	2000	M. K. Tiwari & N. Vidyarthi [55]	√	√									
37	2001	N. Kumar & K. Shanker [39]	√										
38	2001	N. K. Vidyarthi & M. K. Tiwari [64]	√	√									
39	2002	H. Yong & Z. Wu [50]	√										
40	2003	M. Gamila & S. Motavalli [40]							√		√		
41	2004	R. R. Kumar et al. [65]	√	√									
42	2004	T. Sawik [41]						√					
43	2004	R. Swarnkar & M. K. Tiwari [45]	√	√									
44	2005	M. Aldaihani & M. Savsar [46]			√								
45	2006	N. Nagarjunaa et al. [23]	√										
46	2006	M. Goswami & M. Tiwari [24]	√	√									
47	2006	M. K. Tiwari et al. [47]	√	√									
48	2006	A. Kumar et al. [51]		√			√	√		√	√	√	
49	2007	A. Turkcan et al. [52]			√	√							
50	2007	M. K. Tiwari et al. [25]	√	√									
51	2007	S. Biswas & S. Mahapatra [61]	√										
52	2008	A. Prakash et al. [66]	√	√									
53	2008	S. Biswas & S. Mahapatra [62]	√										
54	2008	S. Ponnambalam & L. Kiat [63]	√	√									
55	2008	V. Tyagi & A. Jain [53]	√										
56	2009	M. Yogeswaran et al. [56]	√	√									
57	2010	S. K. Mandal et al. [57]	√	√	√								√
58	2011	M. I. Mgwatua [32]		√	√								
59	2012	V. M. Kumar et al. [58]	√	√									
60	2012	C. Basnet [59]	√										
61	2012	M. Arikan & S. Erol [48]	√		√				√				
62	2012	U. K. Yusof et al. [54]	√	√									
63	2012	D. Kosucuoglu & U. Bilge [60]									√		
64	2012	A. M. Abazari et al. [43]	√										
65	2012	W. F. Mahmudy et al. [77]	√	√									

- Mathematical programming (MP): MP, Linear MP, Non-linear MP, GA based MP
- Swarm Optimization (SO): SO, Particle SO (PSO), Modified PSO
- Queueing network model (QNM): QNM, Single server closed QNM
- Mixed-integer programming (MIP): MIP, GA based MIP
- Branch and bound algorithms (B&BA) : B&BA, New B&BA
- Integer programming (IP): IP, linear IP
- Non-linear programming
- Stochastic model

Table 5: Optimization techniques used for solving machine loading problems in FMS

Sr.	Year	Researcher Name	1	2	3	4	5	6	7	8	9	10	11	12	13	14	15	16	17	18	19	20	21
1	1983	K. E. Stecke & F. B. Talbot [2]	√																				
2	1984	K. E. Stecke & T. L. Morin [67]				√																	
3	1985	K. Shankar & Y. J. Tzen [14]	√																				
4	1986	M. Berrada & K. E. Stecke [26]								√													
5	1986	K. E. Stecke [68]															√						
6	1986	J.G.S. Kumar & K. E. Stecke [69]																				√	
7	1986	T. J. Greene & R. Sadowski [29]						√															
8	1987	S. C. Sarin & C. S. Chen [30]			√																		
9	1988	J. A. Ventura et al. [15]	√																				
10	1989	K. Shankar & A. Srinivasulu [27]	√													√							
11	1990	B. Ram et al. [16]	√																				
12	1990	K. M. Bretthauer et al. [31]									√												
13	1990	H. C. Co et al. [32]						√															
14	1990	M. Liang & S. P. Dutt [33]						√															
15	1992	Y. D. Kim & C. A. Yano [28]							√														
16	1992	S. K. Mukhopadhyay et al. [17]	√																				
17	1993	K. Kato et al. [18]	√																				
18	1993	Ming Liang [34]										√											
19	1993	Y. -D. Kim [70]																			√		
20	1994	Ming Liang [35]										√											
21	1995	E. K. Stecke & F. B. Talbot [19]	√																				
22	1997	M. K. Tiwari et al. [20]	√																				
23	1997	V. N. Hsu & R. D. Matta [36]	√																				

#	Year	Author								
24	1997	H.-K. Roh & Y.-D. Kim [71]								√
25	1997	D. H. Lee et al. [72]						√		
26	1997	Y. D. Kim and C. A. Yano [73]			√					
27	1998	S. K. Mukhopadhyay et al. [44]		√						
28	1998	D.-H. Lee & Y.-D. Kim [74]						√		
29	1998	G. K. Nayak & D. Acharya [21]	√	√						
30	1998	T. J. Sawik [37]					√	√		
31	1999	F. Guerrero et al. [38]				√				
32	1999	J. N. D. Gupta et al. [75]							√	
33	2000	N. Kumar & K. Shanker [49]	√							
34	2000	D.-H. Lee & Y.-D. Kim [22]	√							
35	2000	S. Rahimifard & S. Newman [76]							√	
36	2000	M. K. Tiwari & N. Vidyarthi [55]	√							
37	2001	N. Kumar & K. Shanker [39]				√				
38	2001	N. K. Vidyarthi & M. K. Tiwari [64]	√							
39	2002	H. Yong & Z. Wu [50]	√			√				
40	2003	M. Gamila & S. Motavalli [40]								
41	2004	R. R. Kumar et al. [65]								√
42	2004	T. Sawik [41]				√				
43	2004	R. Swarnkar & M. K. Tiwari [45]	√							
44	2005	M. Aldaihani & M. Savsar [46]					√			
45	2006	N. Nagarjunaa et al. [23]	√							
46	2006	M. Goswami & M. Tiwari [24]	√							
47	2006	M. K. Tiwari et al. [47]		√						
48	2006	A. Kumar et al. [51]	√							
49	2007	A. Turkcan et al. [52]	√							
50	2007	M. K. Tiwari et al. [25]	√							
51	2007	S. Biswas & S. Mahapatra [61]			√					
52	2008	A. Prakash et al. [66]						√		
53	2008	S. Biswas & S. Mahapatra [62]			√					
54	2008	S. Ponnambalam & L. Kiat [63]			√					
55	2008	V. Tyagi & A. Jain [53]	√							
56	2009	M. Yogeswaran et al. [56]		√						
57	2010	S. K. Mandal et al. [57]	√	√	√					
58	2011	M. I. Mgwatua [42]		√						
59	2012	V. M Kumar et al. [58]	√		√					
60	2012	C. Basnet [59]	√							
61	2012	M. Arikan & S. Erol [48]		√						
62	2012	U. K. Yusof et al. [54]	√							
63	2012	D. Kosucuoglu & U. Bilge [60]				√				
64	2012	A. M. Abazari et al. [43]		√						
65	2012	W. F. Mahmudy et al. [77]	√							

- Modified immune algorithm
- Approximative lexicographic approach
- Iterative algorithms
- Hierarchical approach
- Branch & backtrack procedure
- Combined machine loading algorithms
- Dispatching approach
- Due-Date Based Loading methods
- Dynamic approach
- Fuzzy Logic

CONCLUSION ARRIVED ON MACHINE LOADING OBJECTIVES AND OPTIMIZATION TECHNIQUES IN FMS

Detailed study of the machine loading problem is conducted by the authors. The conclusions of the research throttled are divided into three sections as below.

Conclusion on Machine Loading Objectives

On exhaustive study, twenty eight loading objectives are observed in the reviewed literature. Tick marks ($\sqrt{}$) in Table 4 are showing the density for repeatability of the machine loading objectives, which concludes that a research with maximum loading objectives is still required for solving the machine problem. Maximizing expected production rate (throughput) & unbalancing the workload per machine for a system of groups of pooled machines of unequal sizes are the two objectives on which most of the researchers have worked. Balancing of workload on machines for a system of groups of pooled machines of equal sizes is the second most researched loading objective. Minimizing make span is the third most researched loading objective. Minimizing job tardiness is fourth loading objective in the order. Minimizing mean job flow time & minimizing production cost are found at fifth position in the order. Loading objectives observed at sixth rank are maximizing profitability, maximizing the assigned workload, maximizing the part types in each batch, maximizing utilization of system, minimizing subcontracting costs and minimizing the total number of cutting tools required. Material handling time, maximizing routing flexibility of the batches, minimizing earliness, minimizing mean lateness, minimizing mean machine idle time, minimizing overall machining cost, minimizing production time, minimizing the effect of breakdowns, minimizing the maximum workload of the machines, minimizing the number of late jobs, minimizing total flexible manufacturing cell cost per unit of production, minimizing total inter-station transfer time, minimizing total processing time, minimizing part movements and minimisation of the total distance travelled by parts during their production are the loading objectives that are least considered.

Conclusion on Optimization Techniques in FMS

The categorized literature review concludes that the researcher's major emphasis and contribution are towards the use and application of global optimization techniques and with natural optimization techniques, too. Heuristic Algorithms is the mostly used optimization technique by researchers, followed by Genetic Algorithms (GA). Mixed Integer Programming (MIP) & Simulated Annealing (SA) approach are the third mostly used optimization techniques. Linear Mathematical Programming (LMP) is next in the queue succeeded by Integer Programming (IP). At sixth level is Particle Swarm Optimization (PSO) approach. The least used optimization techniques are Tabu search, Swarm Optimization Approach, Branch and backtrack procedure, Branch and bound approach, Combined machine loading (CML) algorithms, Dispatching approach, Due-Date Based methods, Dynamic approach, Fuzzy Logic, Global criterion approach, Hierarchical approach, Artificial immune algorithm, Iterative algorithms, Lexicographic approach, Non-linear programming, Queueing network model and Stochastic model.

Conclusion on validation approaches

A few research problems are solved and the results are compared with previous research results. The results are validated by comparing with literature available results.

Methodologies Findings and Interpretations

A problem when solved for a limited or less number of objectives, it is rather a customized solution for a problem. For general solution, the problem needs to be solved for all possible objectives. On extreme analysis of the machine loading problem and objectives, and on discussion with the academicians and industrialists, the authors emphasise to solve the loading problem for maximization of throughput, part types in a batch, routing flexibility, balancing/ unbalancing of system and workload, and minimization of make-span, delivery dates (covering lateness, tardiness and earliness), part movements, subcontracting costs, machine processing time, tool magazine capacity, number of cutting tools required, breakdowns, non-splitting of jobs, time spend by job on machines in one study. Machine loading problem should be solved for general solution to the problem, for maximum number of objectives. All these objectives are having a common goal of optimizing the production and manufacturing costs.

The literature review reports the application of heuristics, global optimization techniques and some other optimization techniques for solving the loading problem for the listed objectives. Among these approaches, the global optimization techniques were more frequently adopted and the results as founded by the researchers were more accurate and acceptable. Based on regressive analysis of the available literature, and skills and concluding remarks, the authors suggest for the use of natural optimization techniques like swarm optimization for further research. The results of swarm optimization were found more reliable and acceptable as compared to GA, and PSO has attractive characteristics. PSO retains knowledge of all previous particles, which is destroyed in GA when the population changes. PSO is a mechanism of constructive cooperation and information-sharing between particles. Due to the simple concept, ease of implementation, and quick convergence, PSO has gained much attention and has been successfully applied to a wide range of applications.

RESEARCH GAPS AND SCOPE OF RESEARCH IN LOADING OF MACHINES IN FMS

There exists a research gap among the literature available. There are several future scopes that are still not worked out, or still to be worked in a more optimized manner. Based on our observation and exhaustive study such revealed research gap are listed below: Need of integration of loading with other decisions in the neighbourhood of loading (K. Shankar & A. K. Agrawal, 1991); need to reduce excessive computing times (Y. D. Kim & C. A. Yano, 1989); further need of optimization (N. K. Vidyarthi & M. K. Tiwari, 2001, M. K. Tiwari et al., 2007; Amir Musa Abazari et al., 2012); research is required to develop planning softwares (D. H. Lee et al., 1997); PLC controller needs to be enhanced (M. C. Zhou et al., 1993); waiting time for parts and idling time for machines need attention [Mussa I. Mgwatu, 2011]; research by imposing constraints on the availability of resources i.e. jigs, fixtures, pallets, material handling devices needs to be carried out (K. Kato, 1993, N. K. Vidyarthi & M. K. Tiwari, 2001; N. Nagarjuna et al., 2006; Akhilesh Kumar et al., 2006; M. K. Tiwari et al., 2007; Sandhyarani Biswas & S. S. Mahapatra, 2007; Sandhyarani Biswas & S. S. Mahapatra, 2008; Santosh Kumar Mandal et al. 2010; Amir Musa Abazari et al., 2012); new solution methodology needs to be proposed (Santosh Kumar Mandal et al. 2010); need of AI in the field of FMS) Chinyao Low et al., 2006; Sandhyarani Biswas & S. S. Mahapatra, 2008);

Need to use dedicated robot (Majid M. Aldaihani & Mehmet Savsar, 2005); need of simulation studies for FMS (K. Shankar & A. K. Agrawal, 1991; N. K. Vidyarthi & M. K. Tiwari, 2001). Availability of a number of research gaps and that too identified by various eminent researchers from time to time evacuates the need of vast research for solving the observed PPC problems i.e. machine loading problems in FMS.

The authors are working to solve the loading problem with more number of objectives in a single study and for the development of knowledge base system for the machine loading problem. The authors suggest for the development of a knowledge base for all five productions planning problems; part type selection problem, machine grouping problem, production ratio problem, resource allocation problem and loading problem in a single study incorporating the individual objectives of the five individual problems and their respective technological and capacity constraints.

REFERENCES

1. Stecke, K.E. (1983) Formulation and Solution of Nonlinear Integer Production Planning Problems for Flexible Manufacturing Systems. Management Science, 29, 273-288. http://dx.doi.org/10.1287/mnsc.29.3.273

2. Stecke, K.E. and Talbot, F.B. (1983) Heuristic Loading Algorithms for Flexible Manufacturing Systems. Proceedings of the Seventh International Conference on Production Research, Windsor, 22-24 August 1983.

3. Muhlenbein, H. (1992) Parallel Genetic Algorithms in Combinatorial Optimization. In: Balci, O., Sharda, R. and Zenios, S.A., Eds., Computer Science and Operations Research: New Developments in Their Interfaces, Pergamon Press, Oxford, 441-456.

4. Russell, S.J. and Norvig, P. (2002) Artificial Intelligence: A Modern Approach. Second Edition, Prentice Hall, Englewood Cliffs.

5. Rastrigin, L.A. (1963) The Convergence of the Random Search Method in the External Control of Many-Parameter System. Automation and Remote Control, 24, 1337-1342.

6. Metropolis, N., Rosenbluth, A.W., Rosenbluth, M.N., Teller, A.H. and Teller, E. (1953) Equation of State Calculations by Fast Computing Machines. The Journal of Chemical Physics, 21, 1087-1092. http://dx.doi.org/10.1063/1.1699114

7. Glover, F. (1989) Tabu Search—Part I. Operations Research Society of America (ORSA). Journal on Computing, 1, 90-206. http://dx.doi.org/10.1287/ijoc.1.3.190

8.   Moscato, P. (1989) On Evolution, Search, Optimization, Genetic Algorithms and Martial Arts: Towards Memetic Algorithms. Technical Report C3P 826, Caltech Con-Current Computation Program 158-79, California Institute of Technology, Pasadena.

9.   Holland, J.H. (1962) Outline for a Logical Theory of Adaptive Systems. Journal of the ACM, 9, 297-314. http://dx.doi.org/10.1145/321127.321128

10. Holland, J.H. and Reitman, J.S. (1977) Cognitive Systems Based on Adaptive Algorithms. ACM SIGART Bulletin, 63, 49. http://dx.doi.org/10.1145/1045343.1045373

11. Friedberg, R.M. (1958) A Learning Machine: Part I. IBM Journal of Research and Development, 2, 2-13. http://dx.doi.org/10.1147/rd.21.0002

12. Dorigo, M., Maniezzo, V. and Colorni, A. (1996) The Ant System: Optimization by a Colony of Cooperating Agents. IEEE Transactions on Systems, Man, and Cybernetics Part B: Cybernetics, 26, 29-41. http://dx.doi.org/10.1109/3477.484436

13. Eberhart, R.C. and Kennedy, J. (1995) A New Optimizer Using Particle Swarm Theory. Proceedings of the Sixth International Symposium on Micro Machine and Human Science, Nagoya, 4-6 October 1995, 39-43. http://dx.doi.org/10.1109/MHS.1995.494215

14. Shankar, K. and Tzen, Y.J.J. (1985) A Loading and Dispatching Problem in a Random Flexible Manufacturing System. International Journal of Production Research, 23, 579-595. http://dx.doi.org/10.1080/00207548508904730

15. Ventura, J.A., Chen, F.F. and Leonard, M.S. (1988) Loading Tools to Machines in Flexible Manufacturing Systems. Computers & Industrial Engineering, 15, 223-230.

16. Ram, B., Sarin, S. and Chen, C.S. (1990) A Model and Solution Approach for the Machine Loading and Tool Allocation Problem in FMS. International Journal of Production Research, 28, 637-645.

17. Mukhopadhyay, S.K., Midha, S. and Murlikrishna, V. (1992) A Heuristic Procedure for Loading Problem in Flexible Manufacturing Systems. International Journal of Production Research, 30, 2213-2228. http://dx.doi.org/10.1080/00207549208948146

18. Kato, K., Oba, F. and Hashimoto, F. (1993) Loading and Batch Formation in Flexible Manufacturing Systems. Control Engineering Practice, 1, 845-850. http://dx.doi.org/10.1016/0967-0661(93)90252-M

19. Steeke, E.K. and Talbot, F.B. (1995) Heuristics for Loading Flexible Manufacturing Systems, Flexible Manufacturing Systems: Recent Developments. Elsevier Science B.V., Amsterdam, 171-176.

20. Tiwari, M.K., Hazarika, B., Vidyarthi, N.K., Jaggi, P. and Mukhopadhyay, S.K. (1997) A Heuristic Solution Approach to the Machine Loading Problem of FMS and Its Petri Net Model. International Journal of Production Research, 35, 2269-2284. http://dx.doi.org/10.1080/002075497194840

21. Nayak, G.K. and Acharya, A.D. (1998) Part Type Selection, Machine Loading and Part Type Volume Determination in FMS Planning. International Journal of Production Research, 36, 1801-1824. http://dx.doi.org/10.1080/002075498192977

22. Lee, D.H. and Kim, Y.-D. (2000) Loading Algorithms for Flexible Manufacturing Systems with Partially Grouped Machines. IIE Transactions, 32, 33-47.

23. Nagarjuna, N., Mahesh, O. and Rajagopal, K. (2006) A Heuristic Based on Multi-Stage Programming Approach for Machine-Loading Problem in a Flexible Manufacturing System. Robotics and Computer-Integrated Manufacturing, 22, 342-352. http://dx.doi.org/10.1016/j.rcim.2005.07.006

24. Goswami, M. and Tiwari, M.K. (2006) A Reallocation-Based Heuristic to Solve a Machine Loading Problem with Material Handling Constraint in a Flexible Manufacturing System. International Journal of Production Research, 44, 569-588.

25. Tiwari, M.K., Saha, J. and Mukhopadhyay, S.K. (2007) Heuristic Solution Approaches for Combined-Job Sequencing and Machine Loading Problem in Flexible Manufacturing Systems. International Journal of Advanced Manufacturing Technology, 31, 716-730.

26. Berrada, M. and Stecke, K.E. (1986) A Branch and Bound Approach for Machine Load Balancing in Flexible Manufacturing Systems. Management Science, 32, 1316-1335. http://dx.doi.org/10.1287/mnsc.32.10.1316

27. Shankar, K. and Srinivasulu, A. (1989) Some Selection Methodologies for Loading Problems in a Flexible Manufacturing System. International Journal of Production Research, 27, 1019-1034. http://dx.doi.org/10.1080/00207548908942605

28. Kim, Y.-D. and Yano, C.A. (1994) A New Branch and Bound Algorithm for Loading Problems in Flexible Manufacturing Systems. International Journal of Flexible Manufacturing Systems, 6, 361-381. http://dx.doi.org/10.1007/BF01324801

29. Greene, T.J. and Sadowski, R.P. (1986) A Mixed Integer Programming for Loading and Scheduling Multiple Manufacturing Cells. European Journal of Operation Research, 24, 379-386.

30. http://dx.doi.org/10.1016/0377-2217(86)90031-7

31. Sarin, S.C. and Chen, C.S. (1987) The Machine Loading and Tool Allocation Problem in a Flexible Manufacturing System. International Journal of Production Research, 25, 1081-1094. http://dx.doi.org/10.1080/00207548708919897

32. Bretthauer, K.M. and Venkataramanan, M.A. (1990) Machine Loading and Alternate Routing in a Flexible Manufacturing System. Computers and Industrial Engineering, 18, 341-350. http://dx.doi.org/10.1016/0360-8352(90)90056-R

33. Co, H.C., Biermann, J.S. and Chen, S.K. (1990) A Methodical Approach to the Flexible Manufacturing System Batching, Loading and Tool Configuration Problems. International Journal of Production Research, 28, 2171-2186. http://dx.doi.org/10.1080/00207549008942860

34. Liang, M. and Dutt, S.P. (1990) A Mixed-Integer Programming Approach to the Machine Loading and Process Planning Problem in a Process Layout Environment. International Journal of Production Research, 28, 1471-1484. http://dx.doi.org/10.1080/00207549008942806

35. Liang, M. (1993) Part Selection, Machine Loading and Machining Speed Selection in Flexible Manufacturing Systems. Computers and Industrial Engineering, 25, 259-262. http://dx.doi.org/10.1016/0360-8352(93)90270-8

36. Liang, M. (1994) Integrating Machining Speed, Part Selection and Machine Loading Decisions in Flexible Manufacturing Systems. Computers & Industrial Engineering, 26, 599-608.

37. Hsu, V.N. and De Matta, R. (1997) An Efficient Heuristic Approach to Recognize the Infeasibility of a Loading Problem. International Journal of Manufacturing Systems, 9, 31-50.

38. awik, T.J. (1998) A Lexicographic Approach to Bi-Objective Loading of a Flexible Assembly System. European Journal of Operational Research, 107, 656-668. http://dx.doi.org/10.1016/S0377-2217(97)00091-X

39. Guerreore, F., Lozano, S., Koltai, T. and Larraneta, J. (1999) Machine Loading and Part Type Selection in Flexible Manufacturing System. International Journal of Production Research, 37, 1303-1317. http://dx.doi.org/10.1080/002075499191265

40. Kumar, N. and Shanker, K. (2001) Comparing the Effectiveness of Workload Balancing Objectives in FMS Loading. International Journal of Production Research, 39, 843-871.

41. Gamila, M.A. and Motavalli, S. (2003) A Modeling Technique for Loading and Scheduling Problems in FMS. Robotics and Computer

Integrated Manufacturing, 19, 45-54.

42. Sawik, T. (2004) Loading and Scheduling of a Flexible Assembly System by Mixed Integer Programming. European Journal of Operational Research, 154, 1-19. http://dx.doi.org/10.1016/S0377-2217(02)00795-6

43. Mgwatua, M.I. (2011) Interactive Decisions of Part Selection, Machine Loading, Machining Optimisation and Part Scheduling Sub-Problems for Flexible Manufacturing Systems. International Transaction Journal of Engineering, Management, & Applied Sciences & Technologies, 2, 93-109.

44. Abazari, A.M., Solimanpur, M. and Sattari, H. (2012) Optimum Loading of Machines in a Flexible Manufacturing System Using a Mixed-Integer Linear Mathematical Programming Model and Genetic Algorithm. Computers & Industrial Engineering, 62, 469-478. http://dx.doi.org/10.1016/j.cie.2011.10.013

45. Mukhopadhyay, S.K., Singh, M.K. and Srivastava, R. (1998) FMS Loading: A Simulated Annealing Approach. International Journal of Production Research, 36, 1529-1547. http://dx.doi.org/10.1080/002075498193156

46. Swarnkar, R. and Tiwari, M.K. (2004) Modeling Machine Loading Problem of FMSs and Its Solution Methodology Using a Hybrid Tabu Search and Simulated Annealing-Based Heuristic Approach. Robotics and Computer-Integrated Manufacturing, 20, 199-209. http://dx.doi.org/10.1016/j.rcim.2003.09.001

47. Aldaihani, M.M. and Savsar, M. (2005) A Stochastic Model for the Analysis of a Two-Machine Flexible Manufacturing Cell. Computers & Industrial Engineering, 49, 600-610. http://dx.doi.org/10.1016/j.cie.2005.09.002

48. Tiwari, M.K., Kumar, S., Kumar, S., Prakash and Shankar, R. (2006) Solving Part-Type Selection and Operation Allocation Problems in an FMS: An Approach Using Constraints-Based Fast Simulated Annealing Algorithm. IEEE Transactions on Systems, Man, and Cybernetics—Part A: Systems and Humans, 36, 1170-1184.

49. Arikan, M. and Erol, S. (2012) A Hybrid Simulated Annealing-Tabu Search Algorithm for the Part Selection and Machine Loading Problems in Flexible Manufacturing Systems. International Journal of Advanced Manufacturing Technology, 59, 669-679. http://dx.doi.org/10.1007/s00170-011-3506-0

50. Kumar, N. and Shanker, K. (2000) A Genetic Algorithm for FMS Part Type Selection and Machine Loading. International Journal of Production Research, 38, 3861-3887.

51. Yong, H.H. and Wu, Z.M. (2002) GA-Based Integrated Approach to FMS Part Type Selection and Machine Loading Problem. International Journal of Production Research, 40, 4093-4110. http://dx.doi.org/10.1080/00207540210146972

52. Kumar, A., Prakash, Tiwari, M.K., Shankar, R. and Baveja, A. (2006) Solving Machine-Loading Problem of a Flexible Manufacturing System with Constraint-Based Genetic Algorithm. European Journal of Operational Research, 175, 1043-1069. http://dx.doi.org/10.1016/j.ejor.2005.06.025

53. Turkcan, A., Akturk, M.S. and Storer, R.H. (2007) Due Date and Costbased FMS Loading, Scheduling and Tool Management. International Journal of Production Research, 45, 1183-1213.

54. Tyagi, V. and Jain, A. (2008) Assessing the Effectiveness of Flexible Process Plans for Loading and Part Type Selection in FMS. Advances in Production Engineering & Management, 3, 27-44.

55. Yusof, U.K., Budiarto, R. and Deris, S. (2012) Constraint-Chromosome Genetic Algorithm for Flexible Manufacturing System Machine-Loading Problem. International Journal of Innovative Computing, Information and Control, 8, 1591-1609.

56. Tiwari, M.K. and Vidyarthi, N.K. (2000) Solving Machine Loading Problem in Flexible Manufacturing System Using Genetic Algorithm Based Heuristic Approach. International Journal of Production Research, 38, 3357-3384. http://dx.doi.org/10.1080/002075400418298

57. Yogeswaran, M., Ponnambalam, S.G. and Tiwari, M.K. (2009) An Efficient Hybrid Evolutionary Heuristic Using Genetic Algorithm and Simulated Annealing Algorithm to Solve Machine Loading Problem in FMS. International Journal of Production Research, 47, 5421-5448.

58. Mandal, S.K., Pandey, M.K. and Tiwari, M.K. (2010) Incorporating Dynamism in Traditional Machine Loading Problem: An AI-Based Optimization Approach. International Journal of Production Research, 48, 3535-3559. http://dx.doi.org/10.1080/00207540902814306

59. Kumar, V.M., Murthy, A.N.N. and Chandrashekar, K. (2012) A Hybrid Algorithm Optimization Approach for Machine Loading Problem in Flexible Manufacturing System. Journal of Industrial Engineering International, 8, 3. http://dx.doi.org/10.1186/2251-712X-8-3

60. Basnet, C. (2012) A Hybrid Genetic Algorithm for a Loading Problem in Flexible Manufacturing Systems. International Journal of Production Research, 50, 707-718.

61. Kosucuoglu, D. and Bilge, U. (2012) Material Handling Considerations in the FMS Loading Problem with Full Routing Flexibility. International Journal of Production Research, 50, 6530-6552.

62. Biswas, S. and Mahapatra, S.S. (2007) Machine Loading in Flexible Manufacturing System: A Swarm Optimization Approach. Proceedings of the Eighth International Conference on Operations and Quantitative Management, Bangkok, 17-20 October 2007.

63. Biswas, S. and Mahapatra, S.S. (2008) Modified Particle Swarm Optimization for Solving Machine Loading Problems in Flexible Manufacturing Systems. International Journal of Advanced Manufacturing Technology, 39, 931-942.

64. Ponnambalam, S.G. and Kiat, L.S. (2008) Solving Machine Loading Problem in Flexible Manufacturing Systems Using Particle Swarm Optimization. World Academy of Science, Engineering and Technology, 39, 14-19.

65. Vidyarthi, N.K. and Tiwari, M.K. (2001) Machine Loading Problem of FMS: A Fuzzy-Based Heuristic Approach. International Journal of Production Research, 39, 953-979. http://dx.doi.org/10.1080/00207540010010244

66. Kumar, R.R., Singh, A.K. and Tiwari, M.K. (2004) A Fuzzy Based Algorithm to Solve the Machine-Loading Problems of a FMS and Its Neuro Fuzzy Petri Net Model. International Journal of Advanced Manufacturing Technology, 23, 318-341. http://dx.doi.org/10.1007/s00170-002-1499-4

67. Prakash, A., Khilwani, N., Tiwari, M.K. and Cohen, Y. (2008) Modified Immune Algorithm for Job Selection and Operation Allocation Problem in Flexible Manufacturing Systems. Advances in Engineering Software, 39, 219-232. http://dx.doi.org/10.1016/j.advengsoft.2007.01.024

68. Stecke, K.E. and Morin, T.L. (1985) The Optimality of Balancing Workloads in Certain Types of Flexible Manufacturing Systems. European Journal of Operational Research, 20, 68-82.

69. Stecke, K.E. (1986) A Hierarchical Approach to Solving Grouping and Loading Problems of Flexible Manufacturing Systems. European Journal of Operational Research, 24, 369-378. http://dx.doi.org/10.1016/0377-2217(86)90030-5

70. Shanthikumar, J.G. and Stecke, K.E. (1986) Reducing Work in Progress Inventory in Certain Classes of Flexible Manufacturing Systems. European Journal of Operation Research, 26, 266-271. http://dx.doi.org/10.1016/0377-2217(86)90189-X

71. Kim, Y.-D. (1993) A Study on Surrogate Objectives for Loading a Certain Type of Flexible Manufacturing Systems. International Journal of Production Research, 31, 381-392. http://dx.doi.org/10.1016/0377-2217(86)90189-X

72. Roh, H.-K. and Kim, Y.-D. (1997) Due-Date Based Loading and Scheduling Methods for a Flexible Manufacturing System with an Automatic Tool Transporter. International Journal of Production Research, 35, 2989-3004.

73. Lee, D.-H., Lira, S.-K., Lee, G.-C., Jun, H.-B. and Kim, Y.-D. (1997) Multi-Period Part Selection and Loading Problems in Flexible Manufacturing Systems. Computers & Industrial Engineering, 33, 541-544.

74. Kim, Y.D. and Yano, C.A. (1997) Impact of Throughput Based Objective and Machine Grouping Decisions on the Short-Term Performance of Flexible Manufacturing System. International Journal of Production Research, 35, 3303-3322. http://dx.doi.org/10.1080/002075497194084

75. Lee, D.-H. and Kim, Y.-D. (1998) Iterative Procedures for Multi-Period Order Selection and Loading Problems in Flexible Manufacturing Systems. International Journal of Production Research, 36, 2653-2668. http://dx.doi.org/10.1080/002075498192418

76. Gupta, J.N.D. (1999) Part Dispatching and Machine Loading in Flexible Manufacturing System Using Central Queues. International Journal of Production Research, 37, 1427-1435. http://dx.doi.org/10.1080/002075499191337

77. Rahimifard, S. and Newman, S.T. (2000) Machine Loading Algorithms for the Elimination of Tardy Jobs in Flexible Batch Machining Applications. Journal of Materials Processing Technology, 107, 450-458.

78. Mahmudy, W.F., Marian, R.M. and Luong, L.H.S. (2012) Solving Part Type Selection and Loading Problem in Flexible Manufacturing System Using Real Coded Genetic Algorithms—Part II: Optimization. World Academy of Science, Engineering and Technology, 69, 778-782.

Chapter 12

FLEXIBLE MANUFACTURING OF CONTINUOUS PROCESS ENTERPRISES WITH LARGE SCALE AND MULTIPLE PRODUCTS

Yigang Xu, Yifei Du, Yong Zeng, Shiming Li

School of Management and Economics, University of Electronic Science and Technology of China, Chengdu, China

ABSTRACT

Compared with modularized enterprises, continuous Process (CP) enterprises exhibit differing characteristics in manufacturing flexibility. Relative to the vast literature on the flexibility of modularized enterprises, the manufacturing flexibility of CP enterprises has aroused little attention in current literature. In this paper, we develop a new theory to give a reasonable explanation for the flexible manufacturing problem of CP enterprises with multiple products. Based on the analysis of flexible manufacturing process of Jiahua enterprise, we find that CP enterprises can create product diversity through component flexibility, control flexibility and mixed flexibility. The discussion on the permutoid, arising from the product production transformation process of CP enterprises, further reveals the possibility of cost reduction in production transformation and thus efficiency enhancement of manufacturing flexibility. The cost-efficient flexibility of CP enterprises is realized by properly scheduling the production process, shortening the continuous transforming time of the permutoid and reverse development of the permutoid, with exploiting the characteristics of compatibility, continuity and reversibility of the permutoid. Finally, we conclude that the study on the manufacturing flexibility of CP enterprises will contribute to the flexible production transformation in the industries with CP.

INTRODUCTION

The keep-going high price motivates more consideration on energy-saving in the cement industry. Besides developing precalcining kiln technology, utilizing industrial waste heat and using frequency conversation technology largely, the production scale of a single cement production line has been expanded rapidly. In 2000, the average production capacity of a single production line of Chinese cement industry was 85200 tons clinker per year, the average coal consumption of one ton clinker was 156.35 kg standard coal, there were only 11 precalcining kiln production lines whose daily production capacity were above 4000 tons and there was not a single production line with 10,000 tons daily production capacity. Whereas in 2010 the average production capacity of a single production line reached 1 million tons clinker per year, the average coal consumption of one ton clinker was 115 kg standard coal, there were 860 precalcining kiln production lines whose daily production capacity were above 4000 tons and there were 7 precalcining kiln production lines with 10,000 tons daily production capacity[1]. This development phenomenon can also be observed from the industries such as glass industry, metallurgy industry, chemical industry, oil refining industry and so on.

The rapid expansion of the production scale created chances for cement enterprises to reduce energy consumption and made large production lines win cost advantage over small ones. Under the condition of serious overcapacity, the continuous low cement price forced many small common cement enterprises to close their small cement kilns.

It's obvious that the small cement clinker production line is getting more and more uncompetitive on cost. For the special cement production enterprises, they are facing a new challenge, which is how to revolute under the condition of cost pressure and how to keep the former production mode of multiple small production lines with single product in a single production line in such difficult situation. Under the condition of limited order for a single product, expanding the production scale of a single production line means the needs to change the production mode, that is to say the cement enterprise will certainly produce multiple cement clinker products in a single large production line, which is the focus issue of our research.

The production process of cement clinker is a continuous manufacturing process, which is apparently different from the modular production we normally understand. In a continuous production industry, what push flexible manufacturing of multiple products and whether there exist other flexible strategies which differ from the modular production. As the current theoretical models pay less attention to CP enterprise, we discuss and analyze the case of flexible manufacturing of Jiahua enterprise which is the largest special cement production enterprise of China and try to develop a new theoretical model which can describe and explain the multiple products flexible manufacturing of CP enterprise.

The paper contains four parts. The first part is introduction. The second part review related theoretical research, such as mass customization, flexible manufacturing and modularity theory. The third part is specific case analysis and discovery. And the forth part establishes conceptual frameworks for the multiple products flexible manufacturing of CP enterprise based on the case study. Finally, the fifth party is the discussion and conclusion of the paper, which points out the meaning of the research and scope of application, illustrates the further research directions.

RELATED RESEARCH

Mass Customization, Flexible Manufacturing and Modularity

Mass customization (MC) is a production method coming from the diversity of customer demand and development of information technology in 1990s. Joseph Pine systematically studied MC earlier, who thought that the core of MC is the diversity of product types and sharp increase of customization requirements without increasing cost, and the scope is the large scale production of personalized customization products and services and the largest advantage is providing strategic advantage and economic values (Pine, 1992) [1]. Actually MC provides two new insights for us. One is the ability to provide every customization product on the position of customers at the time, place and in a mode customers anticipate. The other one is producing multiple products rapidly with the cost of large scale production through flexible organization, equipments and processes on the position of enterprise (Hart, 1995) [2].

The multiple and customized product put forward higher requirements for the product producing ability of the enterprise. The traditional rigid production line is specially designed for a single product, which can't satisfy the manufacturing requirement of multiplication and customization. MC requires enterprise to have flexible product producing ability. It mainly constructs and promotes the flexible product producing ability of the enterprise through flexible manufacturing system (FMS), effective integration of networked manufacture and flexible management of the enterprise (Dai & Guo, 2006) [3]. The functions of FMS are mainly embodied on the two aspects: one is from the flexibility in the product manufacturing process, the other is the flexibility that influences and determines the production (Sethi & Sethi, 1990) [4].

The promotion of the enterprise's flexible manufacturing level should be realized by low cost, which is the key for the enterprise to run in MC mode.

The best way to realize MC is to establish a modularity component which can configure multiple products and services with the lowest cost and highest individuation level (Pine, 1992).

Masahi ko Aoki (2003) [5] promotes the definition of the module based on studying the definitions of many modules, who thought that the module is a semi-selfdiscipline subsystem that can consist of systematic and specific independent function. The modules interconnect by certain rules through standard surface structure and other functional semi-self-discipline subsystems to construct a more complicated system or process. The modules should have the following characteristics: 1) independence: that means the module can be designed individually, innovated freely and the design concept will not be influenced by other subsystems; 2) relation: that means all modules follow common design rules which should have stable external interfaces and the modules with the same external interface can be interchanged; 3) structural function: that means the modules can not exist and work alone, they need to connect with each other to establish a more complicated system.

Modularity production has apparent function on reducing cost (Gu & Qi, 2001) [6] because modularity design is to produce large quantity products by composing limited universal modules and standard modules, whereas universal modules and standard modules can realize the production in large scale and low cost (Zhang, 2010) [7]. Therefore, the production problem of customized products is transformed or partly transformed into the mass production problem, which helps enterprise to produce the product of any quantity for a single customer or small scale and multiple products market with the cost and speed of large scale production (Anderson & Pine, 1997) [8].

Similarly, the flexibility of the enterprise represents the adaptability for the enterprise to the external changes, and the strategy established on the modularity basis is the best way to deal with changes (Baldwin & Clark, 1997) [9]. Baldwin and Clark (2000) [10] furthermore studied the ways to create diversified products by modularity design. They promoted that six simple instruction characteristics which are splitting, substituting, augmenting, excluding, inverting and porting can realize the diversity of product design.

Continuous Process

Continuous Process (CP) is a process that the raw materials are put into a station and run through the assembly line in a certain order until the finished products are produced in the end. The technology process of CP is sequent and the physical structure of the product of CP is inseparable. The production process is put in use in some industries, such as metal-refractory industry, papermaking industry and chemical industry all belong to CP (Wang, 2011) [11].

Under the condition of a certain technology process control, as the process of CP from input of the material and output of the product is continual, which involves many kinds of transforming process such as separation, decomposition, recombination and composition of material, the mode of CP is a non-modularity production mode.

It can be concluded following discrepancy between CP and modularity process. First, in CP the production line for the material's input and output is a complete system, the termination of any link in the production chain will block the system and influence the quality and quantity of the output product, whereas the production line system of modularity process is separable and combined. Second, the materials of CP will experience a serious of process of separation, decomposition or combination under the control of a certain technology process, which is difficult to be separated to independent subsystems, whereas modular enterprises actually combine each independent module to form a more complicated system in a certain structural way. Third, in CP, the production transformation of multiple products in a single production line is arranged in time order and can't be parallel to the modular enterprises to implement the production.

As the current theoretical models mainly study the MC and flexible manufacturing problem based on modularity process enterprises, which is difficult to explain and illustrate some key strategies of flexible manufacturing of the production enterprise. In this essay we try to develop a new theoretical model to explain the multiple products flexible manufacturing of CP enterprise, which is different from that of modularity process enterprises.

AN INVESTIGATION ON THE FLEXIBLE MANUFACTURING WITH MULTIPLE PRODUCTS

Settings

Jiahua enterprise is one of the largest special cement enterprises of China at present and one of the fastestgrowing special cement enterprises in recent years. At present Jiahua enterprise owns five factories (See Table 1) which are separately located in Leshan City, Emeishan City, Pengzhou City, Mianning County of Liangshan City and Pingshan County of Shijiazhuang City. As the differences of product types and factory locations and each factory has its independent customer group, except the factory in Mianning County of Liangshan State has two 2500T/D precalcining kiln production lines, other factories only have one production line and each production line should produce multiple products according to the personalized requirement of customer.

Table 1: The distribution chart of sub-factories of Jiahua enterprise

Name of factory	Scale of production line	Location	Production type	Note
Leshan Jiahua Cement Factory	One 1800T/D precalcining kiln production line	Jianong Town, Shawan District, Leshan City, Sichuan Province	All kinds of hydraulic engineering and oil well cement and related customer-specific products of the above cement system	
Emeishan Qianghua Special Cement Company of Limited Liability	One 500T/D precalcining kiln production line	Longchi Town, Emeishan City, Sichuan Province	All kinds of decoration and thermo technical cement and related customer-specific products of the above cement system	
Chengdu Jiahua Special Engineering Material Company of Limited Liability	One 1200T/D precalcining kiln production line	Xiaoyudong Town, Pengzhou, Chengdu City, Sichuan Province	All kinds of oil well and road cement and related customer-specific products of the above cement system	
Sichuan Jiahua Jinping Special Cement Company of Limited Liability	Two 2500T/D precalcining kiln production lines	Lugu Town, Mianning County of Liangshan State, Sichuan Province	All kinds of hydraulic engineering and road cement and related customer-specific products of the above cement system	The second production line was started production in April 2012 and is now in the process of debug and pilot run
Shijiazhuang Jiahua Special Engineering Material Company of Limited Liability	One 1000T/D precalcining kiln production line	Pingshan Town, Pingshan County, Shijiazhuang City, Hebei Province	All kinds of oil well cement and related customer-specific products of the above cement system	

Source: Inside information of Jiahua enterprise

And the actual situation is the second production line of the factory in Mianning County of Liangshan State is now in the process of debug and pilot run, so the factory is actually mainly running in single production line.

The production process of cement clinker mainly depends on industrial long kiln and calciner. The characteristic of CP from the raw material input and the cement clinker output is typical. Generally speaking, the common cement enterprises only produce standard common Portland cement clinker, so there is not the problem of multiple products production in a single production line during the calcining process. However, special cement enterprises are exceptions as the production line should satisfy the requirement of flexible manufacturing for different kinds of clinker of different kinds of special cement in a single production line. Therefore, it can be observed the flexible manufacturing process and characteristic of CP enterprise from Jiahua enterprise.

The early special cement businesses of Jiahua enterprise mainly focused on the production of oil well cement. As the scale of the production line was small and the main production line was a 50,000 tons/year vertical cylinder preheater kiln, the factory usually produced in the way of single product in a single production line. In 2000 the cost of coal only counted to 17% of the cement clinker cost, while the price of coal had been grown higher and higher after 2003 and in 2006 the cost of coal counted to 40% of the cement clinker cost. As the energy consumption of small vertical cylinder preheater was too much and competitiveness pressure, in April 2008 Jiahua enterprise changed the vertical cylinder preheater kiln of Emeishan Qianghua Special Cement Company of Limited Liability to be new dry process precalcining kiln, of which the production capacity was 200,000 tons cement clinker per year. It was the first precalcining kiln production line of the enterprise.

Except improving the production capacity, the development of hydraulic engineering market of Jiahua enterprise promoted the requirement of producing multiple products in the same production line of Emeishan Qianghua Company. At the same time, Jiahua enterprise got the order of the special cement for dam impervious wall from China Guodian Dadu River Hydropower Development Co., Ltd. Pubugou Hydropower Station and moderate heat Portland cement from Sinohydro Corporation Shawan Hydropower Station. Plus the original class G oil well cement, Emeishan Qianghua Company tried to produce three kinds of products in the same production line and made preliminary success.

In May 2009, a new 2000T/D (600,000 tons clinker per year) new dry process precalcining kiln production line in Shawan District established by Jiahua enterprise was put into production. In July 2009, Jiahua enterprise acquired Jinping Special Cement Company in Mianning County Liangshan State which had a 2500T/D (775,000 tons clinker per year) new dry process precalcining kiln production line at that time and the second 2500T/D new dry process precalcining kiln production line was completed in 2012. After that, Jiahua enterprise owned two factories in Pengzhou City, Sichuan Province and Pingshan County, Shijiazhuang City, Hebei Province by technology reformation and acquisition. At the same time, the enterprise phased out all small cement kiln production lines and replaced with precalcining kiln production line so that the scale of production line of Jiahua enterprise improved to a new level.

In recent years the enterprise's business has been extended to the areas such as decoration, nuclear engineering, marine engineering and transportation from oil well industry and hydraulic engineering industry. In December 2009, the general manager of the enterprise clearly pointed out in the annual work report that Jiahua enterprise would be positioned on "the manufacturer of cement-based functional material and supplier of application solutions". In order to adapt to the enterprise's development strategy, the enterprise is also trying to construct a kind of large scale flexible manufacturing mode which can rapidly respond to customer's requirement in manufacturing process.

As a strategic consideration, the enterprise must push flexible manufacturing mode to the systematic production mode. From that time all five factories of the enterprise began to carry out manufacturing flexibility production.

CP in the Special Cement Production

All special cement production lines of Jiahua enterprise use new dry process precalcining process. Usually the special cement is produced by grinding special cement clinker and each kind of special cement has unique corresponding relationship with its clinker. For the convenience of study, the essay only takes the production process of clinker as the process of CP to discuss the case.

The production process of special cement mainly includes all the CP from prepare raw meal burden to calcine at high temperature in kiln until the clinker output, such as preheating, decomposion, calcine, cooling and output process, all of which are integrated controlled by central control system. The burden of raw meal is mainly carried out according to the difference of the product type of special cement. If the product belongs to Portland series, the raw meals are mainly limestone and all kinds of siliceous material, ferruginous material or aluminous calibration material. If the product doesn't belong to Portland series, it needs to add other materials such as sulfide, phosphide or fluoride into limestone and bauxite to form the cement minerals such as aluminates, sulpho-aluminates, phosapholuminates or fluoroaluminates. The control methods of technology process mainly includes the control of preheater, calciner, kiln system (the head, middle part and end of the kiln), the temperature of cooling machine; the air volume, air speed and pressure of the pipes connecting equipment system; the rotary speed and flow rate and speed of the material of rotary kiln system; the place, pressure and coal's flow speed of kiln head; the appearance of the flame in kiln head and so on.

Take the production process of Portland series special cement clinker for example, after the raw meal goes into the kiln, firstly the preheater will heat the raw meal by the waste heat of the waste gas exhausted from rotary kiln and calciner to preheat the raw meal and decompose a part of carbonates; secondly the calciner will decompose the preheated material rapidly at suspended status or fluidized status to make the decomposition rate of raw meal above 90%; thirdly the rotary kiln will calcine the pre-decomposed material to make carbonates decompose more quickly and perform a series of solid status reactions to form the minerals such as C_2S, C_3A and C_4AF in cement clinker. With the temperature of the material increases to approach 1300°C, the minerals such as C_2S, C_3A and C_4AF will become liquid status. At this time C_2S dissolved in the liquid status will react with CaO to form a large number of C_3S. After the clinker is sintered, the cooler of cement clinker will cool the high temperature from rotary kiln and then transform to the clinker storehouse It can be seen from the above technical process (Figure 1) that the production of special cement clinker is finished in the precalcining system. The precalcining kiln is a continuous production system composed by preheater, calciner, and rotary kiln and cooler, which is a closed system. The raw meal experiences a uniform motion process inside the kiln system from being input to output of the kiln, during which time the material experiences a series of physical and chemical reactions inside the system by the large number of energy changes between materials and finally form clinker product.

Central control room is a product by combining calcining technology and computer integration technology, which is established on the basis of the information change between kiln system and outside operators. Usually a complete clinker production line will have 3000 - 5000 data collection points, which will be reflected on the computer screen in central control room after been integrated and analyzed and according to the analysis result of the data the operators in central control room can prejudge the running status of the kiln system and instruct adjusting the method and parameters of process technology control so as to control the system.

Adjustment on Components of Raw Material and Control Methods

The manufacturing flexibility of multiple special cement products of Jiahua enterprise is realized by adjusting components of raw material and changing the control methods of clinker calcinations process. Table 2 listed the components of raw material and the finished clinker of six different kinds of special cement products. The variability of components lead to the variety of products.

Table 3 shows the adjustment of parameter of technology process control promotes the product variety under the condition without changing raw material consituents.

Table 4 listed the method and parameter of process technology control of six kinds of special cement products. The changes of method and parameter of process technology control lead to the variety of products.

Flexible Production Exploration

Development of Product Sequencing Technology

In the practices of product conversions, the technicians of Jiahua enterprise found that the conversions between some kinds of products would be very smooth, while some other kinds of products would be very difficult, even suspended the production as kiln system breakdown during product conversions.

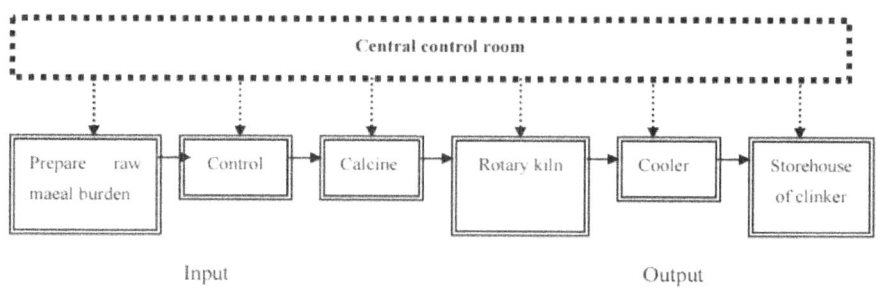

Input   Output

Figure 1: The technology process of special cement clinker production.

Table 2: The components of a part of special cement products of Jiahua enterprise

	Raw meal							Clinker					
Product	Low alkali common use	Class G oil well	Supho-resistant cement	Ordinary low heat	High magnesium and low heat	High magnesium and moderate heat	Product	Low alkali common use	Class G oil well	Supho-resistant cement	Ordinary low heat	High magnesium and low heat	High magnesium and moderate heat
Ratio %							**Mineral composition %**						
Limestone	82.90	82.20	81.20	79.50	69.00	72.20	C_3S	61.42	62.16	49.06	33.81	29.56	50.93
Shale					12.00	11.50	C_2S	14.06	15.95	28.81	45.07	44.18	25.03
Copper slag		6.50	6.50	6.50	6.00	6.30	C_3A	9.12	1.11	1.34	1.26	2.09	1.60
Dolomite					11.00	10.00	C_4AF	10.64	16.29	16.26	16.02	16.39	15.66
Red sand	9.90	11.30	12.30	14.00									
Aluminum tail	7.20	13.55			2.00								
Composition %							**Composition %**						
Loss	34.53	33.33	32.96	32.37	33.34	33.95	Loss	\	\	\	\	\	\
SiO₂	12.71		14.37	15.89	14.68	13.59	SiO₂	20.95	21.80	22.86	24.54	23.12	22.03
Al₂O₃	2.97	1.76	1.83	1.97	2.20	1.78	Al₂O₃	5.68	3.85	3.93	3.85	4.24	3.90
Fe₂O₃	2.16	3.49	3.51	3.50	3.53	3.32	Fe₂O₃	3.50	5.36	5.35	5.27	5.39	5.15
CaO	44.31	44.67	44.14	43.59	40.88	42.66	CaO	64.82	64.22	63.11	62.34	59.33	61.90
MgO	0.85	1.06	1.06	1.05	3.24	3.03	MgO	1.31	1.59	1.58	1.55	4.75	4.46
Value KH	1.086	1.069	0.991	0.879	0.876	1.013	KH	0.924	0.917	0.855	0.788	0.779	0.87
N	2.48	2.58	2.69	2.90	2.56	2.66	N	2.28	2.37	2.46	2.69	2.40	2.43
P	1.38	0.5	0.52	0.56	0.62	0.54	P	1.63	0.72	0.73	0.73	0.79	0.76

Source: Inside information of Jiahua enterprise

Table 3: The adjustment of method and parameter of process technology control changes the performances of a part of products

Product	Standard	The adjustment of method and parameter of process technology control	Purpose	Note
Sulpho-aluminate cement (type A)		Calcine clinker at normal temperature	Satisfy the requirement of standard	Standard product
Sulpho-aluminate cement (type B)	GB 20472-2006	Light burn clinker appropriately	Besides satisfying the requirement of standard, it increases the content of f-CaO to shorten setting time so that the color of clinker gets lighter. It's suitable for European style decoration.	Customized product
Sulpho-aluminate cement (type C)		Over burn clinker appropriately	Besides satisfying the requirement of standard, it reduces the content of f-CaO to lengthen setting time so that the color of clinker gets darker. It's suitable for European engineering grouting.	Customized product
High magnesium and moderate heat Portland cement (type A)	GB 200-2003	Calcine thick clinker at slow rotary speed	Besides satisfying the requirement of standard, in increases early stage hydrated expansion and compensates the concrete shrinkage.	Customized product
High magnesium and moderate heat Portland cement (type B)		Calcine thin clinker at fast rotary speed	Besides satisfying the requirement of standard, in improves MgO activity to increase early stage hydrated expansion furthermore and compensate the concrete shrinkage. It's suitable for arch dam.	Customized product

Source: Inside information of Jiahua enterprise.

Table 4: The process technology control parameters of a part of special cement products of Jiahua

Item / Product	Production per machine (ton)	Primary air (Pa)	Secondary air (Pa)	Tertiary air (Pa)	Temperature of calciner ($^\circ$C)
Low alkali common use cement	68	25,000	5200	120	870
Class G oil well cement	68	25,000	5300	120	870
High magnesium and moderate heat cement	68	27,000	5300	140	880
High magnesium and low heat cement	70	28,000	5400	150	890
Ordinary low heat cement	69	27,000	5300	140	880
Sulfate resistance cement	69	25,000	5200	130	870

Source: Inside information of Jiahua enterprise.

In earlier practices Emeishan Qianghua company produced class G oil well cement and moderate heat Portland cement in the same calcining kiln production line. The factory made success at one time when converting class G oil well cement production to moderate Portland cement production with one time material input, which was because the similarity of the components of two products was very high and the compatibility of "permutoid" was great.

There were also failure cases when the enterprise practiced flexible production. In October 2008, the enterprise closed an ordinary kiln which was used to produce sulpho-aluminate cement. In order to keep customers, Qianghua factory tried to convert to produce sulpho-aluminate cement from producing class G oil well cement and moderate heat Portland cement. As the components of class G oil well cement are high ferric and low aluminium, while the components of sulpho-aluminate cement are high sulfate, high aluminium and low ferric, they are extremely hard to compatible. During the conversions the rotary kiln broke down as the blocked materials in the kiln. The factory spent total five days to clear out the blocked materials and then put into production again. Similarly, during the conversions from sulpho-aluminate cement to class G oil well cement the same broke down happened again and the factory had to stop production, clear out blocked materials and put into production again. Therefore, it can be seen that not all cement products are suitable for the CP in the same production line.

As the pressure of finishing orders, Leshan Jiahua cement factory needs to product 6 kinds of products every month and in peak period the factory will convert production as high as 13 times. After a period of experiences, Leshan Jiahua cement factory found out the regulars of sequencing production conversions, which is similar to the allelopathy theory. For example, the effect of converting class G oil well cement production to low heat Portland cement product would be very poor, while the effect of converting class G oil well cement production to highly sulfate resistance cement first and then to low heat Portland cement would be great. This regular has been applied to other factories of the enterprise too.

See Figure 2 as the product sequencing chartf a part of special cement of Leshan Jiahua factory.

The factory illustrates the following sequence:

1) The KH of low alkali common cement and class G clinker is the same and the control indexes of Fe_2O_3 and Al_2O_3 are different. It just needs to adjust the control indexes of Fe_2O_3, Al_2O_3.

2) The control indexes of Fe_2O_3 and Al_2O_3 of sulpho-resistant clinker and class G clinker are the same and KH of sulpho-resistant clinker is lower than that of class G clinker. It just needs to adjust KH.

3) The control indexes of Fe_2O_3 and Al_2O_3 of sulpho-resistant clinker and low heat clinker are the same and KH of low heat clinker is lower than that of sulpho-resistant clinker. It just needs to adjust KH.

4) The control indexes of KH, Fe_2O_3 and Al_2O_3 of common low heat clinker and high magnesium low heat clinker are the same. It just needs to adjust the content of MgO and alkali.

5) The control indexes of Fe_2O_3, Al_2O_3, MgO and alkali content of high magnesium moderate heat clinker and high magnesium low heat clinker are the same and KH of high magnesium moderate heat clinker is higher

than that of high magnesium low heat clinker. It just needs to adjust KH.

According to illustrations above and the data of Table 2, we find that the smaller the adjustment range of product components is, the easier the product production conversion is.

Improvement of Product Conversion Control Technology

The larger the adjustment range of product components, the product production conversion will be the harder.

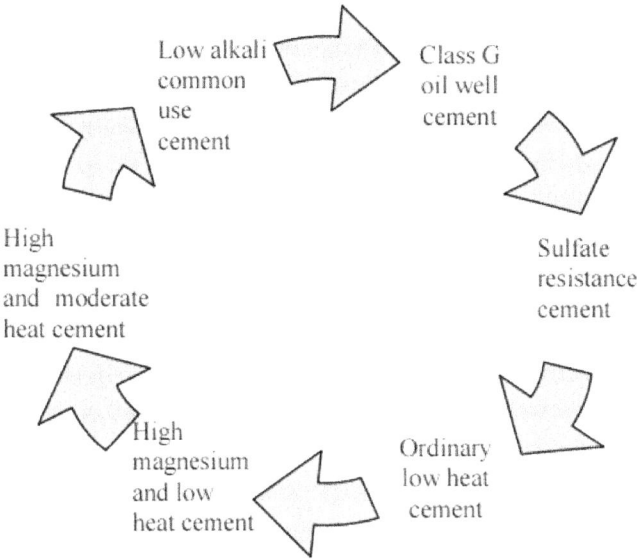

Figure 2: Product sequencing chart of a part of special cement. Source: Inside information of Jiahua enterprise.

It's noticed that it needs to solidify the control parameters and operation methods to a continuous process in order to push the production conversion more smoothly and stably.

Emeishan Qianghua Company experienced a lot of difficulties when converting production of class G oil well cement to special cement for dam impervious wall. The cement is a kind of innovated product from low heat Portland cement and developed jointly by Yangtzi River Commission, owners and Jiahua enterprise. Besides there are obvious differences on the components compared with that of moderate heat Portland cement and class G oil well cement, the calcining temperature is 70˚C lower than that of moderate Portland cement, so the factory didn't produce qualified products for two days after production conversion and got the qualified products in the third day, but the production cost was high, and a lot of alternate materials were wasted.

In the second half year of 2010, Jiahua Jinping Company in Liangshan State always complained that the customized slight expansion moderate heat Portland cement offered to Ertan Hydroelectric Development Corporation Jinping Power Station and Guandi Power Station cost too much as the selection rate of slight expansion moderate heat Portland cement was too low resulting too much wasted alternate materials. In order to solve the selection rate problem, Jiahua enterprise organized related technicians to analyze and found out that the light burning was the main reason for the problem as the variation value of burned temperature was too big, the rotary speed of kiln was faster and the weight per liter of clinker was lower. Finally, Jiahua Jinping Company improved the selection rate of slight expansion moderate heat Portland cement to above 95% just by adjusting a parameter in temperature area of the kiln.

After that Jiahua enterprise carried out a production promotion program, which was led by Engineering Centre. The program systematically summarizes and certifies the operation rules and control parameters of each factory and promoted the time table of modifying and perfecting the flexible manufacturing technology process quality manual and operation rules. In December 2010, all factories of the enterprise finished the modification work of manuals and rules.

Reverse Development on Permutoid

Frequent production conversion will absolutely influence the control of cost, among which the alternate wasted materials is a key factor that influence the cost of multiple products production.

Initially the way to recycle and reuse permutoid is mainly using it as the mixed materials for common cement 32.5. But as the price of common cement is too low, the way can't save too much cost.

Therefore, Jiahua enterprise had been always exploring the better utilization method for permutoid. Leshan Jiahua Cement Factory took the lead in getting that. On one hand, the factory tried to reduce the permutoid and improve the selection rate of qualified products in production conversion. On the other hand, the factory developed permutoid and special cement by adding permutoid into the high components special cement which has complementary relation with it or developing permutoid as the base material of functional mortar of Jiahua enterprise according to the components of permutoid so as to further improve the value of permutoid.

Now the ways Jiahua Cement Factory took have been popularized in the whole enterprise and all factories begin to compare the main economic technology of them with the common cement companies that produce common cement in mass production so as to reduce the production conversion cost to inconspicuous level.

The reverse development technology of permutoid brings value for permutoid which would be wasted materials, which actually is a kind of reverse product development technology conforming to utilization of Industrial wasted resources.

DISCUSSION

Flexible Manufacturing Model in CP Enterprises

By quoting the classification view of economic system, we can view that CP is a closed system, by contrast, modularity process is an open system. The closed system will not have physical interaction with the external world but have energy and information interaction, whereas the open system will not only have energy and information interaction, but also have physical interaction (Zan Ting-quan, 2003) [12].

Therefore, as shown in Figure 3, CP can be illustrated as a process of input and output. In the process of raw material input and output, the system is closed and only has energy and information interaction.

In CP, the materials will have a series of physical and chemical reactions such as separation, decomposition, recombination and composition and finally product under the function of certain energy. The energy functions as a result of the combination of some kind of technology process control method and technology process control parameters, such as the increase or reduction of temperature, increase or reduction of pressure, increase or reduction of air volume, acceleration or deceleration of wind velocity and the transforming mechanical driving force and movement velocity of the material in the internal of the system. As the energy will function on the basis of consuming energy, it can explain why the energy cost in CP will have great influence on the competitiveness of enterprises.

Having entered the closed system, the material will move to the output terminal with appropriate transformation ways in the system and follow the sequence of technology process. The sequence of technology process can be illustrated as following. First, the technology process control method and technology process control parameters that the energy functions are distributed in the transformation spots of the material, for example, the temperature of the system may be lower when the material is input while it may be higher when the material is running in the middle of the system. Second, under the function of energy, the components in the internal of the material have physical or chemical reactions continuously in time order. Third, The material is in different status when it's in different transformation spot and the status of it has different physical or chemical nature in different transformation spot. Four, in the process of input and output of the material, there is some time-lag between the time point of material input and energy function.

The time order of the material inside the system can help us classify the technology process of CP as an unin terrupted production process, however, the relevant production process are not independent and the material inside the system doesn't have a clear physical interface, therefore, the classification of production process of CP is unrelated to modularity.

The control of energy is realized by the information transformation. The higher the automation degree of the closed system is, the more complicated the interaction on the external information will be. Therefore, it can be observed that the computer integrated control system has been applied on a large scale in CP industries such as cement industry, oil refractory industry and chemical industry.

Categories of Flexible Manufacturing in CP Enterprise

The production conversion of the multiple products in a single line of CP enterprise also follows time order. Technologically, there are two factors to drive product conversion, one is the components of raw material and the other is the means and parameter of technology process control when the energy functions. The Table 5 re flects what we think about the conversion method of CP. The vertical axis displays the change of components of raw materials and the conversion of drive products, which makes the production generate flexibility. The horizontal axis displays the means of technology process control and the conversion of adjusting drive products of its parameter, which makes the production generate flexibility.

Table 5: Flexible manufacturing model of CP enterprise

	Unchange of the method and parameter of technology process control	Change of the method and parameter of technology process control
Unchange of the components of raw materials	No flexibility	Control flexibility
Change of the components of raw materials	Component flexibility	Mixed flexibility

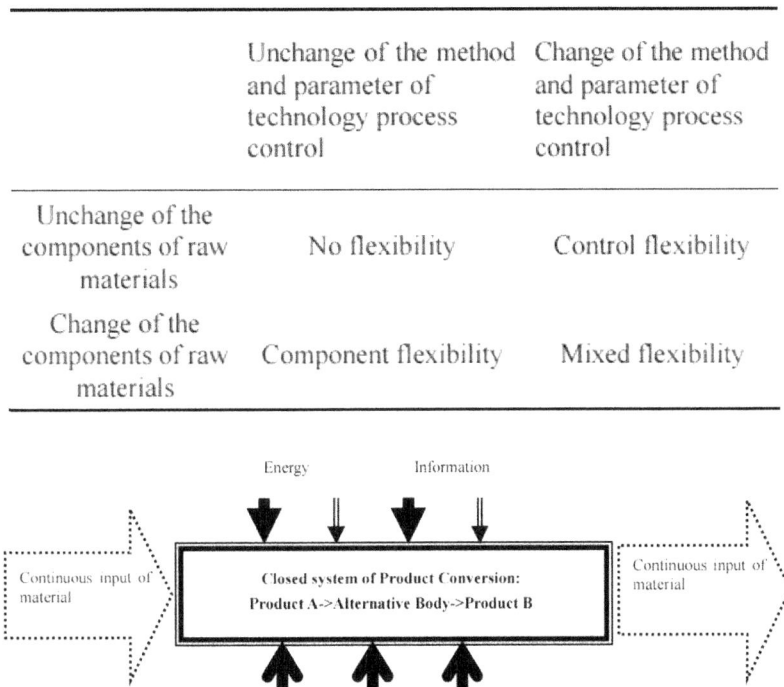

Figure 3: Continuous process of flexible manufacturing.

It can be seen from Table 5 that the flexibility of CP enterprises can function without relying on the group technology of modularity. On the premise that the components of raw materials, energy and parameters never change, the production line runs continuously in a single product production mode. Under the condition that the method and parameter of technology process never change, different products can be produced and the flexibility of manufacture is realized by adjusting the components of raw material , which we call Components Flexibility. Under the condition that raw materials never change, only adjust the method and parameter of technology process such as pressure, temperature, air volume and movement speed of material, it can also produce different products, which we call Control Flexibility.

Besides, in many cases the situation will happen, under which the components of raw materials and the method and parameter of technology process control all need to be adjusted and changed, we classify this kind of flexibility as Mixed Flexibility.

Characteristics of Permutoid for Flexible Manufacturing in CP

All CP enterprises concern the production process of the material input and product output. In order to realize the multiple products production in a single production line, under the condition of continuous production, whether it is component flexibility, control flexibility or mixed flexibility, it will output a product that is neither product A nor product B during the process of the production conversion between product A and product B as a result of the time-lag between material input and energy functions, which we call it Permutoid. Permutoid has the following characteristics:

Compatibility

Permutoid happens under the condition of adjusting the components of raw materials or changing the method and parameter of technology process control. Although it doesn't conform the standard of product A and B, the components between permutoid and product A and B have certain compatibility. Generally speaking, the better the compatibility of the permutoid is, the easier the conversion between two products will be. On the contrary, if the compatibility is very bad, even conflict with each other, the conversion of two products will be hard to realize under the CP. Therefore, the compatibility determines whether the product conversion is feasible or not technologically. Besides, the situation may happen under which the compatibility of the permutoid is worse when product A is converted to product B, but better when it is converted to product C, and the compatibility of the alternate bodies between product C and product B is not bad. Therefore, studying compatibility will help enterprise optimize the product production order in technical level.

Generally speaking, the similarity of the components of two products' raw material can improve the compatibility of the permutoid, the smaller the difference between two products' raw material, the better the components combine. In addition, the good compatibility between two products' raw material can also help to improve the compatibility of alternate bodies even though the components are quite different from each other.

Continuity

Permutoid is an intermedia in the product conversion process of two products. As the production process is always in a continuous state, the permutoid is also in such state, the permutoid will last till product B is output after the production of product A is over. If the conversion is abnormal and the production process never terminates, the permutoid will last a long time. Therefore, the duration of the permutoid and its quantity will directly influence the conversion cost of the product.

Under the condition of the given raw material, the duration and quantity of the permutoid are related to the production control level. Generally, the less the method of technology process control increases or decreases and the smaller the magnitude of every parameter is adjusted, the shorter the duration of the permutoid becomes and the more smoothly the product is converted. Therefore, the smooth degree of conversion negatively corresponds to the adjusted variables of the method and parameter of technology process control.

Reversibility

The permutoid itself is a kind of substandard rejected material. However, it can largely reduce the conversion cost of CP enterprises if recycling, furtherdeveloping and reusing the permutoid with appropriate technology. What's more, furtherdevelopment of the permutoid usually produces a new product, which can increase the manufacturing flexibility. Therefore, the reverse development of the permutoid can also be viewed as a kind of flexible technology of CP enterprises.

Strategic Choices for Large Scale Flexible Manufacturing in CP Enterprises

According to the internal of enterprises, the key of large scale flexible manufacturing of CP enterprises is still how to solve reducing the production cost under the condition of mult-products production, which is similar to modularity enterprises.

Compared with the mass production in a single production line, all CP enterprises face the problem of product conversion cost for large scale and multiple production lines production. So how to reduce the conversion cost of the product is the focus for CP enterprises to carry out flexible manufacturing.

According to the analysis of the flexibility model and permutoid of CP enterprises, CP enterprises can adopt technical strategies below to reduce the conversion cost of flexible manufacturing, as shown in Table 6.

CONCLUSIONS

The exploration of flexible manufacturing of CP enterprise can help to extend the production methods of MC and flexible manufacturing to CP enterprise, which plays positive role in satisfying diversified and personalized requirement or consumers and customers and pushing more industries to carry out flexible production way.

Different from the strategy that modularized industries realize product manufacturing flexibility by standard and universal model design and systematic combinations, CP enterprises mainly create the variety of product by component flexibility, control flexibility and mixed flexibility. However, from the view of enterprises, both modularized enterprises and CP enterprises hope to realize the variety of product by mass production. Therefore, the flexible manufacturing of both modularized enterprises and CP enterprises is based on low cost.

Of course the key strategy for modularized enterprises and CP enterprises reverse to use to reduce the cost is different. CP enterprises pay more attention on permutoid, which arrange product production order appropriately by studying the compatibility of permutoid, shorten the show-up time of permutoid by integrating technology process control and promote the permutoid to be an available reverse product by redoveloping it so as to reduce the production conversion cost.

Table 6: The technical strategy of reducing the conversion cost of CP enterprises

Technical strategy	The way to reduce the conversion cost
Product sequencing technology	According to the compatibility of permutoid, CP enterprises can push conversion, reduce the quantity of permutoid, promote conversion efficiency and reduce conversion cost by optimizing product production sequencing.
Transform control technology	According to the continuity of permutoid, CP enterprises can push conversion, shorten the occurence time of permutoid, promote conversion efficiency and reduce conversion cost by optimizing process control technology.
Reverse product development technology	CP enterpriss can reduce conversion cost by recycling, redevelopment and reusing the permutoid.

While modularized enterprises realize low cost production by standard and universal model mass production.

The paper mainly discusses about flexible manufacturing technology of CP enterprise. Actually the study on the production organization management system related to CP flexible manufacturing is still a new subject. Besides, the writers also are thinking if there is any unique feature in product variety and customized requirement compared with modularized enterprises when considering from the view of customer. Lastly, there may be some new problems for us to pay close attention if we put flexible manufacturing of CP enterprises into supply chain relations.

The practice study in this paper can be extended furthermore. Logically, the flexible manufacturing model, permutoid theory and flexible strategy of CP enterprises can be extended to other CP industries such as glass industry, metallurgy industry, papermaking industry, chemical industry, pharmacy industry and so on from special cement industry. It can be judged whether the flexible strategy in special cement industry can explain the flexible strategy in those industries by further study.

REFERENCES

- J. Pine II, "Mass Customization: New Frontier of Enterprise Competition," China Renmin UP, 2000.

- W. Christopher and L. Hart, "Mass Customization: Conceptual Underpinnings, Opportunities and Limits," International Journal of Service Industry Management, Vol. 6, No. 2, 1995, pp. 36-45. doi:10.1108/09564239510084932

- K.-X. Dai and H.-X. Guo, "The Study of Core Competence of MC Enterprise," Modern Management Science, Vol. 25, No. 3, 2006, pp. 66-67.

- A. K. Sethi and S. P. Sethi, "Flexibility in Manufacturing: A Survey," International Journal of Flexible Manufacturing Systems, Vol. 2, No. 4, 1990, pp. 289-328.doi:10.1007/BF00186471

- M. ko Aoki and A. Haruhiko, "Modularization Times—The Essence of New Industry Structure," Shanghai Far East Press, Shanghai, 2003.

- G. U. Xin-jian and Q. I. Guo-Ning, "Quantitative Analysis Method of Mass Customization," China Mechanical Engineering, Vol. 12, No. 6, 2001, pp. 312-315.

- Y.-H. Zhang, "Mass Customization Strategy for Customer Requirement," Tsinghua University Press, Beijing, 2010.

- D. M. Anderson and J. Pine II, "New Frontier for Business Competition in the 21st Century—Agile Product Development under MC," China Machine Press, Beijing, 1999.

- C. Y. Baldwin and K. B.Clark, "Managing in an Age of Modularity," Harvard Business Review, Vol. 75, No. 5, 1997, pp. 84-93.
- C. Y. Baldwin and B. Kim, "Clark Design Rules: The Power of Modularity," MIT Press, Cambridge, 2000.
- J. Wang, "Production Operation Management," Tsinghua University Press, Beijing, 2011.
- T.-Q. Zan, "Systems Economy: The Essence of New Economy—On the Theory of Modularity," China Industrial Economy, Vol. 20, No. 9, 2003, pp. 23-29.

Chapter 12

ROADMAP FOR LEAN IMPLEMENTATION IN INDIAN AUTOMOTIVE COMPONENT MANUFACTURING INDUSTRY: COMPARATIVE STUDY OF UNIDO MODEL AND ISM MODEL

J. R. Jadhav[1], S. S. Mantha[2], S. B. Rane[1]

[1]Department of Mechanical Engineering, Sardar Patel College of Engineering, Bhavan's Campus, Munshi Nagar, Andheri (West), Mumbai 400 058, India

[2]All India Council for Technical Education (AICTE), 7th Floor, Chanderlok Building, Janpath, New Delhi 110 001, India

ABSTRACT

The demands for automobiles increased drastically in last two and half decades in India. Many global automobile manufacturers and Tier-1 suppliers have already set up research, development and manufacturing facilities in India. The Indian automotive component industry started implementing Lean practices to fulfill the demand of these customers. United Nations Industrial Development Organization (UNIDO) has taken proactive approach in association with Automotive Component Manufacturers Association of India (ACMA) and the Government of India to assist Indian SMEs in various clusters since 1999 to make them globally competitive. The primary objectives of this research are to study the UNIDO–ACMA Model as well as ISM Model of Lean implementation and validate the ISM Model by comparing with UNIDO–ACMA Model. It also aims at presenting a roadmap for Lean implementation in Indian automotive component industry. This paper is based on secondary data which include the research articles, web articles, doctoral thesis, survey reports and books on automotive industry in the field of Lean, JIT and ISM. ISM Model for Lean practice bundles was developed by authors in consultation with Lean practitioners. The UNIDO–ACMA Model has six stages whereas ISM Model has eight phases for Lean implementation. The

ISM-based Lean implementation model is validated through high degree of similarity with UNIDO–ACMA Model. The major contribution of this paper is the proposed ISM Model for sustainable Lean implementation. The ISM-based Lean implementation framework presents greater insight of implementation process at more microlevel as compared to UNIDO– ACMA Model.

INTRODUCTION

India has become manufacturing hub for many global automobile 'Original Equipment Manufacturers' (OEM) and Tier-1 suppliers. They aggressively demand cost-effective quality products with in time on line delivery from their vendors. Small medium enterprise (SME) is facing the pressure from its competitors; mainly large companies as they could provide products of greater value with lower cost as compared to SMEs (Jie et al. 2014). Challenges in today's global competition have forced or required manufacturing firms to look for appropriate manufacturing management strategies in order to enhance their efficiency and competitiveness (Lila 2012). To meet the challenge of offering high standards of quality, cost and delivery (QCD) to these multinational OEMs, Indian manufacturing SMEs must implement effective approaches, such as Lean manufacturing, to continually and systematically improve their operations (Saboo et al. 2014). To remain competitive in fierce global competition, the Indian automotive component industry needs to have paradigm shift in their thinking to improve its production capabilities, productivity, quality and scalability. Implementation of Lean systems has become inevitable to survive in the market. According to Panizzolo et al. (2012), the powerful Lean manufacturing approach that has proved successful as an operations' model in developed economies, as well as in some large Indian companies, is now increasingly being recognized by the small- and medium-sized enterprises (SMEs). In the last 10 years, even the manufacturers located in the developing countries such as China and India are also working to transform their manufacturing base from traditional low-cost, labourintensive 'Fordist' production to higher value, more flexible and more productive 'Lean' manufacturing systems (Panizzolo et al. 2012). Currently, in India about 150 companies in the automobile industry use Lean manufacturing, but it is yet to permeate other areas (Mehta et al. 2012).

There are certain obstacles in the implementation of Lean manufacturing practices. But they can be overcome by successful planning (Mehta et al. 2012). To overcome the challenges in dynamic global market and for propagation of Lean practices, the partnership program was designed and developed by United Nations Industrial Development Organization (UNIDO) in collaboration with Automotive Component Manufacturers Association of India (ACMA) and the

Government of India to support small and medium enterprises (SMEs) in the automotive component sector. The formation of regional clusters consists of 8–10 component manufacturers; each is the focal constituent of this program. Many Lean practices were implemented in more than 130 component manufacturers in three project phases (1999–2010) under the guidance of independent and external UNIDO-trained counselor. Many engineers and upgrading experts of UNIDO, staff of participating company, and members of support organizations have been trained in Lean practices to carry out shop-floor interventions. The progress of shop- floor interventions was monitored and tracked through monthly peer review meetings directed by the counselor at the company level on the basis of key performance indicators (KPI) and mutual learning with other cluster companies through the benchmarking. Many automotive component manufacturing companies were in position to enhance their capacity/capability and develop their skills to participate in a comprehensive continuous improvement program. Benchmarking the progress with peer companies in the same geographical zone empowered participating companies to develop and execute their own vision and mission, including better capacity utilization, better employee participation and healthier labour–management relationships.

The authors have developed a framework for sustainable Lean implementation based on interpretive structural modeling (ISM) methodology. Original theoretical development of ISM is credited to J.W. Warfield. Sage and Smith (1977), Sage (1977) and Farris and Sage (1975) have contributed to the development and application of the ISM methodology for a variety of purposes—especially those concerned with decision analysis and worth assessment in large-scale systems. Interpretive structural modeling is a tool which permits identification of structure within a system. The system may be large or small in terms of numbers of elements; and it is the larger, complex systems which benefit the most from interpretive structural modeling (Farris and Sage 1975). The ISM process transforms unclear, poorly articulated mental models of systems into visible, well-defined models useful for many purposes (Mishra et al. 2012; Ahuja et al. 2009). Interpretive structural modeling (ISM) can be used for identifying and summarizing relationships among specific variables, which define a problem or an issue (Sage 1977; Warfield 1974). It provides us a means by which order can be imposed on the complexity of such variables (Jharkharia and Shankar 2005; Mandal and Deshmukh 1994).

ISM has been used by researchers for understanding direct and indirect relationships among various variables in different industries. Satapathy and Mishra (2013) applied ISM to find interrelationship between design requirements for service quality of electricity utility sector. Kumar et al. (2013) used ISM to model variables for effective customer involvement in green concept implementation in supply chain management. Diabat et al. (2012) analyses the various risks involved in a food supply chain with the help of ISM. Faisal et al. (2007) used ISM to analyses the enablers for supply chain agility. ISM is a well-known technique, which can be applied in various fields. ISM has been applied by a number of researchers to develop a better understanding of the systems under consideration (Mudgal et al. 2010). ISM-based criteria model for the selection of reverse logistics provider was developed by Kannan et al. (2009). Ravi and Shankar (2005) analyzed the barriers of the reverse logistics in automobile industries using the ISM methodology, which shows the interrelationships of the barriers and their levels. Khurana et al. (2010) used ISM for modeling of information sharing enablers for building trust in Indian manufacturing industry. Interpretive structural modeling of risk sources in a virtual organization was carried out by Alawamleh and Popplewell (2011).

The prime purposes of this paper are to study and compare the UNIDO– ACMA Model of Lean implementation and validate the ISM Model through comparison with UNIDO– ACMA Model. It also aims at offering a framework for Lean implementation in automotive component industry. The research is based on secondary data, which include compilation of research articles, web articles, survey reports, thesis and books, etc. on automotive industry. The main aspect of the paper is the framework for sustainable Lean implementation through interpretive structural modeling. The key contribution of this paper is the structural model for Lean implementation with input variables, the process, the anticipated outcomes and its impact on business.

The salient features of the research are:

1.   It provides the understanding of Lean implementation process model adopted by UNIDO–ACMA.

2.   It represents the collective wisdom of Lean practitioners in the form of interpretive structural Model.

3.   It provides the crystal clear information of necessary input, the process, the expected as well as desired outcomes and its impact on the business.

4.   It offers phase-wise road map for sustainable Lean implementation based on ISM.

5.   It provides greater level of understanding of execution process at further microlevel as compare to UNIDO– ACMA Model.

This paper is further organized as follows. Research objective and research methodology are described in second section. Section three contains overview of India's automotive and auto component manufacturing industry. The UNIDO–ACMA Model approach (the partnership roadmap) is introduced briefly in section four. Section five presents the description of ISM-based model for Lean practice bundles. The comparison of UNIDO–ACMA Model and ISM Model for Lean implementation is discussed in section six. Section seven presents the general conclusions with research findings, implications. Section eight and nine contain the limitation of research study and suggestions for future research, respectively.

RESEARCH OBJECTIVES AND METHODOLOGY

Research Objectives

The main objectives of this paper are:

1. To study the UNIDO–ACMA Model and ISM Model of Lean implementation.
2. To compare UNIDO–ACMA Model and ISM Model of Lean implementation.
3. To validate the ISM Model by comparing with UNIDO–ACMA Model.

Research Methodology

The research brings out the similarities and differences of both models. This work can be considered as a roadmap to develop a deployment strategy for successful Lean implementation. First the relevant literature is reviewed. The literature survey was augmented using online computerized database like Taylor and Francis, Science Direct, Google Scholar, Bing, etc. using primary keywords such as just-intime (JIT), Lean, Lean supply, JIT manufacturing, ISM, automotive, automotive components, etc. and secondary key words like practices, benchmarking, modeling, framework, etc. This research is based on secondary data, which comprise of the research articles, web articles, doctoral thesis, survey reports and books on automotive industry in the domain of Lean, JIT and ISM. ISM Model for Lean practice bundles was developed by authors in consultation with four Lean practitioners based in Maharashtra state of India. The development of ISM Model is not presented as it is out of scope of this paper. The flowchart of research methodology used in this study is shown in Fig. 1. Final ISM Model for Lean practice bundles was developed as shown in Fig. 11 based on the input from the team members comprising of Lean/JIT practitioners and experts using ISM technique. ISM Model requires examination of direct and indirect relationships between the Lean practice bundles rather than considering these Lean practice bundles in separation. This analysis will help the managers to devise the strategy for rolling out the Lean/JIT in their organization. The ISM Model was compared with UNIDO–ACMA Model. Both models were analyzed. The discussion highlights the vast similarities and very little difference between two models.

OVERVIEW OF INDIA'S AUTOMOTIVE INDUSTRY

The automotive industry includes the OEM and auto component manufacturers. The automotive industry plays significant role in the development and economic growth of any country. Automobile industry in India grows with leap and bounds since 1991 due to change in industrial polices specifically liberalization and de-licensing of many segments by Government of India. Many foreign automobile companies also started joint ventures with prevailing Indian companies. The Indian automotive industry is witnessing a phase of rapid transformation and growth, mainly driven by stable economic growth and infrastructure development (Wipro 2013). Today India has become the manufacturing hub for global automotive giant companies like Toyota, Honda, Suzuki, Nissan, Hyundai, General Motors, Ford, Fiat, Renault, Mitsubishi, BMW, Daimler etc. The automobile manufacturers of two wheelers, three wheelers, passenger cars, passenger vehicles, multi-purpose vehicles (MPV), commercial vehicles and auto components have created vast opportunities of employment and growth.

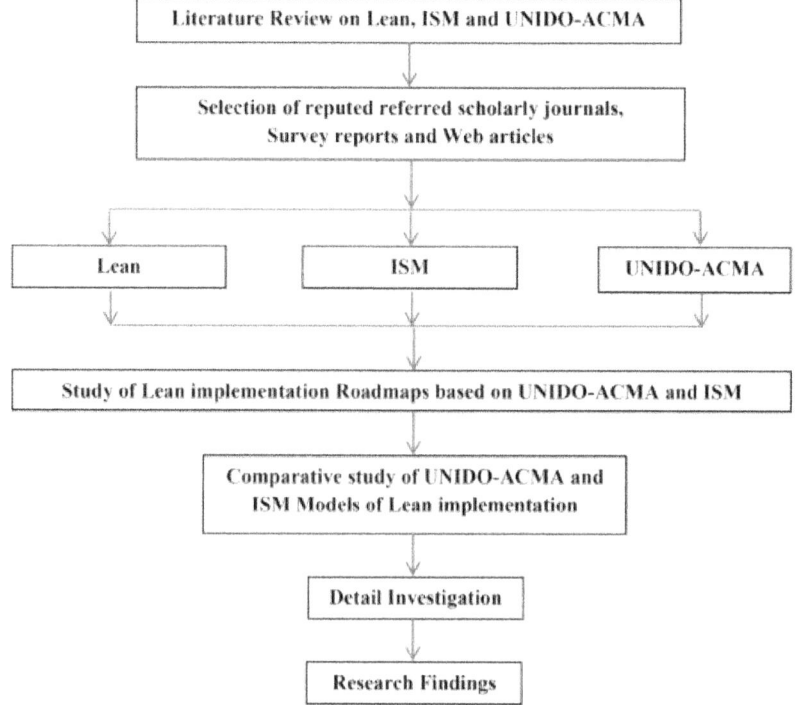

Figure 1: Flowchart for research methodology.

The automobile segment product categories are shown in Fig. 2. Automobile segments are mainly classified into four segment namely passenger vehicles, commercial vehicles, three wheelers and two wheelers. Passenger vehicles include cars, utility vehicles and multi-purpose vehicles. Commercial vehicles include light, medium and heavy commercial vehicles. Three wheelers comprise of passenger carriers and goods carriers. Two wheelers include motorcycles, scooters, mopeds and electric vehicles. Figure 3 shows the production and sales of 4/6 wheelers and 2/3 wheelers industry for year 2011–2012 and 2012–2013. In 4/6 wheelers industry, the contribution of passenger vehicles (car, MPVs and UVs) is much higher than other categories of 4/6 wheelers. In 2/3 wheelers industry, the contribution of motorcycles/step-throughs is much higher than other categories of 2/3 wheelers.

CII SR quarterly update (2009) reported that two wheelers (2Ws), being the most popular means of personal transport, alone account for about 77 % of the total automobile production in India, while passenger vehicles (PVs) account for over 15 % of the sales. However, owing to their lower sales realizations, two wheelers account for only around 32 % of the sales in terms of value while PVs account for around 62 % of the same.

Figure 4 shows Indian automobile industry exports from year 2007 to 2013. A continual increase in Indian automobile industry exports is observed. According to India Brand Equity Foundation (IBEF) (2014), the overall automobile exports grew by 2.03 percent during April–August 2013. According to IBEF (2014), India is set to be among top five automobile producers by 2015. The drivers for this rapid growth are high demand due to expanding middle class population with rising income, young population base, availability of skilled manpower, propagation of modern/advance technology. The statistics related to Indian Automotive Industry mentioned here are published by IBEF. According to the data published by ACMA, the production of passenger vehicles in India was recorded at 3.23 million in 2012–2013 and is estimated to grow at a compound annual growth rate (CAGR) of 13 percent during 2012–2021. The overall automobile exports grew by 2.03 percent during April–August 2013. The Indian automobile industry produced a total 1.69 million vehicles including passenger vehicles, commercial vehicles, three wheelers and two wheelers in August 2013 as against 1.56 million in August 2012, registering a growth of 8.18 percent over the same month last year. The cumulative foreign direct investment (FDI) inflow into the Indian automobile industry during April 2000 to July 2013 was recorded at US$ 8,932 million, amounting to 4.5 percent of the total FDI inflows (in terms of US$), as per data published by Department of Industrial Policy and Promotion (DIPP), Ministry of Commerce. According to a research report by Espirito Santo Securities, the

Indian auto and auto components Industry can be expected to surpass China's growth path by 2021 (Barreto 2013).

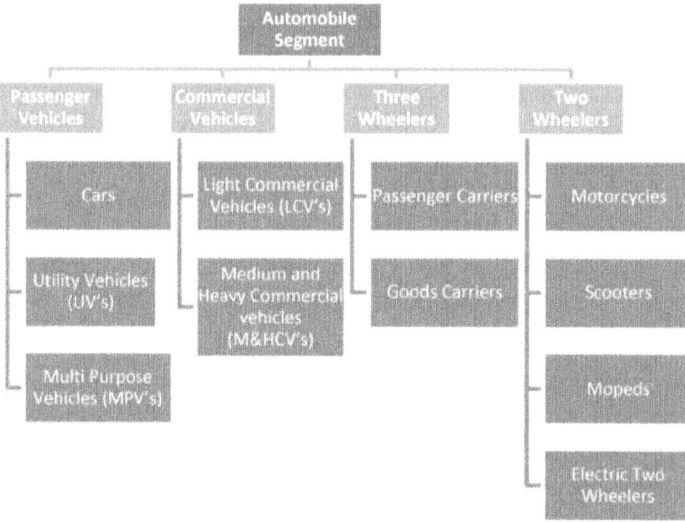

Figure 2: Automobile segment product categories (sources: SIAM, ImaCS and CII SR quarterly Update 2009). SIAM society of Indian automobile manufacturers, IMaCS ICRA management consulting services limited, CII confederation of Indian industries.

Automotive Component Manufacturers Association of INDIA (ACMA)

The All India Automobile & Ancillary Industries Association (AIA & AIA) was established in 1959. It was renamed as ACMA of India in 1982. It is the apex body of the Indian automotive component manufacturing industry. It has more than 650 member companies mainly from auto component manufacturers in India in the organized sector. ACMA member companies contribute more than 85 % of the total auto component production. ACMA member companies cater the needs of domestic vehicle manufacturers as—as OEM, to tier-one vendors, to state transport undertakings, railways, defence organizations and even to the spare parts market. A variety of auto components are also being exported to OEMs and after-markets worldwide. ACMA is actively involved in strengthening the automotive supply chain and enhancing competitiveness of the auto component industry in India through its technology division—ACMA Center for Technology (ACT). ACMA endorses Indian automotive component industry for trade promotion, technology up-gradation, quality improvement, collection and transmission of information. Each one of these is a significant reagent for development and growth of industry.

Automotive Industry Clusters in INDIA

Demand for two wheelers and four wheelers have been rising since 1990 in India. As a result, subcontracting propagates between automobile assemblers (i.e., OEM) and tier-one vendors and between tier-one and tier-two vendors. Small and medium enterprises (SMEs) moved into the rapidly growing market of auto components manufacturing in clusters. Although large enterprise with foreign and technical collaboration with foreign companies has maintained major share of critical components, SMEs which started business in the 1980s and 1990s, have grown. Some of them graduated from SMEs and became large enterprises (Uchikawa and Roy 2013). Today component suppliers offer their own design solutions, do value engineering, and instead of supplying only components, they are beginning to supply automotive systems (Wipro 2013).

The Indian and global OEM's established in the cluster zones are depicted in Fig. 5. At present auto component industry has developed in three major and one minor automotive cluster zones viz. North India, West India, South India and East India. North India zone is established in Haryana–Delhi National Capital Region (NCR) (Delhi, Gurgaon, Manesar, Faridabad, Rewari, Gautam Budh Nagar and Ghaziabad districts). West India Cluster is established in Maharashtra (Mumbai, Pune, Nasik and Aurangabad districts). New hub is being developed at Sanand, Gujarat. South India zone is established in Tamil Nadu (Chennai, Kanchipuram, Krishnagiri, Coimbatore and Tiruvallur districts).

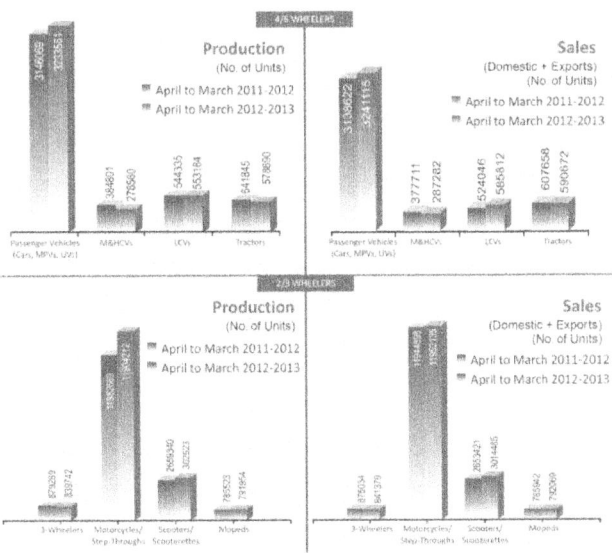

Figure 3: Automobile industry update (sources: SIAM; ACMA and The Industry 2012–2013).

East India Cluster is established in West Bengal (Kolkata and Jamshedpur). The auto and auto components manufacturers (Auto Components, September Auto components 2009) located in various cluster zones are mentioned below.

Major Cluster:

1. North/Central zone: Some of the companies established in this zone are Hero Honda, Honda SIEL, Honda Motorcycle & Scooter India, Maruti Suzuki, Tata Motors, Delphi Auto, JBM, Sona Koyo, Asahi India, Denso India, Lumax and Johnson Matthey.

2. West zone: Some of the companies established in this zone are Daimler, GM, Skoda, Bajaj Auto, Mahindra & Mahindra, Tata Motors, Volkswagen, Bharat Forge, DGP Hinoday, Kirloskar Brothers, TACO Group, SKF Bearings, Supreme Industries, Bright Brothers, Bosch Chassis, Tata Johnson and NRB Bearings.

3. South zone: Some of the companies established in this zone are Ashok Leyland, Ford, TVS Motors, Hyundai, Toyota Kirloskar Motor, Royal Enfield, Volvo, Brake India, Fenner, Rane Group, Visteon, Sundaram Fasteners, Delphi TVS, India Nippon, TI Group, LucasTVS and Ucal Fuel Systems.

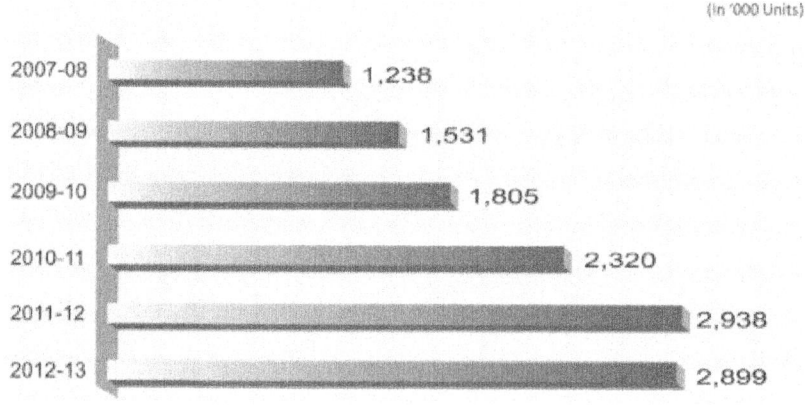

Figure 4: Indian automobile industry exports (sources: SIAM; ACMA and The Industry 2012–2013).

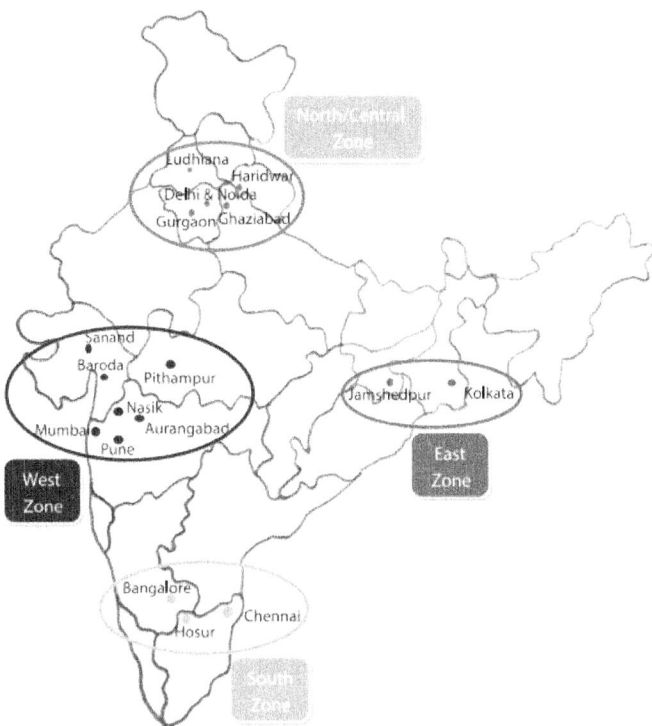

Figure 5: Automotive clusters in India (source: Auto components, September 2009, www.ibef.org).

Minor Cluster:

1.   East zone: Some of the companies established in this zone are Tata Motors, Hindustan Motors, Simpson & Co, International Auto Forgings, Ramkrishna Forgings, JMT and Exide.

The auto components market is split into six product segments namely engine and engine parts, transmission and steering parts, suspension and braking parts, equipment, electrical parts and others. Figure 6 provides the detail classification of each segment. Figure 7 depicts auto component industry update. The turnover of Indian auto component industry is almost consistent from 2010–2013. The highest growth in Indian auto component industry was observed in 2010–2011. However, today the auto component industry is experiencing a sharp negative growth. The auto component industry product range is shown in Fig. 8. Engine parts manufacturing is the largest contributor in component manufacturing. The export destination for 2012 is depicted in

Fig. 9. The largest export destination is Europe contributing almost one-third of auto component exports.

United Nations Industrial Development Organization (UNIDO)

United Nations Industrial Development Organization is the dedicated wing of the United Nations that endorses industrial development for poverty reduction, comprehensive globalization and environmental sustainability. The mandate of the UNIDO is to promote and accelerate sustainable industrial development in developing countries and economies in transition (http://www.unido.org). Many global automobile OEM companies have set up manufacturing facilities in India. These companies are facing global competition. Localized vendor base capable of supplying quality component is need of these OEM's to alleviate the pressure of global competition. The transformation of existing auto component industry into a tier structure, necessity to acquire world class levels in terms of—quality, productivity and reliability are some of the prerequisites for meeting the requirements of the global automobile OEM's. Moreover, the level of outsourcing for product design, development and manufacturing of various components, sub-assemblies and integrated systems by OEM's has increased many folds. In India, automotive industry is one of the sector in which UNIDO is highly active. UNIDO in collaboration with the Government of India and ACMA launched the UNIDO–ACMA partnership programme for the Indian auto component industry in November 1998. The purpose is to help Indian auto component manufacturers to alleviate the effects of the abovementioned challenges.

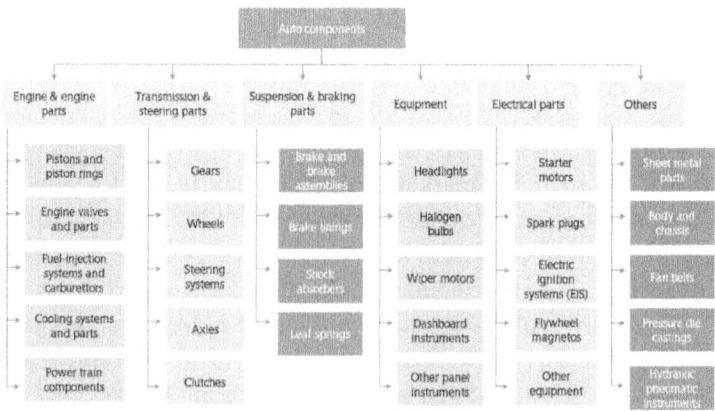

Figure 6: Product segments of auto components (source: Auto components', November 2011).

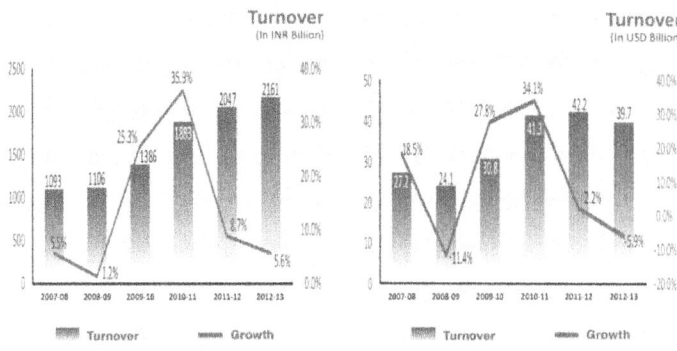

Figure 7: Auto component industry update (source: ACMA and The Industry 2012–2013).

THE UNIDO–ACMA MODEL APPROACH (THE PARTNERSHIP ROADMAP)

"The UNIDO partnership programme was given its shape through the definition of a 'Road Map'— envisaging specific training inputs to firms that, through a process of application at the company level and continuous counseling, would achieve tangible results in terms of improved performance and profits—ultimately leading to a paradigm shift across the automotive component industry to overcome the challenges pointed out above and to take advantage of the manifold business opportunities offered in both the national and global contexts" (UNIDO–ACMA Partnership Programme 2010).

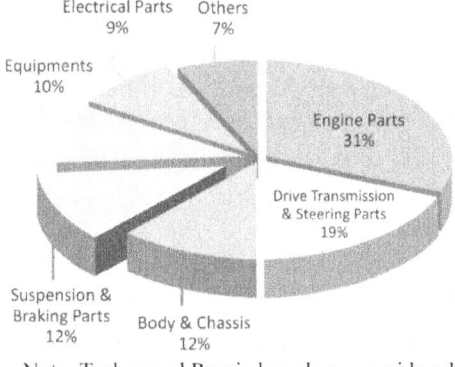

Note: Turkey and Russia have been considered part of Europe

Figure 8: The auto component industry product range (source: ACMA and The Industry 2012–2013).

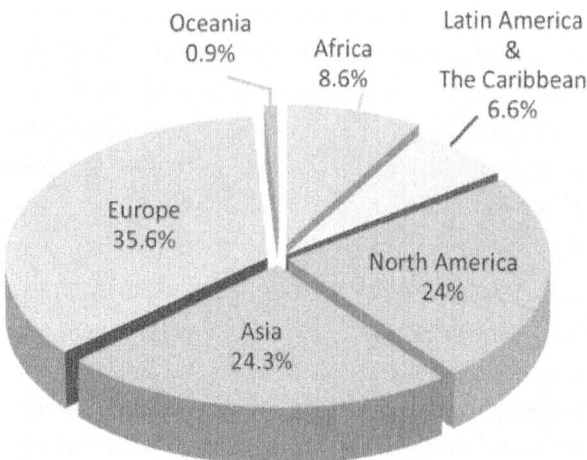

Figure 9: The export destination for 2012 (source: ACMA and The Industry 2012–2013).

Figure 10 provides the roadmap for the UNIDO–ACMA Model. It describes the model in terms of change process, the input required, the expected results and its impact on business. The UNIDO–ACMA Model is described in details as follows:

Employee Involvement

Total employee involvement (TEI) is necessary to create Lean organization. To bring about Lean transformation in organization culture, UNIDO–ACMA introduces intensive classroom as well as practical training on Small Group Activity (SCA), Kaizen (Quality Circle), Suggestion Schemes and safety as a first stage of Lean implementation under the guidance of the counselor. Exploring the improvement opportunities, enhancing employee motivation, improving employee involvement, productivity and reducing absenteeism rate are the primary objectives of first stage. The expected results are zero accidents, operator ownership and better involvement of employees in the form of higher number suggestions. These practices lay the foundation for business excellence through improvements in terms of reduced cost, better quality and on–time in-full delivery.

UNIDO–ACMA programme has strong focus on Kaizen on quality improvement, productivity enhancement, cost reductions and safety. Increased employee involvement is essential for improvements on the shop floor and successful implementation of Kaizen. A strong emphasis on the customer and bringing in innovative business ideas based on total employee involvement has rewarded many companies with a steady 'compound annual growth rate' (CAGR).

UNIDO–ACMA Model suggests implementation of safety practices in first stage. Every activity is supposed to evaluate with an emphasis on safety including well-documented safety policy, formation of safety committee, permanent safety measures and safety audits. The combined expert team work approach to identify, define and solve the shop-floor problems helps in rapid development of a "culture of continuous improvement". Development of "continuous improvement culture" is possible with the applications of input techniques like Five S, My machine Campaign (Help in enhancing safety) and waste elimination (3 M) on shop floor and in the organization. The expected outputs are daily management discipline, waste elimination and operator ownership.

5S- Workplace Management

Workplace organization (5S) and waste elimination (3 M) are combined in second stage of UNIDO–ACMA model. Rigorous and religious 5S implementation naturally laid the foundation for waste elimination. 5S is a Japanese concept of workplace organization and improvement. It helps in building a better working environment and a consistently high-quality process. 5S is a process for building and maintaining a workplace which is clean, sanitary, well organized, pleasant, safe and leads to high performance. Effort should be put to avoid all forms of wastages (Muda), reduce or eliminate strains (Muri) and do the things first time right and every time right (Mura or errors).

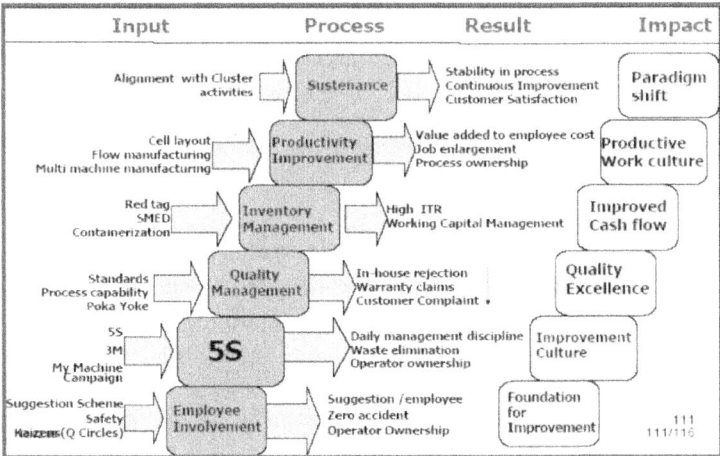

Figure 10: The UNIDO–ACMA Model approach (UNIDO–ACMA partnership programme 2010).

Quality Management

Achieving excellent quality with minimum utilization of resources and cost is ultimate purpose of quality management. Reducing variation and keeping mean on desired target are very important to maintain and improve the quality. Quality excellence can be achieved through the process of quality management by the application of process capability enhancement, standards and Poka Yoke in third stage of implementation of UNIDO–ACMA Model. The desirable outputs are reduction in in-house rejection, warranty claims and customer claims.

Inventory Management

At fourth stage of UNIDO–ACMA Model, inventory management is very important aspect of any business. High inventory blocks the most important resource, i.e., fund. Minimum inventory or high inventory turn ratio (ITR) is desirable to improved cash flow in the business. Costeffective inventory management is possible through application of use of red tags, quick change over, i.e., single-minute exchange of dies (SMED), small batch size, single-piece flow and containerization. This will help in minimizing the working capital to run the business.

Productivity Improvement

Establishment and sustenance of productive work culture are the building block of productivity improvement. This can be achieved through application of Lean techniques such as cell layout, flow manufacturing, multi-machine manufacturing and autonomous maintenance. The expected results are value added to employee cost, job enlargement and process ownership. Final goal of any improvement efforts is the cost reduction. UNIDO–ACMA Model considered inbuilt "Low cost practice".

Sustenance

Adoption of Lean means embracing change in cultural and attitude. To maintain the momentum of improvement is a challenging task. Many organizations implementing Lean fail to continue the Lean practices by back tracking to the old ways of doing the things. The ultimate expected impact of Lean implementation is paradigm shift in UNIDO–ACMA Model. This can be achieved through process of sustenance of Lean practices. The desirable results are stability in process, continuous improvement and customer satisfaction by aligning with cluster activities.

ISM-BASED MODEL FOR LEAN PRACTICE BUNDLES

Figure 11 depicts the ISM-based model for Lean practice bundles suggesting the priority order (in phase-wise manner) of Lean practice bundle implementation. The ISMbased model for Lean practice bundles is described in sequence of its implementation in this section.

Human Resource Management Practice Bundle

According to McNamara (2014), Lean manufacturing has failed to consider human aspects in the past, resulting in undesirable working conditions that can negatively affect commitment, and goes on to identify the role human behaviour plays in the performance of operating systems. Hasle et al. (2012) propose a strong focus on the human side, arguing that Lean must be regarded not only as tools but as an integrated part of a broader sociotechnical system (Oudhuis and Tengblad 2013). HR practices can be conceptualized as the glue that holds the other Lean production practices together (Longoni et al. 2013; de Treville and Antonakis 2006; Cua et al. 2001). The core of Lean adoption is through people, to have workers buy into the ideas and be part of the overall initiative (Tan et al. 2012). Knowledge of individual workers depends on level of team interaction, where the ability to capture and share rapid cycles of learning supports the cornerstone 'expert engineering workforce' (Ringen and Holtskog 2011; Kennedy 2003). Employees must be aware, motivated and trained systematically to make a smooth transition to Lean culture. Lean environment specifically JIT is very dynamic in nature due to pull mechanism. The specific focus must be on HRM practices having impact on cost, quality, delivery and flexibility. The practices in HRM bundle include Quality Circle, communication of goals, effective employee development programs, creating a culture of Lean improvement, rewards and recognition and effective labor management relations.

Creativity and Innovation Practice Bundle

In today's process manufacturing environment, innovation is viewed as critical to sustainable growth and business profitability (Gecevska et al. 2012). According to de Haan et al. (2012), a Lean system continually challenges workers to creatively use their talents, skills, and experience to signal anything that may be identified as waste and to remove impediments to a job well done, thereby improving process control and output quality.

The second phase of ISM Model developed by author advocates use of creativity and innovation practice bundle which consists of Kaizen (continuous improvement) and creative applications of advance technologies.

Organizations need to be proactive to delight the customers by satisfying their stated and implied needs. 'Early birds catch more worms' means the manufacture that brings more innovative products early in the market will capture significant market share. This can be achieved by supplying innovative products at affordable price. To tap and convert implicit needs of customer (internal as well as external) into innovative products requires lot of creative thinking. Application of Lean thinking is essential to bring innovative value added products or process early in the market. It was found that the openness, creativeness, and the challenging mentality were found positively influencing the Kaizen transfer (Yokozawa et al. 2010). Innovative applications of advance technologies like enterprise resource planning (ERP) softwares (for examples SAP and Baan), information technology (IT), radio frequency identification (RFID), etc. have brought about tremendous transparency, flexibility, rapid communication, etc. in systems of the organizations.

Health and Safety Practice Bundle

According to Longoni et al. (2013), when Lean is done right it need not be mean, rather Lean should continue to be considered a best practice, not just for its potential to improve operational outcomes but also because of its potential to improve the health and safety of the workers who run the system. Organizations must promote a healthy working environment that is reformed to suit employee's different physical and psychological requirements and capacity. A working environment without risks for health and safety should be a given at every workplace (Johansson and Abrahamsson 2007). Appropriate working environment has to be created keeping health and safety of employees in mind to enhance the productivity. The practices in Health and Safety bundle consist of Five S, Six S, Poka Yoke (mistake proofing) for safety, visual management, standardized work, ergonomic workstation or cell design and total productive maintenance (TPM).

Waste Elimination Practice Bundle

Lean operations are characterized by the elimination of wastes occurring in the manufacturing process, thereby facilitating cost reduction (Vinodh and Chintha 2011; Serrano et al. 2008). Lean manufacturing has been mandated by higher level management as a tool to be used in waste reduction (Green et al. 2010). Waste is nothing but the non-value added activity. Wastes are any procedures, materials, equipment, tools or activities that do not add value and can be eliminated or simplified (Al-Tahat and Eteir 2010). The activity which involves detrimental transformation in shape, size, dimension, taste, color, place or any other desirable characteristic from perception of the customer is called as non-value added activity. Usually wastes are hidden and inflate to the product cost. Understandably, consumers are not willing to pay for waste. So it is very important to detect the sources of waste and eliminate them. Muda, Muri and Mura (3 M) can be eliminated through TPM, standardized work and ergonomic workstation or cell design. Application of these techniques reduces the fatigue and stress on employees thereby improving the health of employees. Obviously, these improvements ultimately lead to enhanced productivity.

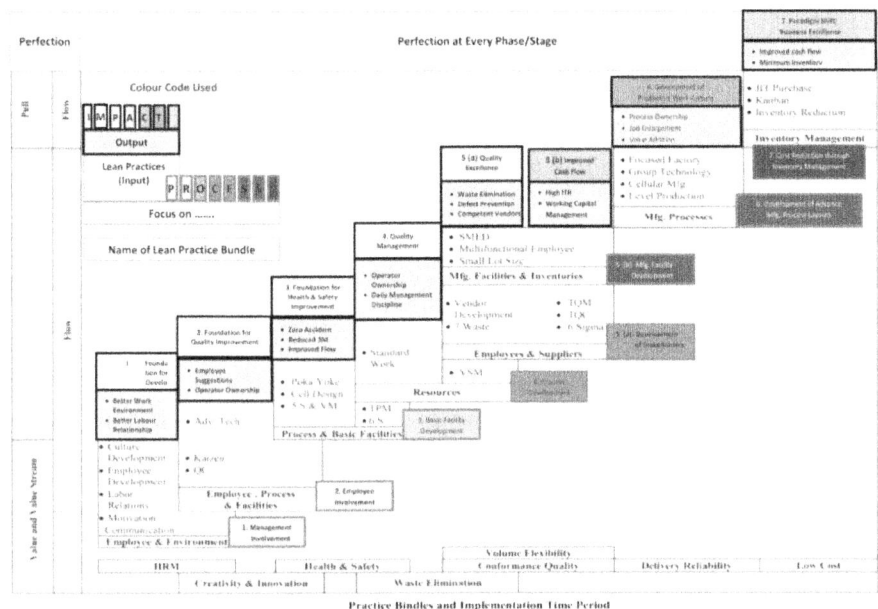

Figure 11: ISM Model—road map for Lean implementation.

Conformance Quality Practice Bundle

Conformance quality practice bundle is used in fifth phase of ISM Model. It emphasizes on the quality excellence through total quality management (TQM) as well as total quality control (TQC) based on statistical quality control (SQC), statistical process control (SPC) and Six Sigma. Total quality control emphasizes on improving internal quality aspect of business. Naturally, if any organization wishes to improve the quality of product, it can be achieved in isolation without involvement of suppliers. The focal company must develop it limited quality supply base (preferably single source) through supplier management or vendor development programs. Vendor development focuses on enhancing external quality aspect of business. These practices stress on quality improvement in all functions in the organization and in the entire supply chain. Supplier must be treated as seamless extension of the focal company.

Volume Flexibility Practice Bundle

Application of volume flexibility practice bundle is suggested at fifth level in ISM Model along with conformance quality practice bundle. Flexibility is the ability of a manufacturing system to cope with changes in the nature, mix, volume or timing of its activities (Garg et al. 2001). If the production changes to meet a peak demand, it must use a flexible process that can meet peak demands and still work efficiently during slacker times (Gupta 2011). The production setup must be capable of coping with dynamic demands in short time. Multi-skilled workers and flexible machines capabilities both are equally crucial in Lean manufacturing. The volume flexibility practice bundle consists of setup time reduction, multifunction employees and small lot size/single-piece flow.

Delivery Reliability Practice Bundle

The customers demand on-time in-full delivery at right place, in exact quantity with specified quality as per the term and conditions of mutual agreement. To fulfill this demand, the manufacturing (supplying) company has to maintain the delivery reliability. Delivery reliability of the organization can be realized with improved flow of material and information throughout the supply chain. The organization's ability to reliably deliver products to the customer may enhance by implementation of focused factory, Group Technology (cellular manufacturing) and uniform workload (production smoothing/Heijunka).

Focused factory attempts to reduce the complexities of the manufacturing process. This may include any or all of the following: simplifying the organizational structure, reducing the numbers of products or processes, and minimizing the complexities of the physical constraints (White et al. 2010). The fundamental idea of Group Technology is to subsume items (e.g., parts, processes, equipment or tools) into families according to their similarity and to take advantage of these groups in order to increase productivity in manufacturing (Bohnen et al. 2011; Shunk 1985). Production smoothing is very critical to make pull production work. More specifically, different variants of a model produced on a shared production line should be distributed over the sequence as evenly as possible (Yavuz 2013). Reduction in batch size and creating the balanced singlepiece flow of the items in entire system are the objectives of production smoothing.

Low-cost Practice Bundle

In the face of current economic crunch, companies are looking for the ways to cope with the situation by opting for cost reduction and quality products at the same time (Qureshi et al. 2013). It is well known that profit earned is equal to the difference of market price and cost price of product. Profit can be increased by reducing cost price of product (Singh and Khanduja 2010). The ISM Model developed by authors considered it as a separate entity at seventh phase/level. It considers the avenues of inventory reduction for radical cost reduction as against incremental cost reduction. JIT implementation demands very high degree of synchronization, communication, commitment, discipline, etc. among supply chain partners. Hence, the focal organization has to attain high level of maturity to implement JIT. The cost reduction is dependent on many successful initiatives of quality improvement programs. That is why it is placed at the final level/phase in ISM Model. The best practices for cost reduction are inventory reduction, Kanban and just-in-time purchasing.

"Sustainability is not a method or a tool, it is the state of a company in which the efficiency of resources is maximized, customers are satisfied to a great extent, an improved condition is long-lasting, success is maintained and competitive advantage is sustained" (Hilton 2013; AlSagheer 2011). Sustainability as the embedment of the practice or 'the way we do things round here' (Radnor 2011; Feldman and Pentland 2003). ISM Model advocates perfection at all phases of Lean implementation. After achieving certain degree of success at each level, it suggests continual improvement at next higher level. Continual improvement acts as a driving force to maintain the zest for improvement through Lean.

DISCUSSION

The comparison of UNIDO–ACMA Model and ISM Model for Lean implementation is shown in Table 1. Both, UNIDO–ACMA and ISM approaches are similar in nature. However, ISM approach splits the first-stage interventions (Employee Involvement) of UNIDO–ACMA approach into three phases viz. first phase as human resource management practice bundle, second phase as Kaizen (continuous improvement) as creativity and innovation practice bundle and then partially third health–safety practice bundle. Quality Circle (group-oriented Kaizen) is lower version of intervention as compared to managementoriented Kaizen. Higher version of intervention demands more creative/innovative ideas and systematic thinking to solve the abnormalities or problems. Belongings of process, conducive environment, availability of resources, availability of expert resource persons and systematic problem solving approach are some of the essential prerequisites to implement higher version of interventions.

Numbers of creative ideas are input for an innovative product. New and innovative products/services attract more and more new customers. It also ensures the survival and growth of the organization. High level of creative thinking at enterprise level throughout the supply chain network is necessary to bring more and more innovative business ideas. This can be accomplished through management Kaizen through involvement of intellectual higher cadre employees. High degree of employee involvement is a necessary for shop-floor innovations and successful implementation of management Kaizen. Hence, the authors suggest execution of creativity and innovation practice bundle only after successful installation of first practice bundle, i.e., human resource management which lays the foundation for process and business excellence as well as nurture the organization culture in favor of creative Lean thinking. The top management should devise a system for talent management and knowledge management in the organization. First two phases of ISM Model works on the development of Lean awareness, Lean culture and continuous improvement. This can be achieved through QC, Kaizen and effective human resources practices. Once the employees are aware and trained in creative thinking, then they can be a great asset in the development of accidentfree and healthy workplace by application techniques like Five S, Six S, TPM, Poka Yoke, visual management, standardized work and ergonomic workstation or cell design, safety Kaizen, etc. Hence, health and safety practice bundle is kept at third phase of Lean implementation.

According to Rose and Deros (2010), the organization no matter the sizes, large or small is crucial to eliminate waste, to increase the profit or return on investment (ROI). Lean relentlessly pursues for eradication of various wastages to make manufacturing system more efficient and effective. Effective waste elimination through high level of interventions like Gemba Walk for identification of Fugai (abnormalities) and value stream mapping is possible. The identification of opportunities for seven types of waste is exposed through value stream mapping, brain storming, Gemba walk for identification of Fugai, etc. Hence, waste elimination practice bundle is placed in phase number four of ISM Model.

Both, quality management of UNIDO-AMA Model and conformance quality practice bundle of ISM Model approaches are similar in nature. With respect to Volume flexibility practice bundle, both approaches are almost identical except development of multifunction employees to enhance volume flexibility. Multi-skilled operators create wide choice for manufacturing different products having different processes. Therefore, the company can satisfy the needs of customers with wide range of products with same human resources and other resources required. Thus, the company can reduce the wastage of skill and talent of its employee.

Delivery reliability practice bundle is useful to ensure the on-time in-full delivery of products. In Group Technology, the components with similar characteristics are grouped together and manufactured in a cell with standard processes. As a result, a small minuscule "focused factories" are being formed as independent manufacturing units within large plants. Cellular manufacturing is based on the principles of Group Technology. The cell layout of machines is usually an U-shaped, to enable smooth onepiece flow production. Cellular manufacturing deploys quick setup, simple machine maintenance by operator himself (autonomous Maintenance), multi-process tools, working on multiple processes and efforts of flexible or multi-functional workforce by means of multi-machine working environment which results in reduction of cumulative lead time. Hence, ISM Model (phase 6) and UNIDO–ACMA Model (stage 5) are similar in nature to improve delivery reliability.

Table 1: Comparison of UNIDO–ACMA Model and ISM Model

Sequence of implementation		Lean practices bundle	Process	Input (Lean practices)	Output (Result)	Impact
UNIDO–ACMA Model	ISM Model					
1			Employee involvement	1. Small Group Activity (SGA)	1. Zero accidents	Foundation for Improvement
				2. Kaizen (Quality Circle)	2. Operator ownership	
				3. Suggestion Schemes	3. Employees suggestion	
				4. Safety		
	1	Human resource management practice bundle	Management involvement	1. Cultural development	1. Better work environment	Foundation for development of culture and employee
				2. Employee development	2. Better labor relationship	
				3. Labor relationship		
				4. Motivation		
				5. Communication		
	2	Creativity and innovation practice bundle	Employee involvement	1. Quality Circle	1. Employee suggestions	Foundation for quality improvement
				2. Kaizen	2. Operator ownership	
				3. Advance technology		
2			Workplace organization (5S) and waste elimination (3 M)	1. Five S	1. Daily management discipline	Improvement culture
				2. My machine campaign (Help in enhancing safety)	2. Waste elimination	
					3. Operator ownership	
				3. Waste elimination (3 M)		
	3	Health and safety practice bundle	Basic facility development	1. Five S and Six S	1. Zero accident	Foundation for health & safety improvement
				2. Poka Yoke for Equipment	2. Reduced 3 M	
				3. Visual management	3. Improved flow	
				4. Standardized work		
				5. Ergonomic workstation or cell design		
				6. TPM		

3		Quality management	1. Standards	Reduction in	Quality excellence
			2. Process capability	1. In-house rejection	
			3. Poka-Yoke	2. Warranty claims	
				3. Customer claims	
4	Waste elimination practice bundle	Process development	1. Standard work	1. Operator ownership	Quality management
			2. Value stream mapping (VSM)	2. Daily management discipline	
			3. Brainstorming		
			4. Gemba walk—Jugai identification		
5(a)	Conformance quality practice bundle	Development of stakeholders (employees and suppliers)	1. TQM and TQC	1. Waste elimination	Quality excellence
			2. SQC/SPC	2. Defect prevention	
			3. Supplier management/vendor development	3. Competent vendors	
			4. Six Sigma		
4		Inventory Management	1. Red Tag	1. High inventory turn ratio (ITR)	Improved cash flow
			2. SMED	2. Working capital management	
			3. Containerization		
5(b)	Volume flexibility practice bundle	Mfg. facility development	1. Setup time reduction (SMED)	1. High inventory turn ratio (ITR)	Improved cash flow
			2. Multifunction employees	2. Working capital management	
			3. Small lot size		
			4. Single-piece flow		
			5. Containerization		
5		Productivity Improvement	1. Cell layout	1. Value added to employee cost	Productive work culture
			2. Flow manufacturing	2. Job enlargement	
			3. Multi-machine Manufacturing	3. Process ownership	
			4. Autonomous Maintenance		
6	Delivery reliability practice bundle	Development of advance mfg. process layouts	1. Focused factory	1. Process ownership	Development of productive work culture
			2. Group Technology	2. Job enlargement	
			3. Uniform workload	3. Value addition	
6		Sustenance	Align with cluster activities	1. Stability in Process	Paradigm shift
				2. Continuous Improvement	
				3. Customer Satisfaction	
7	Low-cost practice bundle	Cost reduction through inventory management	1. Inventory reduction	1. Improved cash flow	Paradigm shift: business excellence
			2. Kanban	2. Minimum inventory	
			3. JIT purchasing		
8	Perfection at all levels of Lean implementation	PDCA Cycle			

Low-cost practice bundle is vital to keep the cost of products under control. JIT is not only inventory reduction mechanism. The JIT is based on "zero concept", which aims to achieve zero defects, zero queues, zero breakdown, zero inventories and so on. It ensures the supply of right parts in the right quantity in the right place and at the right time. In fact, inventory reduction is perhaps the most visible result that JIT brings about. JIT in reality is a philosophy of supply chain excellence (Roy and Guin 1999). Initially conceptualized as an approach to reduce lead time and decrease inventory within a manufacturing plant (Schonberger 1982), JIT has expanded to include a broader set of production and purchasing practices (Kaynak and Hartley 2006; Fullerton et al. 2003; White et al. 1999; Schonberger and Ansari 1984; Hahn et al. 1983; Schonberger and Gilbert 1983). The deployment and implementation of JIT through Kanban in entire supply network or chain are very essential. Implementation of JIT should not be limited to the focal company but also must be extended to its supplier and distribution network to create true Lean enterprise. Of course, JIT implementation requires lot of investment in various resources to convert existing infrastructure to JIT compactible infrastructure.

However, phenomenal benefits will be received through JIT.

Sustenance and perfection are equally important as initiation of improvement. Sustenance and perfection play a vital role in reaping the benefits of improvement in long term. Both models are complementary to each other at this junction. To implement Lean successfully, sustenance of Lean practices and perfection at each phase are absolutely essential. The horizon of UNIDO–ACMA is wider than that of ISM Model. UNIDO–ACMA seeks sustenance of Lean practices throughout the cluster as a single family. Discussions on common platform, benchmarking studies, sharing the lessons learn, etc. are some of the initiatives undertaken by ACMA. ISM Model focuses on successful implementation of Lean in a company.

CONCLUSIONS

Many global automobile OEM and Tier-1 suppliers are setting and expanding the manufacturing facilities in India. The Indian automotive component industry started implementing Lean systems to satisfy the demands of costeffective quality auto components. UNIDO, Automotive Component Manufacturers Association of India (ACMA) and the Government of India launched a joint partnership programme for proliferation of Lean practices in Indian SMEs in various clusters since 1999 under the guidance of trained UNIDO counselors. The primary objectives of this research are to study and compare the UNIDO–ACMA Model of Lean implementation and corroborate the ISM Model by comparing it with UNIDO–ACMA Model. The secondary objective is to propose an ISM-based framework developed by authors, for successful and sustainable Lean implementation in Indian automotive component industry. This research is based on compilation of secondary data, which encompass the research articles, web articles, doctoral thesis, survey reports and books on automotive industry in the realm of Lean, JIT and ISM. ISM Model for Lean practice bundles was developed by authors from inputs received from Lean practitioners.

Both, ISM-based Lean Model and UNIDO–ACMA Model were studied and compared to validate ISM-based Lean implementation model in this research. ISM-based Lean Model and UNIDO–ACMA Model approaches are almost similar. The ISM-based Lean implementation model is thus successfully validated through high degree of resemblance with UNIDO–ACMA Model. The significant contribution of this paper is the ISM Model for sustainable Lean implementation. The ISM-based Lean implementation structure offers greater understanding of implementation process at more microlevel as compared to UNIDO– ACMA Model. ISM Model focuses on successful implementation of Lean in a company where as UNIDO–ACMA seeks sustenance of Lean practices throughout the cluster. To implement Lean successfully, sustenance of Lean practices and perfection at each phase are absolutely essential. ISM-based Lean model advocates implementation of eight Lean practice bundles in sequential order. Application of ISM-based Lean implementation framework offers more clear picture.

LIMITATIONS

This paper primarily focused on Lean implementation in manufacturing segment. The Lean implementation issues in other sectors may slightly differ than manufacturing segment. The issues may vary from country to country, work culture of the organization and geographic location within the country.

FUTURE WORK

Lean implementation roadmap based on ISM can be applied in any manufacturing or service industry. The case studies of such efforts can be prepared.

ACKNOWLEDGMENTS

The authors wish to thank the anonymous referees for their valuable feedback and constructive comments which helped to improve the structure and quality of this paper.

REFERENCES

1. AlSagheer A (2011) Applying six sigma to achieve enterprise sustainability: preparations and aftermath of six sigma projects. J Bus Econ Res 9(4):51–58. Available at: http://www.cluteinstitute.com/ojs/index.php/JBER/article/download/4209/4276. Accessed 01 May 2014

2. ACMA and The Industry (2012–2013). ACMA Annual Report. http://acma.in/pdf/ACMA_Annual_Report_2012-13.pdf. Accessed 10 Dec 2013

3. Ahuja V, Yang J, Shankar R (2009) Benefits of collaborative ICT adoption for building project management. Constr Innov 9(3): 323–340

4. Alawamleh Mohammad, Popplewell Keith (2011) Interpretive structural modelling of risk sources in a virtual organisation. Int J Prod Res 49(20):6041–6063

5. Al-Tahat, MD, Eteir, M (2010) Investigation of the potential of implementing Kaizen principles in Jordanian companies. Int J Prod Dev

10(1/2/3):87–100 Auto components (2009) www.ibef.org. Accessed 15 Dec 2013

6. Barreto C (2013) Indian auto components industry could outpace China in the coming years. ET Bureau. http://articles.economic times.indiatimes. com/2013-01-10/news/36258336_1_auto-compo nents-industry-indian-auto-capita-income. Accessed 22 Dec 2013

7. Bohnen F, Maschek T, Deuse J (2011) Leveling of low volume and high mix production based on a group technology approach. CIRP J Manufact Sci Technol 4(3):247–251

8. CII SR quarterly update (2009) 'CII southern region industry and economic update-auto & auto components'. http://www.cii.in/ webcms/ Upload/Auto%20and%20Auto%20Ancillaries% 20December20091. pdf. Accessed 6 Dec 2013

9. Cua KO, McKone KE, Schroeder RG (2001) Relationships between implementation of TQM, JIT, and TPM and manufacturing performance. J Oper Manag 19(6):675–694

10. de Haan J, Naus F, Overboom M (2012) Creative tension in a Lean work environment: implications for logistics firms and workers. Int J Prod Econ 137(1):157–164

11. de Treville S, Antonakis J (2006) Could lean production job design be intrinsically motivating? Contextual, configurational, and levelsof-analysis issues. J Oper Manag 24(2):99–123

12. Diabat A, Govindan K, Panicker V (2012) Supply chain risk management and its mitigation in a food industry. Int J Prod Res 50(11):3039–3050

13. Faisal MN, Banwet DK, Shankar R (2007) Supply chain agility: analysing the enablers. Int J Agile Syst Manag 2(1):76–91

14. Farris DR, Sage AP (1975) On the use of interpretive structural modeling for worth assessment. Comput Electr Eng 2(2–3):149–174

15. Feldman MS, Pentland BT (2003) Reconceptualizing organisational routines as a source of flexibility and change. Adm Sci Q 48:94–118

16. Fullerton RR, McWatters CS, Fawson C (2003) An examination of the relationships between JIT and financial performance. J Oper Manag 21:383–404

17. Garg S, Vrat P, Kanda A (2001) Equipment flexibility vs. inventory: a simulation study of manufacturing systems. Int J Prod Econ 70(2):125–143

18. Gecevska V, Veza I, Cus F, Anisic Z, Stefanic N (2012) Lean PLM— information technology strategy for innovative and sustainable business

environment. Int J Ind Eng Manag 3(1):15–23

19. Green JC, Lee J, Kozman TA (2010) Managing lean manufacturing in material handling operations. Int J Prod Res 48(10):2975–2993

20. Gupta AK (2011) A conceptual JIT model of service quality. Int J Eng Sci Technol 3(3):2214–2227

21. Hahn CK, Pinto PA, Bragg DJ (1983) Just-in-time production and purchasing. J Purch Mater Manag 19(3):2–10

22. Hasle P, Bojesen A, Langaa P, Bramming P (2012) Lean and the working environment: a review of the literature. Int J Oper Prod Manag 32(7):829–849

23. Hilton RJ (2013) Factors critical to a sustainable deployment of lean six sigma in Australian business. Doctoral Thesis in Business Administration, Monash University India Brand Equity Foundation (IBEF) (2014) Automobile Industry in India. http://www.ibef.org/industry/india-automobiles.aspx. Accessed 14 Jan 2014

24. Jharkharia S, Shankar R (2005) IT-enablement of supply chains: understanding the barriers. J Enterp Inf Manag 18(1):11–27

25. Jie, JCR, Kamaruddin, S, Azid, IA (2014) Implementing the Lean six sigma framework in a small medium enterprise (SME)—a case study in a printing company. Proceedings of the 2014 international conference on industrial engineering and operations management Bali, January 7–9. http://iieom.org/ieom2014/ pdfs/86.pdf. Accessed 02 June 2014

26. Johansson J, Abrahamsson L (2007) The good work—an obsolete vision? 5th International conference on work and learning, Cape Town, South Africa

27. Kannan G, Pokharel S, Kumar PS (2009) A hybrid approach using ISM and fuzzy TOPSIS for the selection of reverse logistics provider. Res Conserv Recycl 54(1):28–36

28. Kaynak Hale, Hartley JL (2006) Using replication research for just-intime purchasing construct development. J Oper Manag 24:868–892

29. Kennedy MN (2003) Product development for the lean enterprise, why Toyota's system is four times more productive and how you can implement it. The Oaklea Press, Richmond

30. Khurana MK, Mishra PK, Jain R, Singh AR (2010) Modeling of information sharing enablers for building trust in Indian manufacturing industry: an integrated ISM and Fuzzy MICMAC approach. Int J Eng Sci Technol 2(6):1651–1669

31. Kumar, S., Luthra, S. and Haleem, A. (2013) Customer involvement in greening the supply chain: an interpretive structural modelling methodology. J Ind Eng Int 9(6):1–13. http://www.jiei-tsb.com/ content/ pdf/2251-712X-9-6.pdf. Accessed 2 June 2014

32. Lila B (2012) A survey on implementation of the lean manufacturing in automotive manufacturers in the eastern region of Thailand. 2nd international conference on industrial technology and management (ICITM 2012), IPCSIT vol 49, IACSIT Press. doi:10.7763/IPCSIT.2012. V49.9

33. Longoni A, Pagell M, Johnston D, Veltri A (2013) When does lean hurt?— an exploration of lean practices and worker health and safety outcomes. Int J Prod Res. doi:10.1080/00207543.2013. 765072

34. Mandal A, Deshmukh SG (1994) Vendor selection using interpretive structural modeling (ISM)'. Int J Oper Prod Manag 14(6):52–59

35. McNamara P (2014) Psychological factors affecting the sustainability of 5S lean. Int J Lean Enterp Res 1(1):94–111

36. Mehta, RK, Mehta, D, Mehta, NK (2012) An exploratory study on implementation of lean manufacturing practices (with special reference to automobile sector industry). Yo¨netim ve Ekonomi 19(2):289–299. http:// www2.bayar.edu.tr/yonetimekonomi/dergi/   pdf/C19S22012/289_299. pdf. Accessed 2 June 2014

37. Mishra S, Datta S, Mahapatra SS (2012) Interrelationship of drivers for agile manufacturing: an Indian experience. Int J Serv Oper Manag 11(1):35–48

38. Mudgal RK, Shankar R, Talib P, Raj T (2010) Modelling the barriers of green supply chain practices: an Indian perspective. Int J Logist Syst Manag 7(1):81–107

39. Oudhuis Margareta, Tengblad Stefan (2013) Experiences from Implementation of Lean production: standardization versus self-management: a Swedish case study. Nord J Work Life Stud 3(1):31–48

40. Panizzolo R, Garengo P, Sharma MK, Gore A (2012) Lean manufacturing in developing countries: evidence from Indian SMEs. Prod Plan Control Manag Oper 23(10–11):769–788

41. Qureshi, MI, Iftikhar, M, Bhatti, MN, Shams, T, Zaman, K (2013) Critical elements in implementations of just-in-time management: empirical study of cement industry in Pakistan. SpringerPlus, 2(645):1–14. http:// www.springerplus.com/content/pdf/ 2193-1801-2-645.pdf. Accessed 2 June 2014

42. Radnor Z (2011) Implementing lean in health care: making the link between the approach, readiness and sustainability. Int J Ind Eng Manag 2(1):1–12. http://www.iim.ftn.uns.ac.rs/casopis/volume2/ ijiem_vol2_no1_1.pdf. Accessed 01 May 2014

43. Ravi V, Shankar Ravi (2005) Analysis of interactions among the barriers of reverse logistics. Technol Forecast Soc Chang 72(8): 1011–1029

44. Ringen Geir, Holtskog Halvor (2011) How enablers for lean product development motivate engineers. Int J Comput Integr Manuf. doi:10.108 0/0951192X.2011.593046

45. Rose, AMN, Deros, B, Rahman, MN.Ab (2010) Development of framework for lean manufacturing implementation in SMEs. The 11th Asia pacific industrial engineering and management systems conference. Melaka

46. Roy RN, Guin KK (1999) A proposed model of JIT purchasing in an integrated steel plant. Int J Prod Econ 59(1–3):179–187

47. Saboo A, Garza-Reyes JA, Er A, Kumar V (2014) A VSM improvement-based approach for lean operations in an Indian manufacturing SME. Int J Lean Enterp Res 1(1):41–58

48. Sage AP (1977) Interpretive structural modeling: methodology for large-scale systems. McGraw-Hill, New York, pp 91–164

49. Sage AP, Smith TJ (1977) On group assessment of utility and worth attributes using interpretive structural modelling. Comput Electr Eng 4(3):185–198

50. Satapathy S, Mishra P (2013) A customer oriented systematic framework to extract business strategy in Indian electricity services. J Ind Eng Int, 9(33):1–18. http://www.jiei-tsb.com/ content/pdf/2251-712X-9-33.pdf. Accessed 2 June 2014

51. Schonberger RJ (1982) The transfer of Japanese manufacturing management approaches to US industry. Acad Manag Rev 7:479–487

52. Schonberger RJ, Ansari A (1984) Just-in-time purchasing can improve quality. J Purch Mater Manag 20:2–7

53. Schonberger RJ, Gilbert JP (1983) Just-in-time purchasing: a challenge for US industry. Calif Manag Rev 26:54–68

54. Serrano I, Ochoa C, De Castro R (2008) Evaluation of value stream mapping in manufacturing system redesign. Int J Prod Res 46(16):4409–4430

55. Shunk DL (1985) Group technology provides organized approach to realized benefits of CIMs. Ind Eng 17(April):74–80

56. Singh BJ, Khanduja D (2010) DMAICT: a road map to quick changeovers. Int J Six Sigma Compet Advant 6(1/2):31–52

57. Tan KH, Denton P, Rae R, Chung L (2012) Managing lean capabilities through flexible workforce development: a process and framework. Prod Plan Control Manag Oper. doi:10.1080/ 09537287.2011.646013

58. Uchikawa S, Roy S (2013) 'The development of auto component industry in India'. http://ihdindia.org/Formal-and-InformalEmployment/Paper-6-The-Development-of-Auto-ComponentIndustry-in-India.pdf. Accessed 6 Dec 2013

59. UNIDO–ACMA partnership programme (2010) Project phases I-III (1999–2010). Final Report. https://www.unido.org/fileadmin/ user_media/Services/PSD/AUTOMOTIVE/Final%20Report%20 India_Layouted.pdf. Accessed 6 Dec 2013

60. Vinodh S, Chintha SK (2011) Leanness assessment using multigrade fuzzy approach. Int J Prod Res 49(2):431–445

61. Warfield JW (1974) Developing interconnected matrices in structural modeling. IEEE Trans Syst Men Cybern 4(1):81–87

62. White RE, Pearson JN, Wilson JR (1999) Just-in-time manufacturing: a survey of implementation in small and large US manufacturers. Manage Sci 45(1):1–15

63. White RE, Ojha Divesh, Kuo Ching-Chung (2010) A competitive progression perspective of JIT systems: evidence from early US implementations. Int J Prod Res 48(20):6103–6124

64. Wipro (2013) Step on the gas—steer into the future of the automotive industry in India. Future thought of business, a wipro thought leadership initiative, IND/TMPL/JAN-DEC2013, 5th edn. http:// www.wipro. com/ftob-automotive-isem/downloads/FTOB%20 Automotive%20 Report%20WEB.pdf. Accessed 6 Dec 2013

65. Yavuz Mesut (2013) Iterated beam search for the combined car sequencing and level scheduling problem. Int J Prod Res. doi:10. 1080/00207543.2013.765068

66. Yokozawa K, Steenhuis, Harm-Jan and de Bruijn, Erik J (2010) The influence of national culture on Kaizen transfer: an exploratory study of Japanese subsidiaries in The Netherlands. Proceedings of The 15th annual Cambridge International Manufacturing Symposium: Innovation in global manufacturing- New models for sustainable value capture, Cambridge, 23–24 September. http://www2.ifm.eng.cam.ac.uk/cim/ symposium2010/proceedings/ 24_yokozawa.pdf. Accessed 6 May 2013

CITATION

CHAPTER 1

D. Dolage and A. Sade, "A Frontier Approach to Measuring Impact of Adoption of Flexible Manufacturing Technology on Technical Efficiency of Malaysian Manufacturing Industry," Technology and Investment, Vol. 3 No. 4, 2012, pp. 266-275. doi: 10.4236/ti.2012.34037.

CHAPTER 2

D. Dolage and A. Sade, "The Impact of Adoption of Flexible Manufacturing Technology on Price Cost Margin of Malaysian Manufacturing Industry," Technology and Investment, Vol. 3 No. 1, 2012, pp. 26-35. doi: 10.4236/ti.2012.31005.

CHAPTER 3

Chatterjee, P & Chakraborty, S. (2014). Flexible manufacturing system selection using preference ranking methods : A comparative study.International Journal of Industrial Engineering Computations , 5(2), 315-338.

CHAPTER 4

Jingshan Li and Ningjian Huang, "Quality Evaluation in Flexible Manufacturing Systems: A Markovian Approach," Mathematical Problems in Engineering, vol. 2007, Article ID 57128, 24 pages, 2007. doi:10.1155/2007/57128.

CHAPTER 5

Ranbir Singha, Rajender Singhb and B.K. Khanc, Meta-Hierarchical-Heuristic-Mathematical- Model of Loading Problems In Flexible Manufacturing System For Development Of An Intelligent Approach, doi: 10.5267/j.ijiec.2015.11.003

CHAPTER 6

Singh, S & Prasher, L. (2014). A production inventory model with flexible manufacturing, random machine breakdown and stochastic repair time.International Journal of Industrial Engineering Computations , 5(4), 575-588.

CHAPTER 7

F Zammori, M. Braglia, M. Frosolini, A measurement method of routing flexibility in manufacturing systems, DOI: 10.5267/j.ijiec.2011.03.001.

CHAPTER 8

Mgwatu, M. (2011). Integration of part selection, machine loading and machining optimisation decisions for balanced workload in flexible manufacturing system. International Journal of Industrial Engineering Computations , 2(4), 913-930.

CHAPTER 9

Paul E. Deering, Sufficient conditions for a flexible manufacturing system to be deadlocked, doi: 10.5267/j.ijiec.2011.08.016..

CHAPTER 10

S.R. Singh, Leena Prasher and Neha Saxenaa, A centralized reverse channel structure with flexible manufacturing under the stock out situation, doi: 10.5267/j.ijiec.2013.05.004.

CHAPTER 11

Singh, R. , Singh, R. and Khan, B. (2015) A Critical Review of Machine Loading Problem in Flexible Manufacturing System. World Journal of Engineering and Technology, 3, 271-290. doi: 10.4236/wjet.2015.34028.

CHAPTER 12

Y. Xu, Y. Du, Y. Zeng and S. Li, "Flexible Manufacturing of Continuous Process Enterprises with Large Scale and Multiple Products," Technology and Investment, Vol. 4 No. 1, 2013, pp. 45-56. doi: 10.4236/ti.2013.41006.

CHAPTER 13

J. R. Jadhav, S. S. Mantha, S. B. Rane, Roadmap for Lean implementation in Indian automotive component manufacturing industry: comparative study of UNIDO Model and ISM Model, DOI 10.1007/s40092-014-0074-6.

INDEX